METAL IONS IN BIOLOGICAL SYSTEMS

Studies of Some Biochemical and Environmental Problems

ADVANCES IN EXPERIMENTAL MEDICINE AND BIOLOGY

Volume 13
CHEMISTRY AND BRAIN DEVELOPMENT
Edited by R. Paoletti and A. N. Davison • 1971

Volume 14
MEMBRANE-BOUND ENZYMES
Edited by G. Porcellati and F. di Jeso • 1971

Volume 15
THE RETICULOENDOTHELIAL SYSTEM AND IMMUNE PHENOMENA
Edited by N. R. Di Luzio and K. Flemming • 1971

Volume 16A
THE ARTERY AND THE PROCESS OF ARTERIOSCLEROSIS: Pathogenesis
Edited by Stewart Wolf • 1971

Volume 16B
THE ARTERY AND THE PROCESS OF ARTERIOSCLEROSIS: Measurement and Modification
Edited by Stewart Wolf • 1971

Volume 17
CONTROL OF RENIN SECRETION
Edited by Tatiana A. Assaykeen • 1972

Volume 18
THE DYNAMICS OF MERISTEM CELL POPULATIONS
Edited by Morton W. Miller and Charles C. Kuehnert • 1972

Volume 19
SPHINGOLIPIDS, SPHINGOLIPIDOSES AND ALLIED DISORDERS
Edited by Bruno W. Volk and Stanley M. Aronson • 1972

Volume 20
DRUG ABUSE: Nonmedical Use of Dependence-Producing Drugs
Edited by Simon Btesh • 1972

Volume 21
VASOPEPTIDES: Chemistry, Pharmacology, and Pathophysiology
Edited by N. Back and F. Sicuteri • 1972

Volume 22
COMPARATIVE PATHOPHYSIOLOGY OF CIRCULATORY DISTURBANCES
Edited by Colin M. Bloor • 1972

Volume 23
THE FUNDAMENTAL MECHANISMS OF SHOCK
Edited by Lerner B. Hinshaw and Barbara G. Cox • 1972

Volume 24
THE VISUAL SYSTEM: Neurophysiology, Biophysics, and Their Clinical Applications
Edited by G. B. Arden • 1972

Volume 25
GLYCOLIPIDS, GLYCOPROTEINS, AND MUCOPOLYSACCHARIDES OF THE NERVOUS SYSTEM
Edited by Vittorio Zambotti, Guido Tettamanti, and Mariagrazia Arrigoni • 1972

Volume 26
PHARMACOLOGICAL CONTROL OF LIPID METABOLISM
Edited by William L. Holmes, Rodolfo Paoletti, and David Kritchevsky • 1972

Volume 27
DRUGS AND FETAL DEVELOPMENT
Edited by M. A. Klingberg, A. Abramovici, and J. Chemke • 1973

Volume 28
HEMOGLOBIN AND RED CELL STRUCTURE AND FUNCTION
Edited by George J. Brewer • 1972

Volume 29
MICROENVIRONMENTAL ASPECTS OF IMMUNITY
Edited by Branislav D. Janković and Katarina Isaković • 1972

Volume 30
HUMAN DEVELOPMENT AND THE THYROID GLAND: Relation to Endemic Cretinism
Edited by J. B. Stanbury and R. L. Kroc • 1972

Volume 31
IMMUNITY IN VIRAL AND RICKETTSIAL DISEASES
Edited by A. Kohn and M. A. Klingberg • 1973

Volume 32
FUNCTIONAL AND STRUCTURAL PROTEINS OF THE NERVOUS SYSTEM
Edited by A. N. Davison, P. Mandel, and I. G. Morgan • 1972

Volume 33
NEUROHUMORAL AND METABOLIC ASPECTS OF INJURY
Edited by A. G. B. Kovach, H. B. Stoner, and J. J. Spitzer • 1972

Volume 34
PLATELET FUNCTION AND THROMBOSIS: A Review of Methods
Edited by P. M. Mannucci and S. Gorini • 1972

Volume 35
ALCOHOL INTOXICATION AND WITHDRAWAL: Experimental Studies
Edited by Milton M. Gross • 1973

Volume 36
RECEPTORS FOR REPRODUCTIVE HORMONES
Edited by Bert W. O'Malley and Anthony R. Means • 1973

Volume 37A
OXYGEN TRANSPORT TO TISSUE: Instrumentation, Methods, and Physiology
Edited by Haim I. Bicher and Duane F. Bruley • 1973

Volume 37B
OXYGEN TRANSPORT TO TISSUE: Pharmacology, Mathematical Studies, and Neonatology
Edited by Duane F. Bruley and Haim I. Bicher • 1973

Volume 38
HUMAN HYPERLIPOPROTEINEMIAS: Principles and Methods
Edited by R. Fumagalli, G. Ricci, and S. Gorini • 1973

Volume 39
CURRENT TOPICS IN CORONARY RESEARCH
Edited by Colin M. Bloor and Ray A. Olsson • 1973

Volume 40
METAL IONS IN BIOLOGICAL SYSTEMS: Studies of Some Biochemical and Environmental Problems
Edited by Sanat K. Dhar • 1973

Volume 41A
PURINE METABOLISM IN MAN: Enzymes and Metabolic Pathways
Edited by O. Sperling, A. De Vries, and J. B. Wyngaarden • 1973

Volume 41B
PURINE METABOLISM IN MAN: Biochemistry and Pharmacology of Uric Acid Metabolism
Edited by O. Sperling, A. De Vries, and J. B. Wyngaarden • 1973

METAL IONS IN BIOLOGICAL SYSTEMS

Studies of Some Biochemical and Environmental Problems

Edited by
Sanat K. Dhar
De Paul University
Chicago, Illinois

PLENUM PRESS • NEW YORK - LONDON

Library of Congress Cataloging in Publication Data

Main entry under title:

Metal ions in biological systems: studies of some biochemical and environmental problems.

(Advances in experimental medicine and biology, v. 40)

Proceedings of a conference on the role of metal ions in biological systems held Nov. 20-21, 1972, at Argonne National Laboratory, sponsored by the Central States Universities, Inc., the Center for Educational Affairs, Argonne National Laboratory, and the U. S. Atomic Energy Commission.

Includes bibliographies.

1. Metals in the body—Congresses. 2. Metal ions—Congresses. 3. Pollution—Congresses. I. Dhar, Sanat K., ed. II. Central States Universities, inc. III. United States. Argonne National Laboratory, Lemont, Ill. Center for Educational Affairs. IV. United States. Atomic Energy Commission. [DNLM: 1. Biochemistry—Congresses. 2. Metals—Congresses. W1 AD559 v. 40 1972 / QV290 M584 1972]

QP531.M47 612'.0152 73-15981

DOI 10.1007/978-1-4684-3240-4

Conference Committee: S. S. Danyluk, S. K. Dhar (Chairman), L. W. Dini, A. B. Edmundson, L. I. Katzin, A. Lindenbaum, J. J. Nisbet, D. O. Van Ostenburg, J. Shen-Miller, M. Schiffer.

Sponsors: Central States Universities, Inc., 9700 South Cass Avenue, Argonne, Illinois; Argonne Center for Educational Affairs, Argonne National Laboratory, Argonne, Illinois; U.S. Atomic Energy Commission.

Proceedings of a conference on the Role of Metal Ions in Biological Systems held November 20 and 21, 1972, at Argonne National Laboratory.

MyCopy version of the original edition 1973

A Division of Plenum Publishing Corporation
227 West 17th Street, New York, N.Y. 10011

United Kingdom edition published by Plenum Press, London
A Division of Plenum Publishing Company, Ltd.
Davis House (4th Floor), 8, Scrubs Lane, Harlesden, London, NW 10 6 SE, England

PREFACE

The articles published in this volume are based on the papers delivered at a conference on the Role of Metal Ions in Biological Systems held November 20 and 21, 1972, at Argonne National Laboratory. The purpose of the conference was to present to an interdisciplinary audience of physical scientists some recent developments illustrating the chemical and environmental participation of the heavy metal ions in the biological system. The invited speakers at the conference are specialists in the fields they describe, and the articles presented here are at a level of interest to readers with backgrounds in physical sciences who are not necessarily doing research in the areas described. The articles are referenced through 1972, and in some cases early 1973, and thus should also be of value to research workers. It is hoped that the book will be of particular interest to chemists, biologists, workers in the fields of environmental science and public health, as well as graduate and senior undergraduate students in these disciplines.

The conference was sponsored by the Central States Universities, Inc., a consortium of sixteen midwestern universities, the Center for Educational Affairs, Argonne National Laboratory, and the United States Atomic Energy Commission. It is my pleasure to thank the members of the conference committee for their ideas and active help in organizing the conference. I am also indebted to many of my colleagues at various CSUI institutions for their valuable suggestions towards the implementation of the conference. Special thanks are due to Mrs. Dorothy Carlson and her associates at Argonne National Laboratory for their flawless arrangements. I would also like to thank Mrs. Susan O'Brien for her skillful typing of the photocopy of the manuscript. Finally, I wish to thank my family for their willing sacrifice during the period the conference and assembling this volume took much of my time that was their due.

Sanat K. Dhar

LIST OF CONTRIBUTORS

N. A. Berger, Laboratory of Molecular Aging, National Institutes of Health, Gerontology Research Center, Baltimore City Hospitals, Baltimore, Maryland

B. G. Blaylock, Environmental Sciences Division, Oak Ridge National Laboratory, Oak Ridge, Tennessee

J. J. Butzow, Laboratory of Molecular Aging, National Institutes of Health, Gerontology Research Center, Baltimore City Hospitals, Baltimore, Maryland

P. Clark, Laboratory of Molecular Aging, National Institutes of Health, Gerontology Research Center, Baltimore City Hospitals, Baltimore, Maryland

G. L. Eichhorn, Laboratory of Molecular Aging, National Institutes of Health, Gerontology Research Center, Baltimore City Hospitals, Baltimore, Maryland

Karl D. Hardman, Division of Biological and Medical Research, Argonne National Laboratory, Argonne, Illinois

Rolf Hartung, Department of Environmental and Industrial Health, The University of Michigan, Ann Arbor, Michigan

Paul R. Harrison, Department of Environmental Control, City of Chicago, Chicago, Illinois

J. Heim, Laboratory of Molecular Aging, National Institutes of Health, Gerontology Research Center, Baltimore City Hospitals, Baltimore, Maryland

Arthur Lindenbaum, Division of Biological and Medical Research, Argonne National Laboratory, Argonne, Illinois

A. S. Mildvan, The Institute for Cancer Research, Fox Chase, Philadelphia, Pennsylvania

J. B. Neilands, Department of Biochemistry, University of California, Berkeley, California

J. Pitha, Laboratory of Molecular Aging, National Institutes of Health, Gerontology Research Center, Baltimore City Hospitals, Baltimore, Maryland

G. H. Reed, Department of Biophysics and Physical Biochemistry, Johnson Research Foundation, University of Pennsylvania School of Medicine, Philadelphia, Pennsylvania

C. Richardson, Laboratory of Molecular Aging, National Institutes of Health, Gerontology Research Center, Baltimore City Hospitals, Baltimore, Maryland

J. M. Rifkind, Laboratory of Molecular Aging, National Institutes of Health, Gerontology Research Center, Baltimore City Hospitals, Baltimore, Maryland

Jack Schubert, Department of Radiation Health, Graduate School of Public Health, University of Pittsburgh, Pittsburgh, Pennsylvania

Michael C. Scrutton, Department of Biochemistry, Temple University School of Medicine, Philadelphia, Pennsylvania

Y. Shin, Laboratory of Molecular Aging, National Institutes of Health, Gerontology Research Center, Baltimore City Hospitals, Baltimore, Maryland

E. Tarien, Laboratory of Molecular Aging, National Institutes of Health, Gerontology Research Center, Baltimore City Hospitals, Baltimore, Maryland

Bert L. Vallee, Biophysics Research Laboratory, Department of Biological Chemistry, Harvard Medical School, and the Division of Medical Biology, Peter Bent Brigham Hospital, Boston, Massachusetts

CONTENTS

COBALT SUBSTITUTED ZINC METALLOENZYMES*

BERT L. VALLEE

Biophysics Research Laboratory, Department of Biological Chemistry, Harvard Medical School, and the Division of Medical Biology, Peter Bent Brigham Hospital, Boston, MA.

It has taken a long time, indeed, for biochemical research on metals to become respectable. Substantial experimental difficulties have beset the field for a long time and overenthusiastic claims, at times based on uncertain facts, were not altogether reassuring to the rank and file of interested biochemists.

The details of the manner in which metals manifest their important biological roles are still relatively unknown. It is appreciated that they function in catalysis, in the synthesis and the stabilization of the structure of proteins and in transport, all processes to which they entail great specificity; but little is known of the chemical details, or the mechanisms by which these processes take place. The essentiality of many metals in biology is now appreciated, but the manner by which they exercise their roles is understood less well. How many aspects of biological specificity may be understood and described in terms of the detailed chemistry of metal interactions with macromolecules? We have been concerned for some time with the mechanism of action of metals in metalloenzymes which we have considered as models for the study of the mechanism of action of enzymes in general.

The chemical basis of biological specificity in general and of enzyme action in particular are fundamental questions. The inactivation of non-metalloenzymes by site-specific reagents can be quite unique, in a manner not observed in simple peptides or

---

*This work was supported by Grant-in-Aid GM-15003 from the National Institute of Health, of the Department of Health, Education, and Welfare.

nonenzymatic proteins. Thus, the inactivation of certain seryl enzymes by diisopropylfluorophosphate or that of sulfhydryl, tyrosyl, histidyl, and lysyl enzymes by other suitable site-specific organic reagents under mild conditions has no facsimile in other fields of chemistry. Questions regarding the properties of enzymes that render certain of their amino acid side chains to be chemically hyperreactive have been raised because the basis of such reactivity is currently unknown, and these features might provide links to their biological functions.

Metal ions, as a class, exhibit properties and reactivities which vary greatly from those of amino acid side chains of proteins. Some metals, when involved directly in activity can serve as excellent labels of active enzyme sites because of the visible spectra of some of their complexes. Thus, blue copper salts and red or brownish iron salts change color on oxidoreduction. The basic characteristics of the spectra might be expected to be preserved in biological systems and yet reveal additional features typical of active sites of metalloenzymes.

The physicochemical properties of many of the metals which are essential to the function of metalloenzymes, mainly those of the first transition series, can, indeed, probe their environments. Basically, physicochemical characteristics of metal complex ions and of metalloenzymes arise from three sources: properties of the metal altered by the ligands, properties of the ligands altered by the metal, and specific _de novo_ properties characteristic only of the resultant complex. Various parameters such as stability constants, the effects of different donor groups upon their formation, redox potentials, spectroscopic, optical rotatory and magnetic properties characterize metal complex ions (Vallee and Wacker, 1970).

The characterization of these parameters has been enhanced by the highly sensitive instrumental methods recently available. These are certain to have even greater impact on the mechanistic features of metals in catalysis just as earlier analytical instrumental developments resulted in the assignment of their biological roles. The conceptual innovations which have accompanied the introduction of nuclear magnetic resonance, electron paramagnetic resonance, optical rotatory dispersion, circular dichroism, magneto circular dichroism, infrared, fluorescence, and other forms of spectroscopy of biological studies are no less profound than those which led to the advances of the previous two decades.

When studied by these means, the pertinent characteristics of metalloenzymes are quite unusual in comparison with those of metal complex ions. This is particularly apparent from the inspection of absorption, optical rotatory dispersion and circular dichroism, and electron paramagnetic resonance spectra of certain iron, copper, and cobalt enzymes. It has been difficult to assign the origin of

Table 1. Spectral parameters of Co(II) complex ions and Co(II) substituted metalloenzymes.

Compound	Absorption λ, nm (Absorptivity, M^{-1} cm^{-1})			Ellipticity λ, nm ($[\theta]^{25} \times 10^{-3}$)		Notes
1. Simple Co(II) complexes						
$[Co(H_2O)_6]^{2+}$	510 (∿5)		1200 (∿2)			Octahedral
$[CoCl_4]^{2-}$	685 (700), complex		1700 (100), complex			Tetrahedral
$[Co(OH)_4]^{2-}$	600 (∿150), complex		1400 (∿50), complex			Tetrahedral
$[Co(Et_4dien)]Cl_2$[a/]	520	660	950			Distortion from trigonal bipyramidal
$[Co(Tren\ Me)_6Cl]Cl$[b/]	500 (120)	625 (80) doublet	800 (∿30) 1750 (∿30)			
2. Co(II) enzymes						
Co(II) carbonic anhydrase	510 (280) 550 (380)	615 (300) 640 (280)	900 (30) 1250 (90)	460 (∿3) 550 (3)	610 (3)	Low symmetry
Co(II) alkaline phosphatase	510 (280) 555 (350)	610 (210) sh 640 (250)		470 (2) 520 (2)	575 (-0.8)	Low symmetry

Table I. Continued.

Compound	Absorption λ, nm (Absorptivity, M^{-1} cm^{-1})			Ellipticity λ, nm ($[\theta]^{25} \times 10^{-3}$)		Notes
Co(II) carboxypeptidase	500 sh	555 (~150) 572 (~150)	950 (~25)	500 sh	538 (-0.5)	Distorted tetrahedral
Co(II) yeast alcohol dehydrogenase[c]	620 (~800) 657 (~800) 710 (~600)					Low symmetry
Co(II) neutral protease	475 sh 525 (sh)	555 (50-100)		500 sh	550 (-0.8)	Low symmetry
Co(II) yeast aldolase	490 sh	530 (150)		485 (2.4)	530 (3)	Low symmetry
Co(II) yeast enolase[d]	490 (sh)	540 (30-40) 575 (sh)		490 sh	540 (-0.5)	Low symmetry

[a] In organic solvents.
[b] In organic solvents.
[c] Enzyme contained both cobalt and zinc.
[d] Metal-enzyme complex.

these spectra which seem to differ from those of most metal complex ions (Vallee and Williams, 1968) suggesting low symmetry of the coordination environment of metals in metalloenzymes. Yet, it is not yet clear, however, what may account for these unusual spectral properties of metalloenzymes. However, the unusual spectral properties of metalloenzymes and their modification by addition of substrates, coenzymes, or inhibitors have been inferred to signal catalytic potential reflecting biological capacity in some manner. On this basis physical properties might be expected to be influenced by conditions which alter function, a phenomenon which has been observed (Lindskog and Ehrenberg, 1967; Kasper, 1968; Simpson and Vallee, 1968; Vallee and Latt, 1970). Data basic to such deductions have been reviewed (Vallee and Williams, 1968; Vallee and Riordan, 1969; Vallee and Latt, 1970).

When enzymes contain a chromophoric metal such studies are readily performed, while the presence of a metal such as zinc would seem to preclude them. Zinc is a poor probe both owing to its colorless complexes and diamagnetism. Cobalt, however, gives rise to characteristic spectra and is paramagnetic, qualities which are desirable when probing the environment of active enzymatic sites. Fortunately, cobalt can be substituted for the native zinc atom(s) of many enzymes while activity is retained. In this manner, the active sites of zinc enzymes can be examined using a cobalt probe. Since the resultant spectra are quite revealing, some of these are summarized in Table I.

Cobalt alkaline phosphatase of *E. coli* serves as an illustration. The enzyme prepared by chromatography on DEAE cellulose contains four zinc atoms per mole of protein. Two of the four are removed rapidly by 8-hydroxyquinoline-5-sulfonic acid and, at the same time, virtually all activity is lost. The other two zinc atoms are apparently not involved directly in catalytic function and, therefore, are removed much more slowly. Restoration of zinc to the apoenzyme confirms that only two of the zinc atoms are required for function. The binding of the first two gram atoms of cobalt by the apoprotein generates a spectraum similar to that of octahedral cobalt complex ions without inducing enzymatic activity. Addition of the second two gram atoms of cobalt generates activity and results to a complex absorption spectrum with maxima at 640 nm (ε = 250), 610 nm (sh 210), 555 nm (350), and 510 nm (280). This spectrum differs from that of octahedral or tetrahedral cobalt complex ions and suggests an unusual coordination environment for the metal at the active site of alkaline phosphatase (Simpson and Vallee, 1968) and, in fact, bears a strong resemblance to that of cobalt carbonic anhydrase (Lindskog and Nyman, 1964) which contains only a single cobalt atom per molecule. The circular dichroism of cobalt phosphatase also suggests an unusual coordination environment, much as do the alterations of the absorption and circular dichroic spectra of the cobalt enzyme when it interacts with phosphate.

The intensity of the spectrum associated with cobalt atoms involved in catalytic activity decreases progressively when titrated over the pH range of 9 to 6. When studied over the same pH range, the degree of restoration of the spectrum associated with these catalytically active cobalt atoms is closely parallel to enzymatic activity of the cobalt enzyme (Simpson and Vallee, 1968).

Quite analogous studies of <u>horse liver alcohol dehydrogenase</u> (LADH) have proven productive also with cobalt or cadmium and similarly result in enzymatically active enzymes which exhibit specific activities and spectral properties characteristic of each of these metals (Drum and Vallee, 1970). Co-LADH has a broad absorption band in the near UV region centered at 340 nm (ε = 6500) associated with two positive and three negative circular dichroic bands between 300 nm and 450 nm. Absorption maxima also occur at 655 nm (1330), 730 nm (800), and in the near infrared between 1000 and 1800 nm (270-540). A small negative ellipticity band is centered at 620 nm (Figs. 1 and 2).

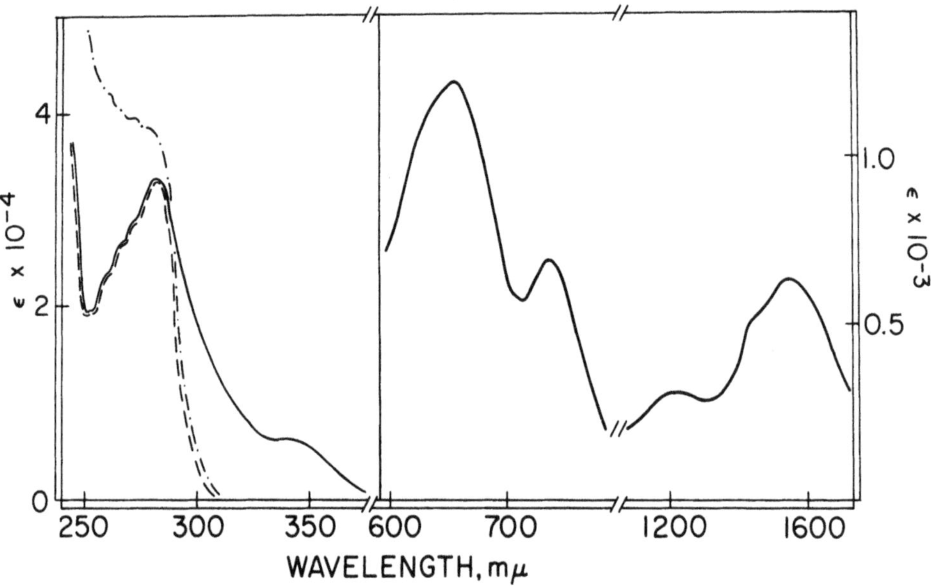

Figure 1. Absorption spectra of the metallodehydrogenases. (---), Zn-LADH; (——), Co-LADH; (—·—·—·). Cd-LADH. 0.1 M tris acetate, pH 7.0 (Drum and Vallee, 1970).

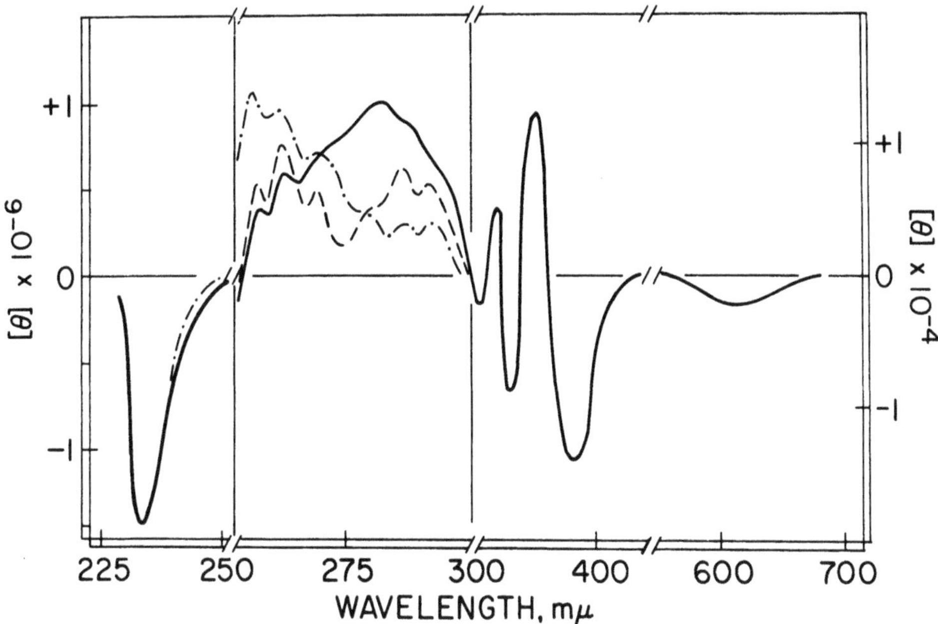

Figure 2. Circular dichroic spectra of the metallodehydrogenases. (---), Zn-LADH; (——), Co-LADH; (—·—·—·), Cd-LADH. The symbol [Θ] denotes molecular ellipticity based on the concentration and molecular weight of LADH. 0.1 M tris acetate, pH 7.0 (Drum and Vallee, 1970).

The intense absorption band centered at 245 nm of Cd-LADH may assist in identifying the ligands of the proteins which bind the metal, since its molar absorptivity ε = 10,200 is close to that of 14,000 reported for the cadmium mercaptide chromophores of metallothionein.

The circular dichroic spectrum of Co-LADH is richer in detail than that of any comparable cobalt enzyme studied to date. It can be analyzed in terms of protein structure, as reflected in the <u>intrinsic</u> and <u>side-chain</u> Cotton effects of the protein, as well as the superimposed <u>extrinsic</u> Cotton effects which result from either binding of NADH or of 1,10-phenanthroline (OP).

The side-chain Cotton effects of LADH are also perturbed by the metal substitutions. The circular dichroic spectra of these metallodehydrogenases exhibit remarkable fine structure between 250 and 300 nm characteristic of each. Below 270 nm the circular dichroic difference spectrum of Cd-LADH versus Zn-LADH exhibits a positive ellipticity band which may correspond to the maximum seen in the absorption spectrum at 245 nm. The magnitude of the ellipticity band of Co-LADH exceeds that of both Zn- and Cd-LADH in the region of maximum absorption due to aromatic amino acid residues. The properties of asymmetrically bound extrinsic chromophores also reflect the effects of metal substitutions in LADH. Distinctive extrinsic Cotton effects with magnitudes related quantitatively to binary complex formation are generated by the reduced coenzyme, NADH, and 1,10-phenanthroline (OP) when bound to Zn-, Co-, and Cd-LADH.

The circular dichroic band of the LADH-Cd OP complex which is centered at 271 nm has an amplitude which markedly exceeds that of LADH-Zn OP, while that of the LADH-Co OP complex is considerably smaller. Thus, sensitive probes of the identity of the metal substituent at the active site are afforded by the rotational strengths of these mixed complexes.

Such studies have shown that in certain metalloenzymes some of the metal atoms are critical to the catalytic step while the remainder seem to be involved largely in maintaining protein structure (Drum, et al., 1967). Cobalt or cadmium when substituted in Zn-LADH appear to confirm the discrete roles for the two metal atoms. Only two of the metal atoms in cobalt and cadmium LADH seem to be accessible to OP, since they bind only 2 moles of OP, as in the native zinc enzyme. Previous criteria can now be augmented by the employment of the unusual absorption and circular dichroic properties to document the selective replacement either of the "functional" or the "structural" zinc atoms with other metals resulting in "metal hybrid enzymes."

The probe properties of metals have also served to study proteolytic enzymes, a number of which are known to contain a metal, most commonly zinc (Vallee and Wacker, 1970; Vallee and Latt, 1970). Analogous studies in yet other, similar systems have become possible. It is now feasible to replace the zinc of both thermolysin and neutral protease by cobalt and other metals to obtain active enzymes which exhibit characteristic spectra (Vallee and Latt, 1970).

Certain patterns are becoming obvious despite the still relatively small number of cobalt enzymes examined in this manner: the absorption spectra of many substituted zinc enzymes are quite similar to one another and many have optically active bands of similar magnitude and position. It is tempting, though premature, to interpret the spectral details and to speculate that existent similarities might indicate common functional characteristics. Similarities and differences of spectra could conceivably imply specific mechanisms.

With the advent of sufficiently discriminating physicochemical methods to discern chemical details of enzyme structure, it is now possible to consider that these details may be related to catalytic potential as recently observed in the carboxypeptidase-procarboxypeptidase system (Behnke and Vallee, 1972).

In carboxypeptidase A zinc can be replaced by cobalt and the spectra of cobalt carboxypeptidase are also indicative of asymmetry around the cobalt atom involved in catalysis (Latt and Vallee, 1971). There is a shoulder at about 500 nm and maxima at 555 and 572 nm with absorptivities, ε, slightly greater than 150 (Fig. 3). Two

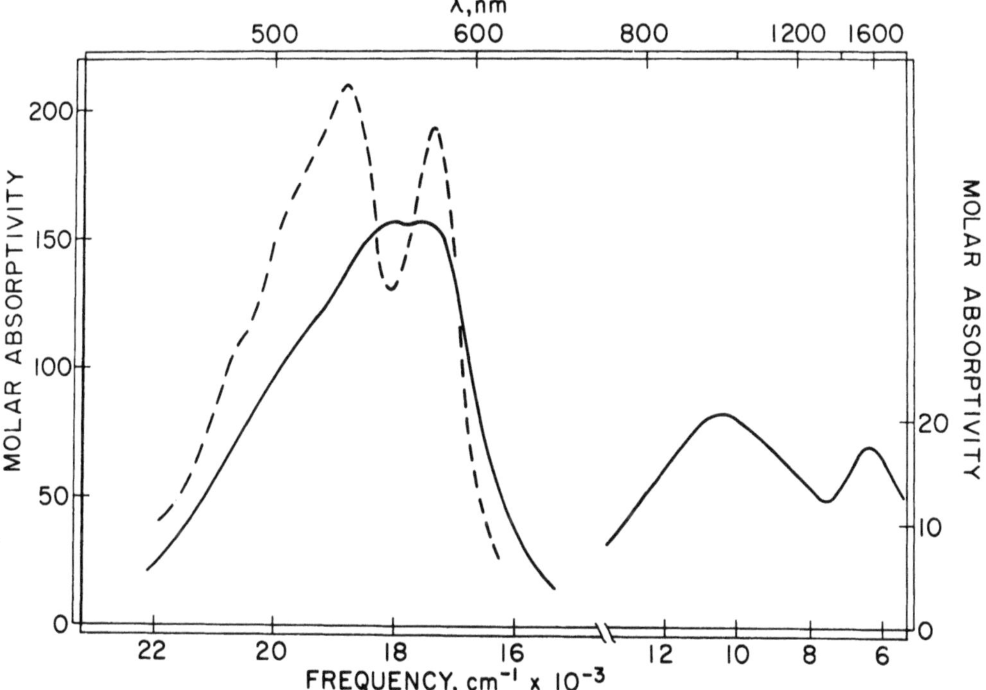

Figure 3. Absorption spectra of cobalt carboxypeptidase for (a) visible region. The enzyme, 1mM, was dissolved in 1 M NaCl, 0.005 M Tris-Cl, pH 7.1, 20° (—— ——). Another enzyme solution, about 3 mM, was diluted with glycerol to 45% v/v and cooled to 4.2°K (- - - -) for spectral measurements. (b) Near-infrared region. Apocarboxypeptidase, 1.5 mM, was dissolved in 1 M NaCl, 0.005 M [D] Tris-Cl, D_2O, pH 7.2. The sample cuvet contained 1.5 mM enzyme plus $Co(SO_4)$ in D_2O buffer to yield a final total cobalt concentration of 2.0 mM, and, hence, a 0.5 mM excess of free Co(II) ions; the reference cell contained 1.5 mM apoenzyme, brought to volume with buffer.

infrared bands are centered at 940 nm and 1570 nm (ε = 20). The location of the high energy band at 940 nm is distinctive, since the near infrared absorption bands of divalent cobalt complex ions are usually found at longer wavelengths bearing out the suggestion that interaction of cobalt with its ligands is stronger than that in cobalt complex ions. The position of these bands in cobalt spectra can often probe the metal environment more effectively than can the visible bands. Overall, the geometry of coordination of the cobalt atom in cobalt carboxypeptidase is irregular.

At liquid helium temperature the absorption measurements show an increase in spectral resolution. There are two new shoulders and the maximum at 555 nm is shifted to 532, while that at 572 nm is maintained at this wavelength. There is no significant reduction in absorptivity; this would argue against participation of vibronic interactions in the generation of spectral intensity. The spectral changes at 4.2°K are reversed by a gradual increase of the temperature to 25°C. The absorption band at 572 nm is rendered optically active by a magnetic field of 47,000 gauss. Circular dichroic spectra reflect the geometry of the metal binding site and may include information on the influence of vicinal factors on the cobalt atom in the enzyme. Enhanced in the spectrum observed at 4.2°K are a negative band at 538 nm and the shoulder near 500 nm which probably correspond to the low wavelength maximum and shoulder of the absorption band. The spectral detail suggests the existence of distinct transitions with different polarizations.

Agents which affect enzymatic activity would be expected to alter the spectra if these were to reflect catalytic potential. Indeed, major shifts in the absorption and circular dichroic spectra of cobalt carboxypeptidase are brought about by a large number of inhibitors (Fig. 4). The patterns of the spectral shifts, observed as a function of the structure and concentration of added inhibitors, are characteristic for inhibitors having different functional groups (Latt and Vallee, 1971).

The effects of substrates on the circular dichroic spectrum of cobalt carboxypeptidase are particularly interesting. The glycyl-L-tyrosine-cobalt carboxypeptidase complex differs from that both of the enzyme itself and from that of most of the enzyme-inhibitor complexes. The extremum of the negative Cotton effect of the cobalt enzyme at 537 nm shifts to 555 nm with an inversion of sign, while the molar ellipticity increases from -500 to +2000, a four-fold change in magnitude (Fig. 4). Major electronic rearrangements around the cobalt atom upon interaction with substrates are suggested by both the inversion of the sign and the marked enhancement of the spectrum. This impression is reinforced by the fact that in the absorption spectrum the band at 940 nm splits into two, located at ∿850 nm and ∿1150 nm, respectively, while that at 1570 nm is shifted to 1420 nm.

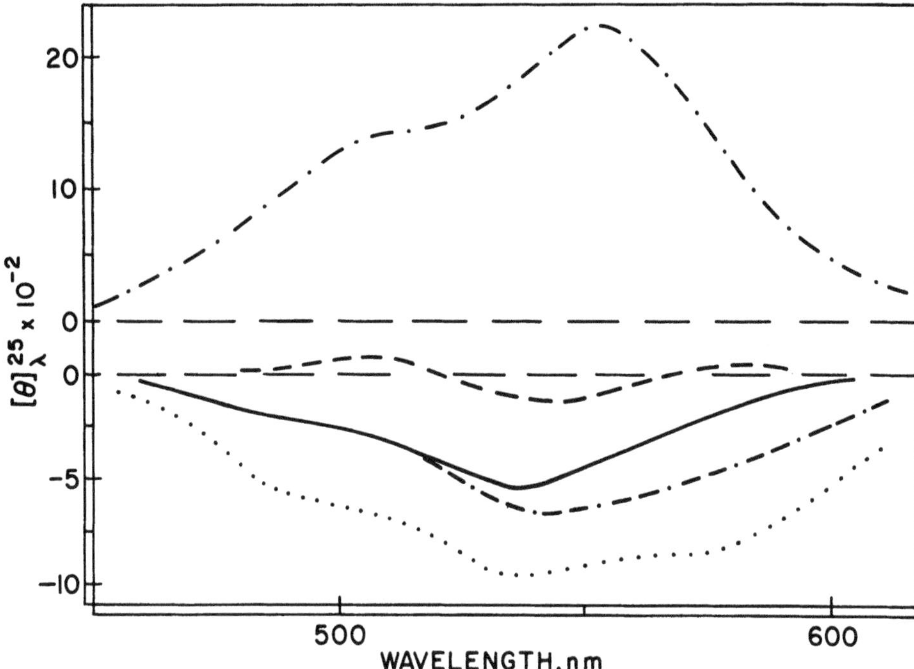

Figure 4. Effect of glycyl-L-tyrosine, β-phenylpropionate, and L-phenylalanine on the circular dichroic spectrum of cobalt carboxypeptidase (——). The enzyme was dissolved in 1 M NaCl, 0.005 M Tris-Cl, pH 7.1. Spectra of the enzyme were also obtained in the presence of glycyl-L-tyrosine, 10 mM (-·-·-·), L-phenylalanine, 9 mM (— — —), of β-phenylpropionate, 2 mM (--·--·), or 9 mM (····).

Jointly the absorption and circular dichroic spectra of cobalt carboxypeptidase indicate that the enzymatically active metal occupies an irregular, asymmetric tetrahedral site in the enzyme. In this example, cobalt serves both as an integral part of the catalytic apparatus and as a spectral probe which is, perhaps, capable of providing new insight into enzyme action (Vallee and Latt, 1970; Latt and Vallee, 1971).

The spectral features of cobalt procarboxypeptidase and their perturbation by substrates and inhibitors are virtually identical with those of cobalt carboxypeptidase and its gly-tyr complex (Figs. 3 and 4), which, in turn, are consistent with distorted tetrahedral geometry of the cobalt atom. If this entatic spectrum is, indeed, an expression of catalytic potential, the zymogen would be expected to exhibit enzymatic activity. Synthesis of a series of acylamino acids has permitted systematic investigation of this hypothesis with the result that the zymogen turns out to be a highly active peptidase toward such substrates. The cobalt spectrum of procarboxypeptidase seems to be so far unique in having predicted the recognition of catalytic activity for a protein hitherto thought to be enzymatically inert.

Thus, the developments in physical-organic and physical chemistry have become increasingly applicable to macromolecules such as enzymes and these developments offer new opportunities which give promise that the recognition of the chemistry underlying the exceptional biological activity and specificity of enzymes may consolidate the now diversified area of study. This hypothesis would seem to be particularly suited for investigations based on the properties of various proteolytic enzymes.

REFERENCES

Behnke, W. D. and Vallee, B. L. (1972), Proc. Nat. Acad. Sci. U. S., 69, 2442.

Drum, D. E., Harrison, J. H., IV, Li, T.-K., Bethune, J. L., and Vallee, B. L. (1967), Proc. Nat. Acad. Sci. U. S., 57, 1434.

Drum, D. E. and Vallee, B. L. (1970), Biochem. Biophys. Res. Comm., 41, 33.

Kasper, C. B. (1968), J. Biol. Chem., 243, 3218.

Latt, S. A. and Vallee, B. L. (1971), Biochemistry, 10, 4263.

Lindskog, S. and Ehrenberg, A. (1967), J. Mol. Biol., 24, 133.

Lindskog, S. and Nyman, P. O. (1964), Biochem. Biophys. Acta, 85, 462.

Simpson, R. T. and Vallee, B. L. (1968), Biochemistry, 7, 4343.

Vallee, B. L. and Latt, S. A. (1970), Structure-Function Relationships of Proteolytic Enzymes (P. Desnuelle, H. Neurath, and M. Ottesen, eds.), Copenhagen: Munksgaard, p. 144.

Vallee, B. L. and Riordan, J. F. (1969), Ann. Rev. Biochem., 38, 539.

Vallee, B. L. and Wacker, W. E. C. (1970), Metalloproteins, The Proteins, 2nd ed. (H. Neurath, ed.) 5, New York: Academic Press.

Vallee, B. L. and Williams, R. J. P. (1968), Proc. Nat. Acad. Sci. U. S., 59, 498.

CHEMISTRY OF IRON IN BIOLOGICAL SYSTEMS

J. B. NEILANDS

Department of Biochemistry
University of California
Berkeley, California 94720

Iron compounds in biology can be classified in various ways. An organization of these substances based on source--whether microbial, plant, or animal--is certain to suffer from considerable redundancy. Thus, the iron porphyrins, and even more so, the ferredoxins are widely distributed on a phylogenetic basis. I have elected to present this material on the themes of structure and function. First, some relevant properties of iron will be recalled to memory. Then the diversified roles of iron in living systems will be enumerated. Finally, the structures of the lower and higher molecular weight compounds of iron will be discussed.

The iron compounds of microorganisms have been touched upon in a monograph titled *Microbial Iron Metabolism* (Neilands, 1973a) as well as in recent reviews (Hutner, 1972; Neilands, 1973b). A general and very lucid summary of the biochemistry of iron has been prepared by Harrison (1971). The problems involved in human iron deficiency have been treated in a book edited by Hallberg, *et al*. (1970). The books by Bowen (1966) and Underwood (1971) are probably the single best sources of information on trace elements in the plant and animal worlds, respectively. Two other monographs worthy of mention are *Bioinorganic Chemistry*, edited by Dessy, *et al*. (1971) and *Inorganic Biochemistry*, edited by Eichhorn (1973).

I. DISTRIBUTION AND PROPERTIES OF IRON

Iron is one of the most common elements on Earth and throughout the solar system. In contrast to its trace concentration in living cells, the element is the most abundant one in the whole planet Earth, the core of which is believed to consist of iron admixed with nickel. On the surface of Earth iron is still quantitatively an important element although it here ranks just below aluminum, the most abundant metal. Common igneous rocks may contain over 5% iron; the average concentration in soil is about 3.8% (Bowen, 1966).

At ca. 0.2×10^{-6} gram atoms per liter, ocean water is a poor source of iron for proliferating cells. This may be an important factor in establishing the fertility of the seas. Phytoplankton exhibit truly fantastic capacities to concentrate this iron and so bring it into the food chains. The concentration factor for ocean plankton and brown algae may be as high as 100,000 times, but the molecular mechanism for this very efficient system remains obscure. Depending on the season, river water may range up to several parts per million in iron; in any case, the concentration is typically higher than in ocean water.

Spectroscopic probes provide convincing evidence that iron is present in intrastellar dust in significant quantities and that the element enjoys general distribution throughout the solar system. The abundance of elements decreases exponentially in the progression light → heavy, but the trend is broken by a peak centered at mass 56 (Burbridge, _et al_., 1957). This is ascribed to an accumulation of the stable ^{56}Fe nucleus in the high temperature fusion reactions in the stars.

It has already been noted that the abundance of iron in the inorganic world stands in stark contrast to its sparing concentration in living cells. Iron is truly a trace substance in biology, albeit its concentration is higher than that of several of its transition element relatives. More iron is required in animals because in this case the element has been assigned the additional task of transporting oxygen to the tissues. The human body contains about 5 grams of iron, of which around 4 grams is in hemoglobin. The remaining 1 gram is virtually all in ferritin and myoglobin. This leaves something of the order of tenths of a gram for the several dozen iron enzymes known to be essential for the proper functioning of the living body.

Plants and microorganisms do not need hemoglobin, although in some of these species oxygen binding pigments are nonetheless present, and iron levels of approximately 0.1 milligram per gram dry weight can be expected in all cells. Anaerobic species, such as the _Clostridia_, are rich sources of the iron sulfur proteins and it has been known for many years that such organisms may contain as much, if not more, iron than their aerobic counterparts (Curran, _et al_., 1943). Microbes, which are renowned for the plasticity of their metabolism, doubtless have means of coping with nutritional deficits in iron. Both glycolysis and the tricarboxylic acid-electron transport chain are widely distributed in aerobic cells, but only the latter requires a battery of iron enzymes. At low iron both aerobic and facultative aerobic cells can partially satisfy their energy demands by glycolysis, form modest yields of cells, and live more-or-less independently of iron.

Probably the most crucial inorganic property of iron, from the point of view of the nutrition of living cells, is the profound insolubility of the ferrous and ferric ions. The solubility product constants for these ions are estimated to be 10^{-15} and 10^{-37}, respectively. The free hexahydrated ions exist only at low values of pH. Polynuclear complexes of more-or-less undefined composition are formed upon neutralization of ferric chloride. The Fe(II) and Fe(III) salts of strong acid anions are soluble, but the phosphates, carbonates, and especially the phytates, are insoluble. Ferrous ion can act as an energy source for the iron-oxidizing bacteria which, perforce, must be able to survive at pH values < 4. Ferric hydrolysates containing oxides and hydroxides of iron may be regular in composition and display a microcrystalline state.

In the 4.5 billion years since the formation of Earth, iron has undergone extensive geochemical transformations. Like the other elements, it is assumed to have been present on the primitive earth in its utmost reduced form. Water brought up from subterranean sources can be shown to contain the iron in the reduced state, and a slice into the Earth's crust will reveal the iron as progressively more reduced at lower depths.

Single cell, procaryotic species are thought to have arisen in an evolutionary sequence starting with the strictly anaerobic bacteria and leading, perhaps after a billion years, to the blue-green algae. Unlike their immediate precursors, the photosynthetic bacteria, the blue-green algae evolve oxygen gas. The appearance of O_2 in the atmosphere must have promptly switched the surface iron of Earth predominantly to the ferric, and highly insoluble, form. This event, in turn, precipitated a demand for the elaboration of special complexing agents to keep iron in solution during its period of transport and insertion into biological loci. Probably all aerobic and facultative aerobic microorganisms produce or require iron transport compounds, collectively termed siderochromes (Neilands, 1973b). In addition, special mechanisms have been devised to dissolve and transport iron in animal organisms, and probably also in plants.

Various fungi have been shown capable of the dissolution and solubilization of iron-containing minerals such as granodiorite, augite, and biotite. One of the active species was shown to be *Aspergillus niger* (Arrieta and Grez, 1971), an organism which synthesizes polyhydroxamate type siderochromes such as ferrichrome. Lichen products have also been shown to promote the chemical weathering of silicates (Iskandar and Syers, 1972). Biological processes such as these keep the iron in the biosphere in a state of continuous transformation.

The two stable valence states of iron, Fe(II) and Fe(III), form complexes and coordination compounds with all types of weak and strong field ligands. A considerable amount of structural information can be obtained by measuring the number of unpaired electrons in those derivatives which happen to have a finite level of paramagnetism. Weak field ligands, such as water and halide ions, allow a maximum number of parallel unpaired spins, as in the free ions. Such compounds are designated "high spin." In the other type, the "low spin" coordination compound, the ligands have moved into two of the 3d orbitals causing spin pairing in the remaining three orbitals and consequent reduction in spin state. In this manner the paramagnetism of Fe(II) compounds can be eliminated. Cyanide is a notoriously strong field ligand, tending to give complexes of low spin. The hemoglobin family of compounds illustrates the relationship between d electron structure and spin state. In deoxyhemoglobin the sixth ligand is water, the complex has four unpaired electrons and is high spin. Oxyhemoglobin, on the other hand, has zero unpaired electrons and is low spin. When the iron is oxidized, as in the methemoglobin series, fluoride binding gives, as expected, a high spin complex (5 unpaired electrons), while cyanide binding reduces the d electron configuration to a single unpaired electron. Magneto-chemical classification of natural iron compounds has been hampered by the lack of suitably sensitive instrumentation. This has limited quantitative measurements of paramagnetism to only the readily available substances, such as hemoglobin. A new magnetometer, based on the theory of superconductivity, should facilitate measurements of paramagnetism in very small amounts of enzymes and proteins which may contain less than 0.01% iron (Dawson, *et al.*, 1972).

Electron spin resonance, a very convenient probe for paramagnetism, has been applied extensively for studies on the environment of the metal in the iron-sulfur proteins. The latter display in the reduced state a characteristic signal at $g = 1.94$. In bacterial systems it is possible to replace both the ^{32}S and the ^{56}Fe of the iron sulfur proteins by the heavier isotopes. A comparison of the electron spin resosnance spectra of the heavy and light proteins confirms the presence of iron sulfur bonding. Electron spin resonance spectra measured as a function of temperature can resolve ligands attached to iron in a completely rhombic field (Peisach, *et al.*, 1971). Also, replacement of ^{56}Fe by ^{57}Fe enables the measurement of Mössbauer spectra, a technique that is quite useful in comparing the equivalence of the environments of iron when several atoms are present per mole of protein.

The sensitivity and precision of light absorption measurement are such that this technique remains the single most useful probe for preliminary classification of natural iron compounds. All heme derivatives can be immediately identified through the intensely absorbing Soret band centered at ca. 400 nm. Ultimate characterization of the particular substitution in the heme nucleus requires that the iron be removed and the spectral peaks of the porphyrin, or porphyrin ester, be recorded (Falk, 1964). The four-banded, "stepladder" spectra between 500 nm and about 600 nm is highly characteristic of the porphyrins, as is their intense orange fluorescence under ultra violet illumination. The latter is sufficiently potent to enable the quantitative determination of the porphyrin content of a single red blood cell.

Other iron compounds possess less spectacular electronic spectra than the hemes but, nonetheless, useful structural information has been gleaned in nearly every case that has been carefully investigated. The polyhydroxamate siderochromes are a case in point. The 3:1 chelates with ferric ion are tea colored in concentrated solution and yellow in dilute solution. The absorbancy maximum is near 430 nm and the molar absorbancy coefficient is ca. 3000/Fe(III). The 1:1 chelates, in contrast, are purple in color, they have an absorbancy maximum near 500 nm, and the absorbancy coefficient drops to about 1000. These bathochromic and hipsochromic shifts can be used to distinguish between the ferric derivative of a polyhydroxamate and a 3:1 monohydroxamate, both of which have identical spectral properties at neutral pH. On acidification the latter 3:1 complex will degrade successively through the 2:1 to the 1:1 complex (or chelate), the latter being at pH 2 purple in color and possessing only 1/3 of the molar absorbancy per iron atom. The typical polyhydroxamate is much more resistant to degradation by acid and may have spectral characteristics at pH 2 and 7 which are virtually identical.

In simple complexes and chelates both ferric and ferrous ions commonly display hexacoordination with a preferred octahedral geometry. Ferric ethylenediaminetetraacetic acid is, however, seven coordinate. Natural coordination compounds of iron are frequently characterized by a departure from a strict regularity of shape. This will be the situation in most iron-proteins. But even in hemin crystallography shows the iron displaced from the plane of the porphyrin ring so that the geometry is most accurately described as square pyramidal (Hoard, 1971). The polyhydroxamate siderochromes of the ferrichrome family contain iron which approaches octahedral symmetry but in the iron sulfur proteins fourcoordinate iron seems to be the common rule.

The ratio of Fe(III)/Fe(II) stability will be largely influenced by the nature of the atoms in the coordination sphere of the central metal ion. Oxygen, nitrogen, and sulfur--the three commonly small, electronegative atoms--will be those most frequently involved in binding Fe(II) or Fe(III) in chelates. Phenolates, enolates, hydroxamates, and all such ligands contributing exclusively oxygen will show a pronounced preference for Fe(III). Ferric complexes tend to be the more stable. An all nitrogen ligand system, as in the ferroins, will form more stable complexes and chelates with ferrous iron. These properties are exploited for colorimetric assessment of valency, a measurement which is beyond the scope of atomic absorption spectroscopy. If a mixture of ferric and ferrous iron is added to a solution of a ferroin type ligand, ferrous iron will accumulate because of its stronger attachment to the organic molecule. Examples of two very common reagents for ferrous and ferric ions, respectively, are bathophenanthroline (1) and tiron (2).

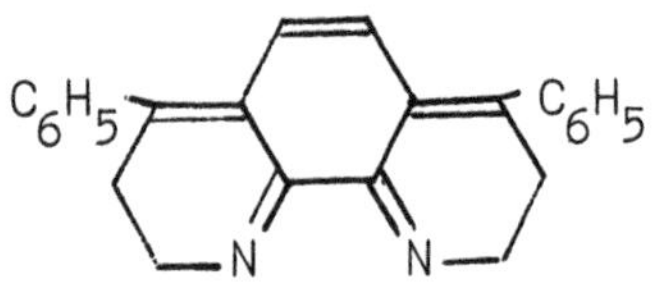

C_6H_5 C_6H_5 N N (1) HSO_3 HSO_3 OH OH (2) N OH (3)

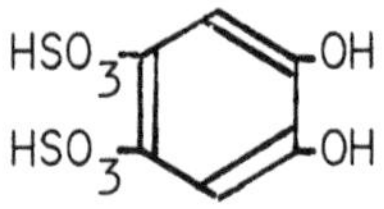

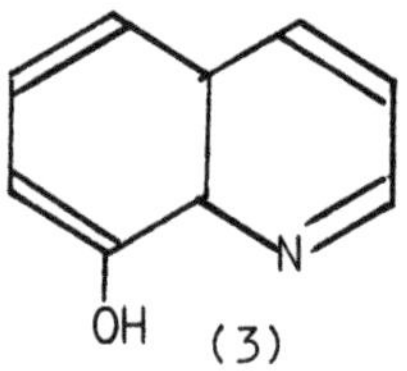

The former is a typical ferroin type reagent for iron. Terosite, a terpyridine, affords a ferrous complex with a very high absorbancy (Zak, *et al.*, 1971). The molar absorbancy for ferrous ferroins typically ranges between 20,000 and 30,000. A reagent which shows high affinity for both Fe(II) and Fe(III) is 8-hydroxyquinoline (3).

2. FUNCTION OF IRON IN BIOLOGY

A. Introduction

Prior to the work of Otto Warburg (1883-1970) and David Keilin (1887-1963), the main significance of iron in biology was in connection with its presence in hemoglobin. The idea that the latter was an iron compound became current in the first part of the last century and in 1886 Zinoffsky estimated, correctly, that horse hemoglobin contains 0.335 per cent iron (Underwood, 1971). The contribution of Warburg and Keilin centered on their demonstration that, in addition to carrying oxygen to the tissues, iron was involved in the reduction of this gas to water. The latter reaction was shown to occur in aerobic cells as part of the process of chemical conservation of energy. This immediately broadened the function of iron from the rather specialized task of O_2 transport to the much more general one of respiration. We now recognize that iron is also involved, as iron sulfur proteins, in electron transfer reactions below the redox level of the pyridine nucleotide coenzymes. The ferredoxins have opened up a new dimension in the metabolic role of iron. Furthermore, the efficient operation of nitrogenase, the enzyme system reducing N_2 to ammonia, requires the participation of four enzyme proteins, all of which contain iron. Iron is most commonly the transition element encountered from

Escherichia coli to man in ribotide reductase, a multi-component collection of enzymes reducing ribotides to deoxyribotides. The only well established role for the latter is in DNA synthesis. Finally, certain iron porphyrins not involved in electron transfer are essential for the metabolism of H_2O_2 and the detoxification of drugs. Thus, the hydroperoxidases, catalase and peroxidase, play roles which are recognized as very important in animal and plant tissues, respectively.

Some of the significant iron compounds in biology are enumerated in Table I. This is a provisional list since new iron-containing

Table I. Some biological iron compounds classified by function.

Function	Compound
Iron transport	Transferrin
	Phosvitin
	Siderochromes
	Hydroxamate type
	Catechol type
	Citrate type
Iron storage	Ferritin
	Hemosiderin
	Phosvitin
	Inorganic polyphosphates
Oxygen transport	Hemoglobin
	Erythrocruorin
	Chlorocruorin
	Hemerythrin (non-heme type)
Oxygen storage	Myoglobin
Oxygenases	Tryptophane dioxygenase (heme type)
	Non-heme iron oxygenases
Hydroperoxidases	Catalase
	Peroxidase
Electron transfer	Cytochromes
	Iron sulfur proteins
	Nitrogenase
	Iron flavoproteins
	Ribotide reductase
Dehydration and isomerization	Aconitase
Unknown	Leghemoglobin
	Pyrimine
	Ferroverdin

substances are constantly being discovered. In addition, only a few of the large number of iron flavoproteins are recorded in Table I.

B. Transport

Transport forms of iron have been positively identified in microorganisms and animals and tentatively identified in plants.

The largest single group of natural iron containing substances belong to the siderochromes, defined as microbial ferric ion transport agents. A very significant feature of these molecules is the repression of their biosynthesis by iron. The addition of only a few milligrams of iron per liter of growth medium completely eliminates the extracellular excretion of the deferri-siderochromes although in most cases small amounts are still formed intracellularly. The universal repression of siderochrome biosynthesis was one of the factors suggesting that these agents must function in iron transport (Neilands, 1957), although this could not be proven until suitable mutants became available (see below).

The biological role of the siderochromes in iron transport was firmly established through the use of mutants selected by the powerful techniques of microbial genetics. In the late 1960's, B. Ames, at the University of California at Berkeley, discovered a group of mutants ("iron mutants") of Salmonella typhimurium which failed to grow on simple media containing citrate. The iron mutants grew normally on complex media or on the simple medium (containing citrate) supplemented with massive quantities of inorganic iron salts. The first clue to the metabolic defect(s) in the iron mutants was gained from the observation that revertent-to-wild type colonies excreted a factor which fed the surrounding cells. This material was isolated from low-iron cultures of the wild type strain as a crystalline, ether-soluble substance. It was characterized as the cyclic triester of 2,3-dihydroxybenzoyl-L-serine and named enterobactin (Pollack and Neilands, 1970). An identical compound, designated enterochelin, has been obtained from E. coli (O'Brien and Gibson, 1970).

The iron mutants of S. typhimurium could be grouped into two classes, one of which excreted copious amounts of a catechol identified as 2,3-dihydroxybenzoic acid. The other class of mutants excreted no catechols but grew on citrate media supplemented with 2,3-dihydroxybenzoic acid. From these data it was concluded that enterobactin is synthesized from chorismate, the two classes of mutants representing blocks before and after 2,3-dihydroxybenzoic acid. Citrate was shown to tie up iron in a chelated form which was unavailable to S. typhimurium; under these conditions the wild type could excrete enterobactin, a much more powerful coordinating agent for iron, and in this way relieve the iron deficiency. The genes for the biosynthesis of enterobactin are closely linked at about 20 minutes on the Salmonella chromosomal map (Pollack, et al., 1970). A similar analysis has been carried out for E. coli, which has a similar high affinity iron transport system, except that the latter organism does not have an inducible uptake system for iron citrate (Rosenberg and Young, 1973.

In addition to enterobactin, very low concentrations of ferrichrome were found to support the growth of iron mutants of S. typhimurium on simple, citrate containing media. This suggested that the organism is equipped with an efficient uptake system for this siderochrome. Strains derived from the iron mutants which were selected on the basis of their resistance to albomycin were found to be defective in the uptake of various siderochromes. Albomycin is a broad spectrum antibiotic which is closely related structurally to many of the siderochromes. It contains a cyclic hexapeptide moiety, a ferric trihydroxamate metal center, and a substituted pyrimidine attached to a seryl side chain. The substituted pyrimidine is apparently the toxic element in the molecule. Albomycin apparently permeates the cell through the uptake system designed for the siderochromes. Once inside the cell, albomycin poisons some vital process which is probably in no way connected with iron metabolism. Many of the permease defective strains map at about 9 minutes on the Salmonella chromosome map. S. typhimurium has not been demonstrated to produce hydroxamate type siderochromes and it is a little mysterious that the organism should be equipped with a specific transport system for these molecules (Luckey, et al., 1972).

Ferrichrome has been shown to function in an iron transport capacity in Ustilago sphaerogena, an organism which has an outsized capacity to produce these compounds and cytochrome c (Emery, 1973). The latter is a mammalian type cytochrome c, but it has a substantially lower isoelectric point (Neilands, 1972; Bitar, et al., 1972).

In animals iron is absorbed mainly in the upper part of the small intestine by a process which remains to be disclosed but which may involve the participation of carrier molecules. Absorption is enhanced in iron deficiency or following hemorrhage or phlebotomy. The primary site of regulation of uptake is at the level of the intestinal mucosa and only about 1 mg of the estimated 15 mg of iron consumed daily is admitted into the human body. The mechanism of this regulatory process is without any doubt the most important unsolved problem in animal iron metabolism.

The circulating plasma contains a ferric ion binding protein, transferrin, which transports iron to the blood stream. Normally only about a third of the total iron binding capacity of serum is saturated with the metal. While reticulocytes will accept iron from a variety of synthetic chelates, the iron donated by transferrin is very rapidly and efficiently delivered to the receptor sites for hemoglobin synthesis. A protein remarkably similar to transferrin, conalbumin, occurs in egg white. Conalbumin is able to bind iron with sufficient avidity to make the metal unavailable to some bacteria, and it may thus serve to protect the yolk from microbial infection. A third related protein, lactoferrin, is found in milk. Its function may be to diminish the availability of iron to the newborn and, in this way, encourage the growth of a lactic flora in the gut. Lactobacilli are unusual in not requiring iron for growth.

Transport forms of iron have been less clearly identified in the plant world, although chelated iron has been used in agriculture for two decades. Plants growing in hydroponic solution at low iron concentrations are believed to excrete specific iron complexing agents. Organic acids occur commonly in plant tissues and citric acid may assist in the translocation of iron.

C. Intracellular Roles

Inspection of Table I suggests that the two main roles of physiologically active iron are oxygen and electron transport. The activation of aconitase by ferrous ion is one case in which iron acts as a metal bridge; however, it is not completely certain that iron is a biologically essential constituent of aconitase.

The oxygen transport and storage function is divided between the heme and the non-heme type of iron protein. The former is widely distributed in lower and higher forms of life while the non-heme type, as represented by hemerythrin (Klotz, 1971), is confined to a relatively few species in four different phlya, namely, the Sipunculids, Polychaete annelids, Priapulids, and the Brachiopods. Leghemoglobin occurs in the nodules of leguminous plants, where its concentration is high during periods of active N_2 fixation. Its function has not been defined, but the molecule resembles very closely the animal counterpart (Ellfolk, 1972). The primary amino acid sequences of apoleghemoglobin and globin have much in common and in both cases the heme group seems to fit into a hydrophobic pocket. The bacteroids of *Rhizobia* perform the heme synthesis while the genetic information for production of the protein is contained in the plant. Free-living *Rhizobia*, as well as other microbial species, synthesize characteristic types of hemoglobins.

Catalase is generally understood to function in the removal of toxic quantities of hydrogen peroxide, the latter generated by the reaction of reduced flavoproteins with oxygen. However, the enzyme is so potent as regards its turnover number and is generally present in such excess that it is suspected of having other functions. It is not completely essential for the human body since there exists an inherited disorder, first discovered in Japan, in which the blood and other tissues are completely devoid of catalase. However, such individuals suffer from severe ulcerations in the oral cavity. It seems likely that catalase is useful in scavenging the hydrogen peroxide generated by bacteria invading the gums through minor punctures of the skin.

In contrast to catalase, which merely decomposes H_2O_2, peroxidases catalyze the oxidation of substrates by means of peroxide. The enzyme from the horse radish root has been the most thoroughly studied.

The cytochrome chain of hemoproteins occurs in all aerobic cells, where their function is the transfer of electrons from oxidizable substrates to O_2 with simultaneous conservation of chemical energy. Thus, three moles of ATP are generated per atom of oxygen reduced to water. The cytochromes were discovered by the British physiologist MacMunn in the middle of the last century, who recognized their distinction from hemoglobin. Later Keilin (1966) rediscovered the pigments and elucidated their role in cellular oxidation. Keilin, working with a low dispersion spectroscope, identified three types of cytochromes, which he designated as a, b, and c, with α and β bands situated at different regions of the visible spectrum. He named the component with bands next to the red as a, and the component with bands shifted furthermost to the blue as c, but the order of electron transfer is b → c → a. Bacterial, plant, and animal tissues have yielded a large number of other cytochromes, in addition to the three basic types described by Keilin and, hence, the multiplicity of names such as a_1, a_3, b_5, etc. The cytochromes concerned with energy conservation are packaged in the mitochondria where they apparently occur in molar proportions.

The respiratory chain contains, in addition to the cytochromes, substantial levels of non-heme iron. The latter is known to function in electron transport.

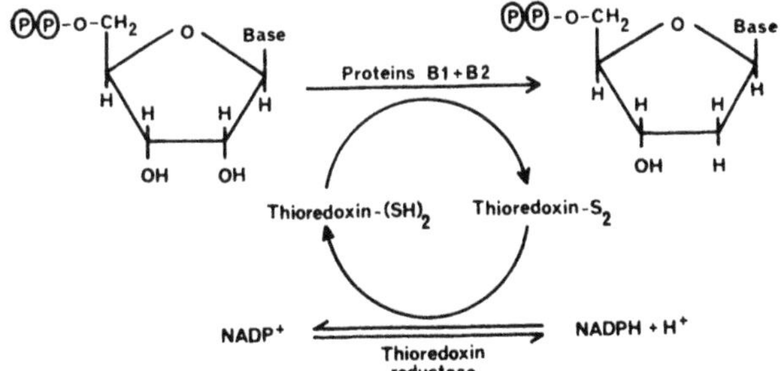

Figure 1. The ribonucleoside diphosphate reductase complex of enzymes showing the role of protein B2, which contains one iron atom per subunit (Karlström, 1972).

The role of iron in nucleotide metabolism is illustrated in Figure 1, which shows a multi-component enzyme system from _E. coli_ thoroughly documented as functioning in the reduction of ribotides to deoxyribotides (Karlström, 1972). Protein B2 is comprised of two identical, or nearly identical, subunits in a particle with molecular weight 78,000 which contains two atoms of non-heme iron per mole. Thioredoxin reductase is a flavoprotein which catalyzes the reduction of a low molecular weight protein containing the couple -S-S-$\rightleftharpoons$ -SH—HS-, named thioredoxin, with electrons derived from $NADP^+$. The fourth protein, B1, is coupled to B2 by Mg^{++} and is concerned in regulating the activity of the system.

The _E. coli_ ribotide reductase system functions in animal tissues but in lactic acid bacteria, and perhaps in a few other species, the cobalt-containing vitamin B_{12} substitutes for the iron-containing protein B2. There are at least one or two other instances known in which iron is connected to nucleic acid metabolism. The growth of _E. coli_ at low iron levels leads to a structural change in a hypermodified base (N^6-(Δ^2-isopentenyl)-2-thioadenosine) situated adjacent to the anticodon loop in tyrosine tRNA, and other tRNA's, in which the thiol group at position 2 is no longer methylated (Rosenberg and Gefter, 1969). In an animal cell culture, the HeLa line, the addition of just that amount of chelating agent which blocks ^{55}Fe uptake inhibits DNA synthesis (Robbins, _et al_., 1972). The cytoplasmic iron was found to be bound to a glucose-containing polysaccharide, nucleolar iron was attached to a single protein and nuclear iron was distributed among several proteins. This study underlines the fact that, in addition to the "mainline" storage, transfer, and hemoprotein forms of iron, many additional iron compounds remain to be characterized in living cells.

The iron sulfur proteins are distributed throughout the microbial, plant, and animal worlds where they function in nitrogen fixation, metabolism of carbon substrates, hydroxylations, and photosynthesis. The universal presence of these proteins in microorganisms--ranging from the strictly anaerobic _Clostridia_, the photosynthetic bacteria, the blue-green algae--as well as in higher plants and animals renders them ideal candidates for studies of protein evolution (Hall, _et al_., 1971).

Iron is intimately involved in nitrogen fixation, widely accepted as one of the basic chemical events crucial to the existence of the animal species, including man. The enzymology of this process has been clarified in recent years and iron is known to be functional at several sites, namely, as ferredoxin, in the iron protein, in the iron-molybdenum protein, and in glutamate synthase (Benemann and Valentine, 1972; Miller, 1973).

Iron is the most commonly encountered cofactor in the large group of enzymes involved in the insertion of molecular oxygen into organic molecules (Nozaki and Ishimura, 1973). Although copper and flavin are also cofactors for the oxygenases, of the 100 members of this class described to date about half contain bound iron or require addition of this metal for activity. The most thoroughly studied member of the group is tryptophane oxygenase, which is somewhat atypical since it is the only one which contains heme as the prosthetic group. One member of the cytochromes, P450, acts as a general hydroxylation enzyme in bacteria, fungi, plants, and animals. It contains iron protoporphyrin which, in the reduced form, binds CO to give a characteristic spectrum with an absorbancy maximum at 450 nm.

Finally, iron is found associated with a large number of flavoproteins. Some of the better known members of this class are succinic dehydrogenase, xanthine oxidase, NADH-cytochrome *c* reductase, and dihydroorotate dehydrogenase.

D. Storage

The iron storage role of ferritin, a protein found in different animal tissues, in plants, and in the fungus *Phycomyces blakesleeanus*, is well established. Its synthesis is stimulated by iron in some unknown manner. It has not been decided if the protein is built around the iron, or if the iron is inserted into the fully formed apoferritin. Ascorbic acid may be implicated in ferritin formation *in vivo* at the expense of hemosiderin (Harrison, 1971), the latter a seemingly somewhat metabolically less active storage form of iron.

Phosvitin, a protein found in the yolk of eggs, has been suggested to function as an iron carrier during embryonic development (Greengard, *et al.*, 1964). Recently, Donella, *et al.* (1972) proposed that a phosvitin-like protein of rat cytosol transports iron into the mitochondria where the iron is liberated by action of a specific phosphatase. In the cytosol the protein is phosphorylated by a specific kinase. Phosvitin and related proteins are relatively low in molecular weight and may contain up to one atom of iron per two atoms of phosphorus. Donella, *et al.* (1972) reported the binding of up to 14.2 moles of iron per 10,000 daltons. This calculates to 1 mole of iron per 700 daltons and, while it is less efficient than ferritin (where the ratio is 1/100), it does seem to suggest than an iron transport role for phosvitin is plausible.

Inorganic polyphosphates occur in microorganisms where they are suspected of acting as iron storage depots, but this has not been proven by direct experimentation.

3. STRUCTURE OF NATURAL COORDINATION COMPOUNDS OF IRON

A. Low Molecular Weight Substances

Most of the low molecular weight iron compounds discovered thus far belong to the siderochrome family (see below). Two exceptions are pyrimine and ferroverdin which, in contrast to the siderochromes, show a strong preferential binding of ferrous ion.

Pyrimine (4) is a ferroin-type ligand isolated from a strain

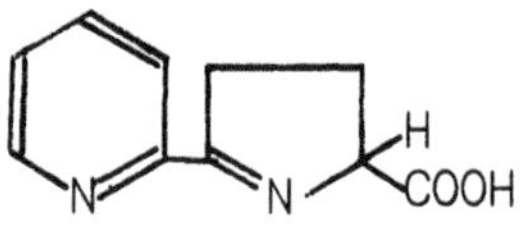

(4) 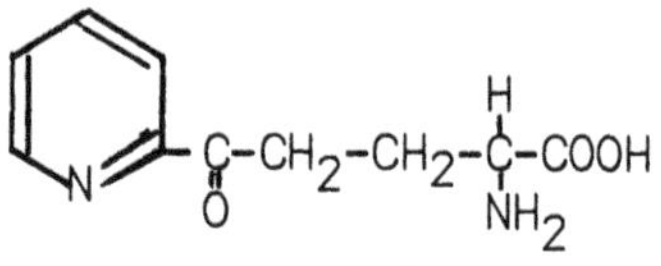 (5)

of Pseudomonas and detected as a brilliant magenta colored ferrous complex. At very low values of pH it exists in the open chain form and can be crystallized as the dihydrochloride (5). At all values of pH approaching the physiological range it ring-closes to give the conjugated diimine structure which shows a high binding affinity for Fe(II). The function of pyrimine is unknown; it would be an ideal substance in which to carry out experiments on the origin of the heterocyclic ring by use of the modern techniques of nuclear magnetic resonance and mass spectrometry.

Ferroverdin (6) is a green-colored complex from a Streptomyces

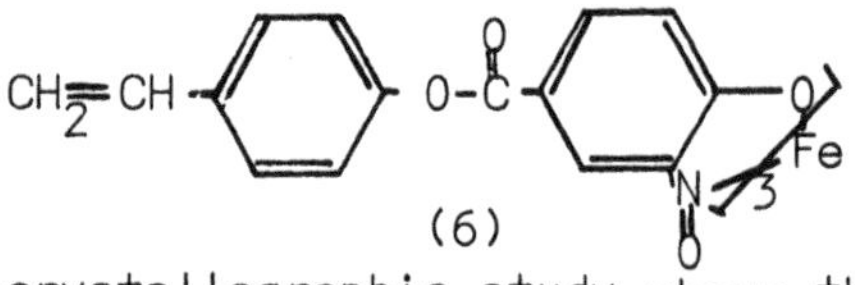

(6)

sp. A crystallographic study shows the ferrous ion coordinated to the nitroso nitrogen and the phenolate oxygen (Candeloro, et al., 1969).

The siderochromes may be roughly classified chemically as either catechols (or phenolates) or as hydroxamates. The distinction on the basis of functional groups is not a perfect one since one siderochrome, mycobactin, contains both types of ligands. Furthermore, some siderochromes contain bound citric acid which affords the opportunity for carboxylate and hydroxylate complexing to iron.

As the biosynthesis of all genuine siderochromes is repressed by iron, it is a relatively simple matter to produce them in quantity by simply withholding iron from the growth medium. A second reason for the wealth of structural information on the siderochromes is their relatively low molecular weight, generally < 1000, and the existence of a special technology for their characterization.

The catechol siderochromes give the Arnow reaction and show an intense blue fluorescence under ultraviolet light. They all contain 2,3-dihydroxybenzoic acid, which is acylated to the amino group of an α-amino acid such as glycine, lysine, or serine. One member of the series, itoic acid (7), contains glycine and is chemically defined as 2,3-dihydroxy-N-benzoylglycine. Another is 2,3-dihydroxy-N-benzoyl-L-serine (8), and a third, in chronological order of discovery, is the bis-2,3-dihydroxybenzoyl derivative of L-lysine (9). Enterobactin (enterochelin) (10) is the cyclic trimer of

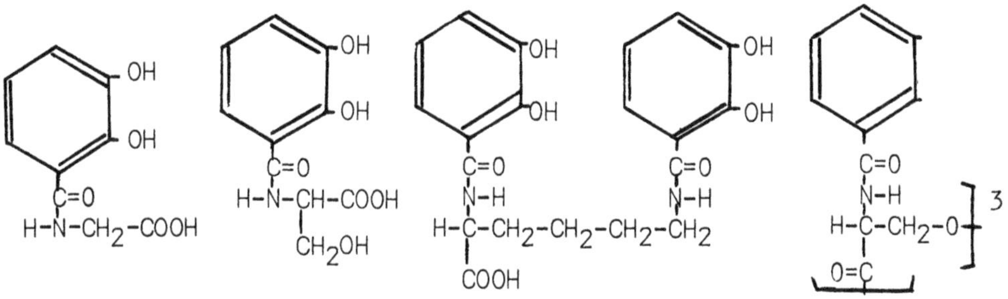

OH OH C=0 H-N-CH$_2$-COOH

(7)

OH OH C=0 H-N-CH-COOH CH$_2$OH

(8)

OH OH C=0 N-H H-C-CH$_2$-CH$_2$-CH$_2$-CH$_2$ COOH; OH OH C=0 N-H

(9)

C=0 N-H H-C-CH$_2$-0- 0=C]3

(10)

2,3-dihydroxybenzoylserine. In the bis-lysine compound and in enterobactin exactly the same number of atoms separate the catechol groups, hence, all of these compounds are probably destined for a role in iron metabolism.

Ferric ion complexes to enterobactin with the displacement of six protons and the resulting chelate has three negative charges per mole. The ferric chelate has limited life at neutral pH owing to its ready oxidation to quinones which, in turn, polymerize into black degradation products. Ferric enterobactin shows an optically active chromophore in the region of the light absorbancy maximum and inspection of molecular models suggests that the ligand atom pairs assume the sense of a left-hand propeller around the central metal ion. This conformation has not been verified by experimental measurements.

The metal-free hydroxamate siderochromes are rapidly oxidized by periodate, which is the single most useful reagent for characterization of these molecules. Since the hydroxamate is universally of the secondary variety and carries an R group on the nitrogen, the periodate oxidation product is a cis-nitrosoalkane dimer, a substance with a very sharp and intense absorbancy at about 267 nm. At the same time, the acyl moiety is liberated as a carboxylic acid. Periodate oxidizes the hydroxamate linkage to an active acylating species which is instantly cleaved in water. Many siderochromes contain peptide as well as hydroxamate bonds and periodate provides a gentle method for selective scission of the latter. Also, the acyl portion may contain double bonds, as in the case of mycobactin and ferrichrome A, and there is a possibility of isomerization accompanying more vigorous methods of hydrolysis. Hydroxamate linkages may also be oxidized by performic acid. This yields the carbon residues on each side of the hydroxamate link as carboxyl terminal chains.

The hydroxamate group is notoriously difficult to reduce but if it is not too severely hindered, it can be hydrogenated with Raney nickel at 50 pounds pressure. An alternative approach is reductive hydrolysis with hydriodic acid, which affords the hydroxyamino group as a primary amine. Raney nickel reduction has proven useful for conversion of hydroxamates to more volatile products which can be analyzed by mass spectrometry.

The most reliable method for preparation of the hydroxamate link is via reduction of the precursor nitro compound with zinc dust in ammonium chloride solution followed by acylation with a suitable reagent. The benzyl group is an effective blocking agent for the hydroxylamino oxygen in view of the general resistance of the hydroxamic acid bond to reduction (Isowa, *et al.*, 1972). The free hydroxylamino function can, of course, be reduced to the primary amine by bubbling H_2 in the presence of a catalyst.

The structures of the hydroxamate type siderochromes have been discussed in several reviews (Neilands, 1973; Hutner, 1972). One interesting feature worthy of note is the recent emergence of citric acid as the backbone holding the metal-binding groups in a series of siderochromes of bacterial origin. Three such citrate type, hydroxamate-containing siderochromes are now known, namely, aerobactin (11), schizokinen (12), and arthrobactin (13). The latter is

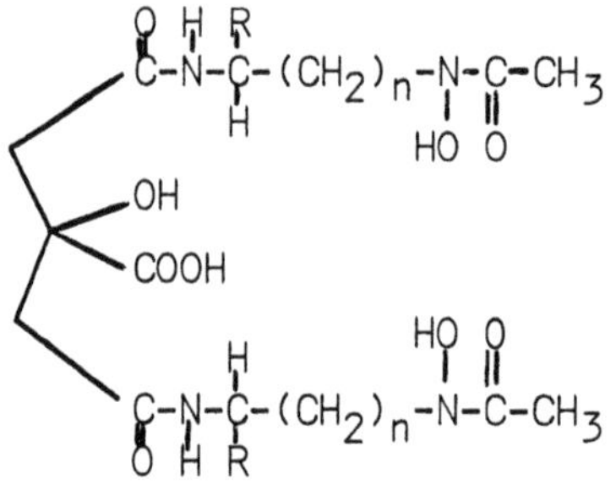

$$\begin{array}{l}
\quad\; \overset{O}{\overset{\|}{C}}\text{-}\overset{H}{N}\text{-}\underset{H}{\overset{R}{C}}\text{-}(CH_2)_n\text{-}\underset{HO}{N}\text{-}\underset{O}{\overset{\|}{C}}\text{-}CH_3 \\
\diagup \\
C\text{-}OH \\
\;\; \text{-}COOH \\
\diagdown \\
\quad\; \underset{O}{\overset{\|}{C}}\text{-}\underset{H}{N}\text{-}\underset{R}{\overset{H}{C}}\text{-}(CH_2)_n\text{-}\overset{HO}{N}\text{-}\overset{O}{\overset{\|}{C}}\text{-}CH_3
\end{array}$$

	R	n
(11)	COOH	4
(12)	H	2
(13)	H	4

the ligand of Terregens Factor, a hydroxamate type siderochrome formed by *Arthobacter* sp. (Linke, *et al.*, 1972). If one knew enough about microbial taxonomy and evolution, it might be possible to say which specimens in the American Type Culture Collection would yield new citrate siderochromes where n = 3, with R = H or COOH. The biological existence of such compounds can be predicted with confidence. The ferric complex of schizokinen is an anion (Mullis, *et al.*, 1971), but details of the conformation of the metal coordination derivatives in this ligand series have not been investigated. No member of this family of siderochromes has yet been obtained by chemical synthesis.

The hydroxamate type siderochrome most thoroughly studied from the standpoint of its biosynthesis is rhodotorulic acid. This cyclic dipeptide is commonly encountered as the siderochrome of yeasts, and strains such as Rhodotorula pilimanae can be induced to form up to 10 grams per liter of crystals. The addition of ^{14}C labeled glutamate, glutamine, proline, arginine, and ornithine led to incorporation of ^{14}C into rhodotorulic acid but afforded little information on the route of biosynthesis. Subsequently, it was found that the organism would grow on media in which the water was replaced by 99.8% D_2O. The addition of H-form precursors to deuterated cultures, followed by examination of the rhodotorulic acid by nuclear magnetic resonance spectroscopy, disclosed the pathway of biogenesis. Ornithine is oxidized to N^{δ}-hydroxyornithine (14) which is acetylated to N^{δ}-acetyl-N^{δ}-hydroxyornithine (16) which, in turn, is cyclized to rhodotorulic acid (17) (Akers and Neilands,

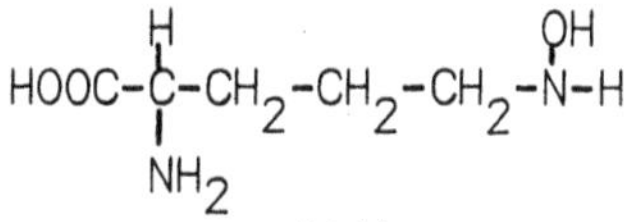

$HOOC\text{-}CH(NH_2)\text{-}CH_2\text{-}CH_2\text{-}CH_2\text{-}N(OH)\text{-}H$ (14)

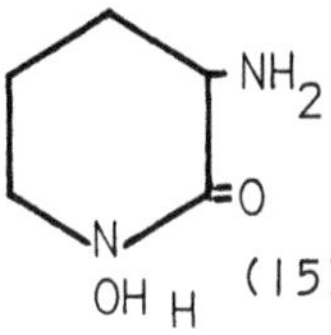

(15)

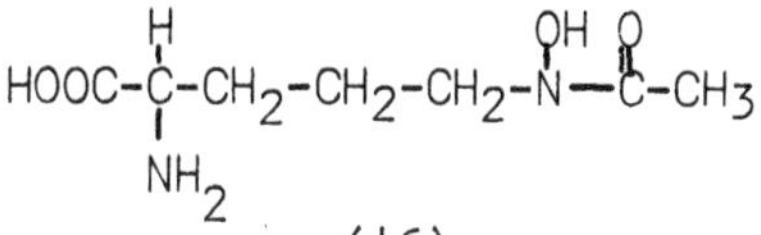

$HOOC\text{-}CH(NH_2)\text{-}CH_2\text{-}CH_2\text{-}CH_2\text{-}N(OH)\text{-}C(O)\text{-}CH_3$ (16)

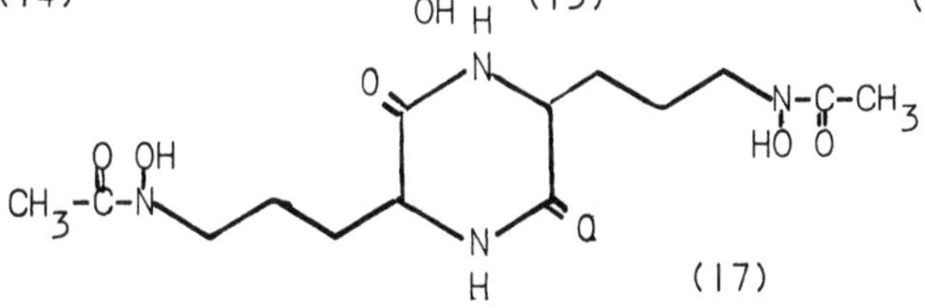

(17)

1973). Growth of the organism at pH 3 was found to interdict the biosynthesis after the first step, resulting in the formation of N^{δ}-hydroxyornithine. At these low values of pH this amino acid is cyclized to 1-hydroxy-3-aminopiperidone (15), which accumulates in the culture. The monomer (16) is dimerized to rhodotorulic acid by the enzyme rhodotorulic acid synthetase (Anke and Diekmann, 1972). The mechanism of this enzyme must be closely related to those participating in the biosynthesis of the cyclic decapeptides gramicidin and tyrocidin. This could be shown by examination of the rhodotorulic acid formed from H-type precursors in deuterated cultures--both light monomers suffer identical replacement of the α-hydrogen when they are activated (probably as thiol esters) on the enzyme. Iron was found to repress both the ornithine oxidase (or oxygenase) step and the rhodotorulic acid synthetase. These data suggest that the genes directing the formation of the enzymes involved in rhodotorulic acid biosynthesis are clustered in an operon (Akers and Neilands, 1973).

Ferrichrome (18), the prototype of the hexapeptide trihydroxamate siderochromes, is structurally one of the most completely characterized molecules in biology. It has been obtained by chemical synthesis and an analog, ferrichrome A, has been subjected to x-ray diffraction analysis. The ferrichromes have also been used to research the differences between the structures in the solid and solution states. The latter is assumed to be biologically the more relevant form.

(18)

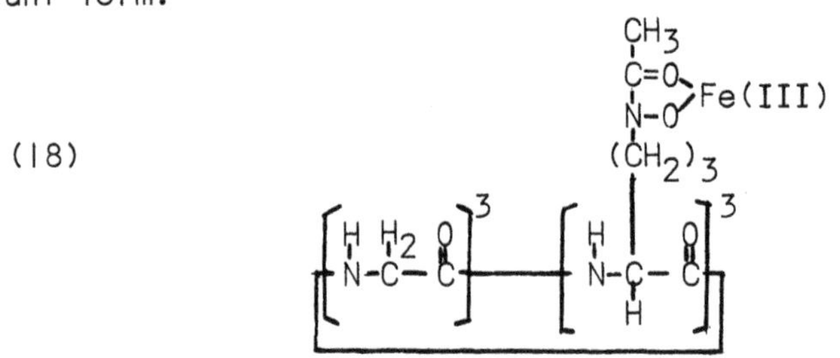

The metal-free ferrichromes exhibit no higher structure orders which impede the rapid exchange of protons at the six amide bonds. In contrast, the metal complexes may have up to four slowly exchanging protons per mole. Inspection of the crystal structure shows that three of these are likely to belong to the N^{δ}-hydroxyornithyl residues. The amide H of one of these residues is "internal" in the sense that it is covered by the side chains holding the metal ion. The amide of the second N^{δ}-hydroxyornithine is bonded to its own side chain and the third residue of N^{δ}-hydroxyornithine has its exchangeable H atom transannularly attached in an H bond to a carbonyl group. The fourth slowly exchanging proton was shown through the use of high resolution nuclear magnetic resonance spectroscopy to belong to the peptide bond in which the NH is contributed by the amino acid residue situated in site 3. In the crystal the distance between the N bearing this proton and the transannular carboxyl oxygen is judged to be approximately 4Å and, hence, no contact is likely over this considerable distance. This subtle, but definite structural difference revealed by nuclear magnetic resonance techniques will now be considered in more detail.

Sharp spectra for the protons in the ferrichromes are difficult to obtain owing to line broadening by the paramagnetic ferric ion. Llinas, _et al_. (1972) solved this problem by removing the iron and replacing it with diamagnetic ions such as aluminum and gallium. Although the ionic radii of Ga^{3+} (r_o = 0.62Å) and Al^{3+} (r_o = 0.53Å) are substantially different, the nuclear magnetic resonance spectra of their siderochrome chelates are, nonetheless, virtually superimposable. Since the ionic radius of iron (r_o = 0.64Å) is very close to that of Ga^{3+}, the two diamagnetic metal ion complexes can be assumed to be satisfactory substitutes for the iron so far as their proton magnetic resonance spectra are concerned.

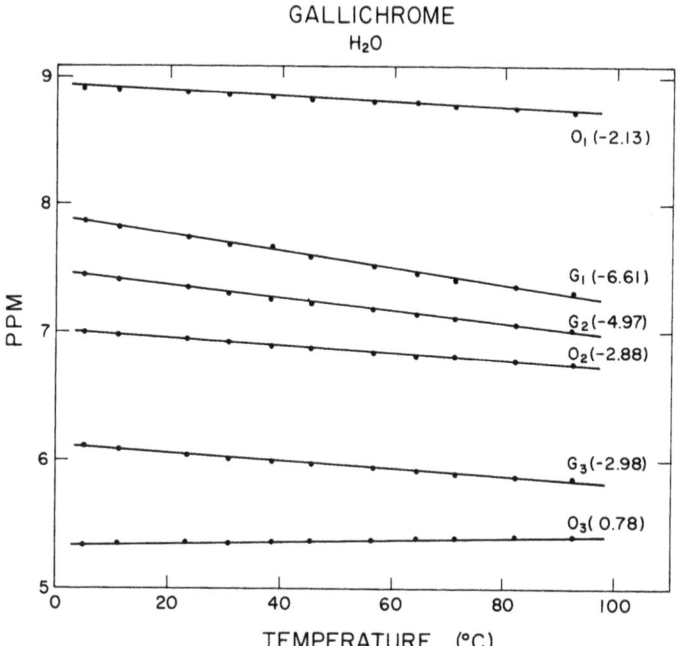

Figure 2. The temperature dependence of the chemical shifts of the amide NH protons of gallichrome at pH 5.14. O and G denote ornithyl and glycyl amide protons, respectively, and the subindex refers to the order of resonance from low to high fields. The line slopes are given in parentheses. Chemical shifts were measured with respect to internal *tertiary* butanol. (Llinas, *et al*., 1972, with permission.)

Figure 2 shows the temperature dependence of the chemical shifts of the size amide protons of gallichrome in water. It will be noted that two residues, designated G_1 and G_2, display slopes which are much larger than the other four. The latter can be shown by exchange in D_2O to be the four which interact more slowly with the solvent. Shifts to higher fields at increasing temperature can be explained as an increased rate of breaking of H bonds, either to the solvent or internally. The latter type should result in a lower slope. The resonances shown in Figure 2 can readily be assigned to glycines or N^{δ}-hydroxyornithines as the former give triplets and the latter doublets, but further assignments within each of the two groups requires reference to a series of ferrichrome analogs of known sequence (Table II).

The amide resonances measured for the analogs listed in Table II are shown in Figure 3. Considering first the non-ornithyl residues, it can be seen that the glycyl triplet centered at 1674.5 Hz moves, in alumicrocin, to a doublet at 1611 Hz. The single serine in alumicrocin is known to be situated at site 2 (Llinas and Neilands, 1973). A further substitution of a glycine in alumichrome

Table II. Amino acid sequences of ferrichrome analogs.

Analog	Amino acid sequence*						Acyl group
	site			site			
	3	2	1	3	2	1	
ferrichrome	gly	gly	gly	orn	orn	orn	acetate
ferricrocin	gly	ser	gly	orn	orn	orn	acetate
ferrichrysin	ser	ser	gly	orn	orn	orn	acetate
ferrichrome A	ser	ser	gly	orn	orn	orn	trans-β-methylglutaconate

*gly = glycyl; ser = seryl; orn = acyl-N^{δ}-hydroxyornithyl. The sites are designated by the convention of Zalkin, et al. (1966).

and alumicrocin by a serine at site 3 removes the triplets at 1263.5 and 1259.3 Hz, respectively, giving the doublet at 1356 Hz in alumichrysin. The single glycine present in all of the peptides and located at site 1 is shifted to just slightly lower fields in the ferrichrome analogues. This single glycine is required to form the cyclohexapeptide ring, otherwise the conformation of the ferrichrome peptide moiety would be radically different.

The ornithine at site 2 is sandwiched between the other two ornithines and its amide is bonded to its own side chain. It only moves significantly in ferrichrome A, where the acyl substituent, trans-β-methylglutaconic acid, presents a double bond conjugated with the hydroxamate link. The amide of the ornithyl residue at site 3 lies next to the peptide bond formed by the nitrogen of the constant glycyl residue and the carboxyl of the residue at site 2. It hence senses substitutions at the latter site, moving from 1471.5 Hz in alumichrome to 1493 Hz in alumicrocin. The remaining ornithyl residue, at site 1, provides the internal proton. It resonates at 1181 Hz in alumichrome and moves to slightly higher fields in the other analogs as the shielding is increased by introducing further seryl residues.

This analysis leads to the structure shown in Figure 4 for the ferrichrome peptides, i.e., there is some kind of contact, perhaps a weak hydrogen bond, between the proton of the residue at site 3 and the carbonyl oxygen of the site 3 ornithyl residue (Llinas, et al., 1972).

The structures of the common naturally occurring iron porphyrins are shown in Table III. Here it should be noted that heme d, which contains a dihydropyrrole ring, belongs to the chlorins rather

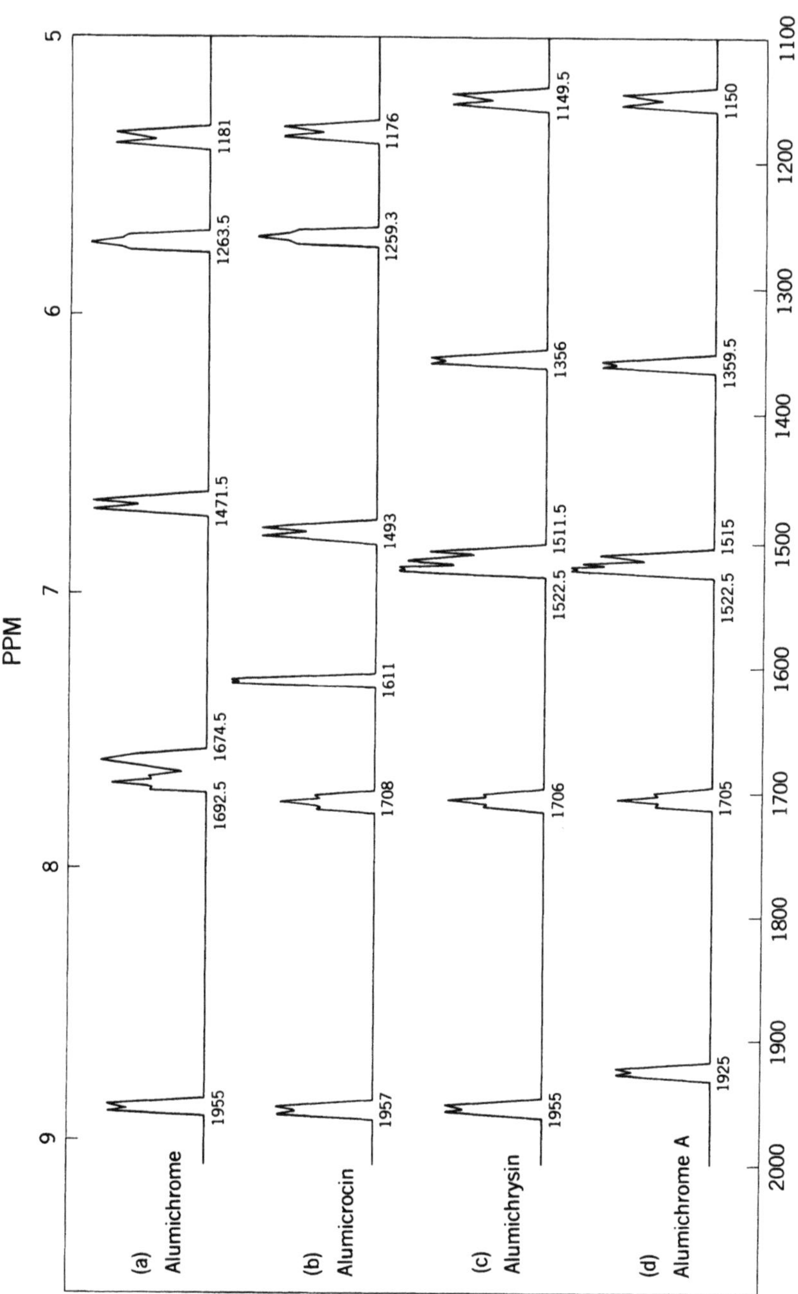

Figure 3. The amide NH resonances of the ferrichrome analogs listed in Table II. The spectra were obtained at 56.5° in deuterated dimethylsulfoxide and are referred to internal tertiary butanol. (Llinas, et al., 1972, with permission.)

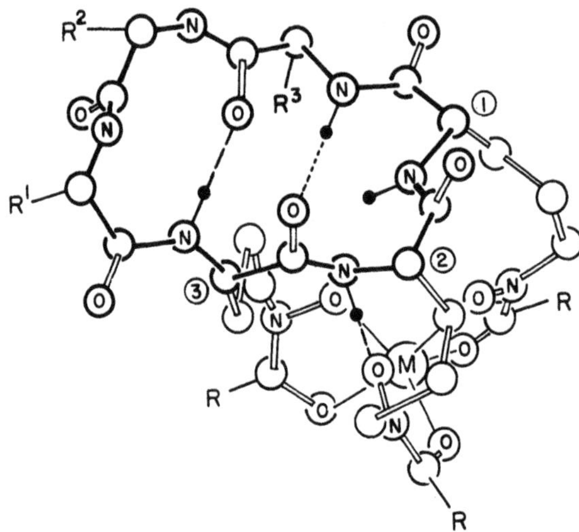

Figure 4. The solution conformation of the ferrichromes, as deduced by crystallography (Zalkin, et al., 1966) and high resolution proton magnetic resonance spectroscopy (Llinas, et al., 1972, with permission).

than to the porphyrins. Recently, Hoard (1971) has reviewed the stereochemistry of hemes and metalloporphyrins, as derived by crystallography. Probably the most significant feature of this work is the realization that the metal atom may be displaced from the plane of the porphyrin ring by as much as 0.8Å. An in-plane/out-of-plane variable location of the metal ion gives a square pyramidal/square planar choice of geometry that seems to be intimately correlated with the spin type. The hole in the porphinato core (19) is apparently too small to accomodate the large radius of the high spin Fe(II). The fluctuation between high spin and low spin forms depends on the nature of the axial ligands and provides a partial explanation for the cooperativity among subunits in the oxygenation of hemoglobin.

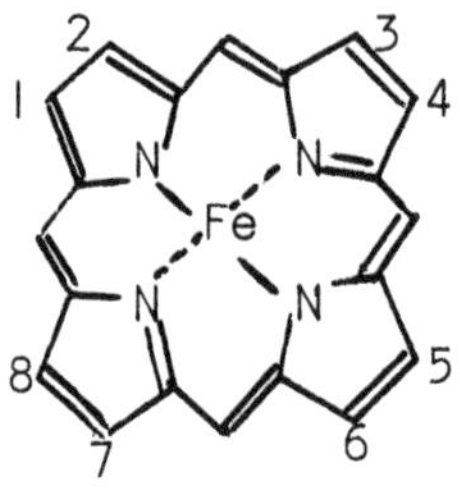

(19)

Table III. Structure of the prosthetic group of some common heme compounds.

Derivative	Distribution	Substituent in Ligand (19)*							
		1	2	3	4	5	6	7	8
Iron protoporphyrin IX	Hemoglobin Myoglobin Hydroperoxidases Cytochromes _b_ Tryptophane oxidase	M	V	M	V	M	CE	CE	M
Heme a	Cytochromes _a_	M	-C(OH)H-[CH_2-CH=C(M)-CH_2]$_3$-H	M	V	M	CE	CE	F
Heme c	Cytochromes _c_	M	-CH(CH_3)-S-R	M	-CH(CH_3)-S-R	M	CE	CE	M
Heme d	Cytochromes _d_	M	-CH(OH)-R_1	M	-CH=C(R_2)(R_3)	M	CE	H, CE	H, M

*CE = carboxyethyl; F = formyl; H = hydrogen; M = methyl; V = vinyl.

See the monograph by Marks (1969) for a more extensive discussion of the chemistry of heme compounds. Caughey (1971) has reviewed the structure-function relationship in cytochrome oxidase. The general chemistry and biochemistry of the cytochromes have been treated _in extenso_ by Lemberg and Barrett (1972).

B. Iron Enzymes and Proteins

In hemoglobin the four-chain tetramer of the type $\alpha_2\beta_2$ is mainly coiled in α-helix form around the heme groups. The exterior of the protein is hydrophylic and the interior hydrophobic--a condition necessary to prevent oxidation of the Fe(II) by O_2. The iron is five-coordinate and bonded to a histidyl residue in deoxyhemoglobin. After attachment of the O_2 the parallel spin electrons in the d orbitals pair, the paramagnetism disappears, and the iron is pulled back into the plane of the porphinato ring. On picking up the first O_2 the change in radius of the ferrous iron atom, some 13%, triggers alterations in the tertiary structure of hemoglobin which affects the binding of successive molecules of O_2 (Perutz, 1970).

Dickerson and his colleagues have determined the crystal structure of both ferri and ferrocytochrome c at resolution below 3Å (Dickerson, 1972). This elegant work reveals the heme sited in a hydrophobic pocket in both oxidation states of the molecule. Some conformational differences have been noted between the two forms. Thus, on reduction a phenylalanine residue goes into the heme pocket and an isoleucine residue moves out into the aqueous surroundings.

The x-ray diffraction work on the hemoproteins, coupled with the experiments of Kassner (1972) with iron mesoporphyrin IX, appear to explain the range of redox potentials observed in this series of electron transfer catalysts (Table IV). It seems that the more

Table IV. Influence of heme milieu on redox potentials*.

Hemoprotein	Redox potential (neutral pH)	Comment
Cytochrome c	+260 mV	native protein
Heme c peptides	-50 mV	small peptides
Mesoheme IX bipyridine	+100 mV	aqueous solution
Mesoheme IX bipyridine dimethyl ester	+400 mV	benzene solution
Cytochrome b	+35 mV	in mitochrondria
Cytochrome b	-340 mV	in aqueous solution
Cytochrome c_2	+320 mV	very hydrophobic pocket
Cytochrome b_5	+20 mV	heme nucleus exposed

*See the papers by Kassner (1972), Dickerson (1972), and Dickerson, et al. (1972) for a more complete discussion of the influence of the dielectric constant of the heme environment on the redox potential.

hydrophobic the environment, the higher the redox potential. As seen in Table IV, there is an almost perfect correlation between the dielectric constant and the value of the standard redox potential in a series of hemes and hemoproteins. Smaller perturbations are achieved by insertion of electron-withdrawing substituents in the periphery of the porphyrin macrocycle.

Serum transferrin has a molecular weight just under 80,000 and binds two atoms of ferric iron (Harrison, 1971). The binding constants for iron are in the range of 10^{30}; other transition metal ions are also attached to the apoprotein but their stabilities are lower. Three tyrosines and two histidines are thought to be linked to the iron, although this has not been demonstrated conclusively; a bicarbonate ion is also involved, but metals can be bound in its absence. A conformational change in the protein accompanies metal binding.

Apoferritin has a molecular weight of almost 0.5 million and binds variable amounts of iron as a crystalline hydrated ferric oxide. The protein consists of 20 to 24 subunits arranged in a hollow sphere or shell which contains the mineral core. The latter is largely independent of the protein coating. Hemosiderin, a secondary stored form of iron, has a less clearly defined chemical constitution.

Considerable advances have been made in the structure of the iron sulfur proteins. The simplest members of this series do not contain inorganic (or labile) sulfide and are represented by rubredoxin (20), where the iron is bound by four cysteinyl residues

(20)

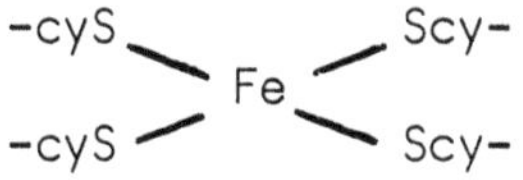

```
-cyS        Scy-
    \      /
      Fe
    /      \
-cyS        Scy-
```

protruding from two adjacent peptide chains (Herriott, et al., 1970). The ferredoxins, which contain H_2S and iron in varying ratios, are all low molecular proteins with a characteristically negative redox potential. The "plant" type, which is also found in microorganisms, has two irons and two labile sulfurs per mole (21) (Tsibris and

(21)

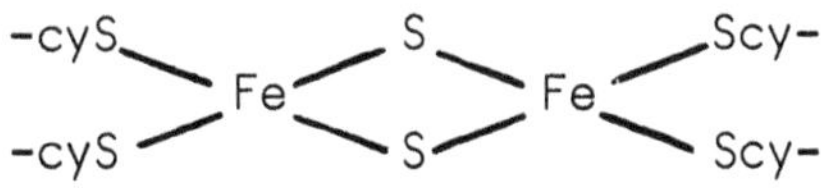

```
-cyS         S         Scy-
    \      /   \      /
      Fe         Fe
    /      \   /      \
-cyS         S         Scy-
```

Woody, 1970). Jensen and his coworkers have prepared an x-ray structure of an 8-iron, 8-sulfide microbial ferredoxin, which apparently consists of two approximately cubic arrangements of the type shown per mole of protein (22) (Sieker, et al., 1972). Herskovitz, et al. (1972) prepared and measured the x-ray and physical properties of an iron-sulfur model compound of the type $[Fe_4S_4(SR)_4]^{2-}$. This

(22)

proved to be an inorganic "cubane analog" resembling closely the active site of *Chromatium* High Potential Iron Protein (HiPIP), and presumably the active site of iron sulfur proteins containing 8 irons per mole. The electrons of the complex are apparently delocalized and not relegated to discrete sites such as 2Fe(II), 2Fe(III), and $4S^{2-}$. HiPIP has been regarded as having an anomalously high redox potential for an iron-sulfur protein. The fact that the ferredoxins and HiPIP are separated by a potential difference of three quarters of a volt and yet contain the same Fe_4S_4 cluster as a prosthetic group is a paradox which has been considered by Carter, *et al.* (1972). These workers conclude that *three* oxidation states occur in the iron-sulfur proteins and that HiPIP and ferredoxin, respectively, oxidize and reduce a common species, designated "C."

4. ENVIRONMENTAL ASPECTS

Although there are important agricultural, nutritional, and medical problems associated with iron, this heavy element does not enjoy the same type of negative notoriety as lead and mercury. None of these inorganic poisons may, in the long run, be as significant as plutonium, the mass production of which in power reactors--especially in the breeder--has been referred to by my colleague John Gofman as "man's most immoral act" (Gofman and Tamplin, 1971).

Iron overload occurs in idiopathic hemochromatosis and is a consequence of the particular supportive therapy in Cooley's anemia. The only effective therapy for the latter disease is repeated transfusion and, since there is no biological means for excretion of iron, the patient eventually succumbs from iron poisoning. Acute iron poisoning is a fairly common accident in young children; some success has been achieved by prompt treatment with Desferal, a trihydroxamate polymer containing two residues of l-amino-5-hydroxyamino pentane, two residues of succinate, and one residue of acetate. The free amino group of Desferal confers high water solubility on the molecule so that the iron complex is rapidly excreted through the kidney.

Iron deficiency anemia is a very prevalent disease; treatment with ferrous iron preparations is quite effective. The recent proposal to double the iron enrichment of bread may have negative effects on the male population and in those individuals tending toward hemochromatosis.

Impaired ability to form cytochrome P450 could have a bearing on the capacity of the individual to tolerate drugs.

The oxidation of ferrous iron by bacteria leads to the formation of acid, which may have a corrosive effect on the immediate environment.

Plants frequently require chelated forms of iron even though high concentrations of the element may be present in the soil. Large amounts of EDTA are used for this purpose in furnishing iron for fruit trees and other produce. It is perhaps well to bear in mind that this practice may, in turn, introduce the problem of toxic residues.

REFERENCES

Akers, H. A. and Neilands, J. B. (1973), Biochemistry, in press.

Anke, T. and Diekmann, H. (1972), FEBS Letters, 27, 259.

Arrieta, L. and Grez, R. (1971), Appl. Microbiol., 22, 487.

Benemann, J. R. and Valentine, R. (1972), Adv. Microbial Physiol., 8, 59.

Bitar, K. G., Vinogradov, S. N., Nolan, C., Weiss, L. J., and Margoliash, E. (1972), Biochem. J., 129, 561.

Bowen, H. J. M. (1966), Trace Elements in Biochemistry, London: Academic Press.

Burbridge, E. M., Burbridge, G. R., Fowler, W. A., and Hoyle, F. (1957), Rev. Mod. Physics, 29, 547.

Candeloro, S., Grdenic, D., Taylor, N., Thompson, B., Viswamitra, M., and Crowfoot-Hodgkin, D. (1969), Nature, 224, 589.

Carter, C. W., Kraut, J., Freer, S. T., Alden, R. A., Sieker, L. C., Adman, E., and Jensen, L. H. (1972), Proc. Nat. Acad. Sci. U. S., 69, 3526.

Caughey, W. S. (1971), Bioinorganic Chemistry, R. Dessy, J. Dillard, and L. Taylor, eds., Washington, D. C.: American Chemical Society.

Curran, H. R., Brunstetter, B. C., and Meyers, A. T. (1943), J. Bacteriol., 45, 485.

Dawson, J. W., Gray, H. B., Rossman, G. R., Schredder, J. M., and Wang, R. H. (1972), Biochemistry, 11, 461.

Dessy, R., Dillard, J., and Taylor, L., eds. (1971), Bioinorganic Chemistry, Washington, D. C.: American Chemical Society.

Dickerson, R. E. (1972), Scient. Amer., April.

Dickerson, R. E., Takano, T., and Kallai, O. B. (1972), Fifth Jerusalem Symposium, B. Pullman and E. D. Bergmann, eds., in press.

Donella, L., Pinna, A., and Moret, V. (1972), FEBS Letters, 26, 249.

Eichhorn, G., ed. (1973), Inorganic Biochemistry, Amsterdam: Elsevier.

Ellfolk, N. (1972), Endeavour, 31, 139.

Emery, T. F. (1973), Microbial Iron Metabolism, J. B. Neilands, ed., New York: Academic Press.

Falk, J. E. (1964), Porphyrins and Metalloporphyrins, Amsterdam: Elsevier.

Gofman, J. W. and Tamplin, A. R. (1971), Poisoned Power, Emmaus, Pa.: Rodale Press.

Greengard, O., Sentenac, A., and Mendelsohn, N. (1964), Biochem. Biophys. Acta, 90, 406.

Hall, D. O., Cammack, R., and Rao, K. K. (1971), Nature, 233, 136.

Hallberg, L., Harwerth, H. G., and Vannotti, A., eds. (1970), Iron Deficiency, New York: Academic Press.

Harrison, P. M. (1971), Clin. Toxicol., 4, 529.

Herriott, J. R., Sieker, L. C., Jensen, L. H., and Lovenberg, W. (1970), J. Mol. Biol., 50, 391.

Herskovitz, T., Averill, B. A., Holm, R. H., Ibers, J. A., Phillips, W. D., and Weiher, J. F. (1972), Proc. Nat. Acad. Sci. U. S., 69, 2437.

Hoard, J. L. (1971), Science, 174, 1295.

Hutner, S. H. (1972), Annual Reviews of Microbiology, vol. 26,

C. E. Clifton, ed., Palo Alto, California: Annual Reviews, Inc., pp. 313-346.

Iskandar, I. K. and Syers, J. K. (1972), J. Soil Sci., 23, 255.

Isowa, Y., Takashima, T., Ohmori, M., Kurita, H., Sato, M., and Mori, K. (1972), Bull. Chem. Soc. Japan, 45, 1461.

Karlström, O. (1972), Doctoral Dissertation, Royal Technical High School, Stockholm, Sweden.

Kassner, R. J. (1972), Proc. Nat. Acad. Sci. U. S., 69, 2263.

Keilin, D. (1966), The History of Cell Respiration and Cytochromes, New York: Cambridge University Press.

Klotz, I. M. (1971), Subunits in Biological Systems, S. N. Timasheff and G. D. Fasman, eds., New York: Marcel Dekker.

Lemberg, R. and Barrett, J. (1972), The Cytochromes, New York: Academic Press.

Linke, W. D., Crueger, A., and Diekmann, H. (1972), Arch. Mikrobiol., 85, 44.

Llinas, M., Klein, M. P., and Neilands, J. B. (1972), J. Mol. Biol., 68, 265.

Llinas, M. and Neilands, J. B. (1973), Bioinorganic Chem., 2, 159-165.

Luckey, M., Pollack, J. R., Wayne, R., Ames, B. N., and Neilands, J. B. (1972), J. Bacteriol., 111, 731.

Marks, G. S. (1969), Heme and Chlorophyll, London: Van Nostrand.

Miller, R. E. (1973), Microbial Iron Metabolism, J. B. Neilands, ed., New York: Academic Press.

Mullis, K. B., Pollack, J. R., and Neilands, J. B. (1971), Biochemistry, 10, 4894.

Neilands, J. B. (1957), Bact. Rev., 21, 101.

Neilands, J. B. (1972), Structure and Bonding, 11, 145.

Neilands, J. B., ed. (1973a), Microbial Iron Metabolism, New York: Academic Press.

Neilands, J. B. (1973b), Inorganic Biochemistry, G. Eichhorn, ed., Amsterdam: Elsevier, pp. 167-202.

Nozaki, M. and Ishimura, Y. (1973), Microbial Iron Metabolism, J. B. Neilands, ed., New York: Academic Press.

O'Brien, I. G. and Gibson, F. (1970), Biochem. Biophys. Acta, 215, 393.

Peisach, J., Blumberg, W. E., Lode, E. T., and Coon, M. J. (1971), J. Biol. Chem., 246, 5877.

Perutz, M. F. (1970), Nature, 228, 726.

Pollack, J. R. and Neilands, J. B. (1970), Biochem. Biophys. Res. Commun., 38, 989.

Pollack, J. R., Ames, B. N., and Neilands, J. B. (1970), J. Bacteriol., 104, 635.

Robbins, E., Fant, J., and Norton, W. (1972), Proc. Nat. Acad. Sci. U. S., 69, 3708.

Rosenberg, A. H. and Gefter, M. L. (1969), J. Mol. Biol., 46, 581.

Rosenberg, H. and Young, I. G. (1973), Microbial Iron Metabolism, J. B. Neilands, ed., New York: Academic Press.

Sieker, L. C., Adman, E., and Jensen, L. H. (1972), Nature, 235, 40.

Tsibris, J. C. M. and Woody, R. W. (1970), Coordination Chem. Rev., 5, 417.

Underwood, E. J. (1971), Trace Elements in Human and Animal Nutrition, New York: Academic Press.

Zak, B., Baginski. E. S., Epstein, E., and Weiner, L. M. (1971), Clin. Toxicol., 4, 621.

Zalkin, A., Forrester, J. D., and Templeton, D. H. (1966), J. Am. Chem. Soc., 88, 1810.

SOME EFFECTS OF METAL IONS ON THE STRUCTURE AND FUNCTION OF NUCLEIC ACIDS

G. L. EICHHORN, N. A. BERGER, J. J. BUTZOW, P. CLARK, J. HEIM, J. PITHA, C. RICHARDSON, J. M. RIFKIND, Y. SHIN, and E. TARIEN

Laboratory of Molecular Aging, National Institutes of Health, Gerontology Research Center, Baltimore City Hospitals, Baltimore, Maryland 21224

The biological process in which nucleic acids are involved generally require the participation of metal ions. The same metal ions which are essential to biological function at one concentration may be harmful at other concentrations. For this reason I shall not restrict the metals to be discussed to those commonly considered toxic, even though the interest of this symposium is focused upon toxicity.

The effects of metal ions on the structure and function of nucleic acids are so numerous that it is not possible to approach a comprehensive coverage of this topic here. Reference may be made to literature reviews on the subject (1-4). For the purpose of this symposium I shall confine myself to some selected interactions that have been studied in our laboratory that illustrate the variety of metal binding sites on nucleic acids as well as the dramatic effects that metal binding exerts on the structure and function of these substances.

To understand the nature and implications of the interactions of metal ions with nucleic acids it is necessary to understand the structure and function of the biological macromolecules themselves. I shall therefore provide a brief introduction for those who are not familiar with the chemistry of nucleic acids.

STRUCTURE AND FUNCTION OF NUCLEIC ACIDS

All nucleic acid molecules consist of a ribose phosphate

backbone (Fig. 1), to which are attached heterocyclic bases such as uracil, guanine, cytosine, and adenine. There are two types of nucleic acids, ribonucleic acid, RNA, which contains the 2'hydroxyl on the ribose group, and deoxyribonucleic acid, DNA, which lacks the 2'hydroxyl group. The monomers of the nucleic acid molecules are called nucleoside (base + ribose) or nucleotide (base + ribose + phosphate).

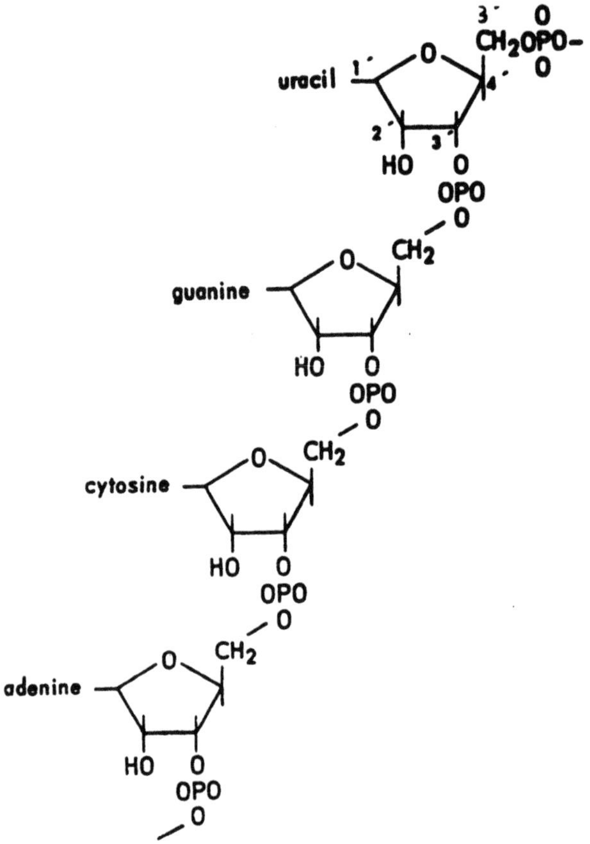

Figure 1. Schematic structure of nucleic acid backbone.

It is evident from an inspection of these molecules that they contain several types of electron donors to which metal ions can attach themselves: phosphate groups, ribose hydroxyl groups (in the case of RNA) and the bases. Metals are devious and will bind to just about every available site, under different conditions. The consequence of binding to many different sites is many different reactions. Figure 2 reveals the structures of adenine, cytosine, guanosine, and uracil bases, including their points of attachment to the ribose to make them nucleosides. The nucleosides are called adenosine (A), cytidine (C), guanosine (G), and uridine (U). Two

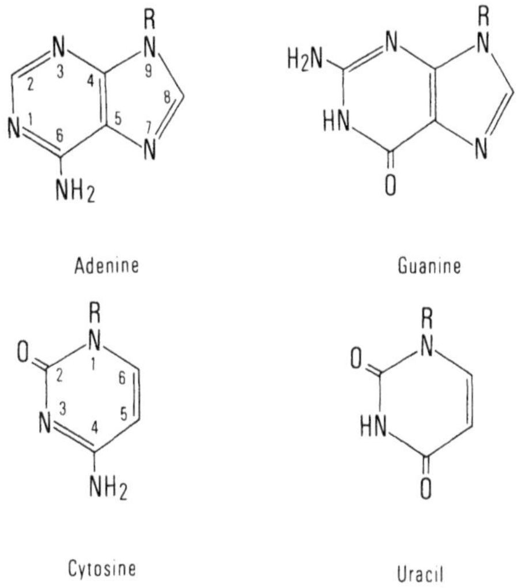

Adenine Guanine

Cytosine Uracil

Figure 2. Structures of some heterocyclic bases found in nucleic acids.

other nucleosides with which we shall be concerned are the guanosine lacking an amino group, which is called inosine, and the 5-methyluracil riboside, which is called thymine.

DNA is, of course, the primary bearer of the genetic information in the nucleus in the form of the sequence of bases on a chain. It is a double helical molecule whose two chains are linked through hydrogen bonding of the bases. In the replication of DNA the double helix unwinds into two chains, each of which forms a second chain around it so that two double helices are obtained from one. Figure 3B is a diagram of the DNA molecule. The ribbons represent the ribose phosphate backbone, and the lines perpendicular to the helical axis represent the bases, which as can be seen, are hydrogen bonded. A characteristic of the DNA molecule is that if there is an adenine on one strand there must be a thymine on the other (Fig. 3A), and if there is a guanine on one strand there must be a cytosine on the other. Therefore, the bases are complementary bases, the two strands contain complementary codes, and both chains of the DNA molecule contain the genetic information.

Figure 4 contains a highly schematic representation of some biological functions of nucleic acids. DNA serves as a template for the formation of messenger RNA, which carries the genetic message from the nucleus into the cytoplasm. This process is often called transcription. In the cytoplasm the messenger RNA becomes attached to particles known as ribosomes, on which message units on the mRNA are recognized by smaller RNA molecules called transfer RNA,

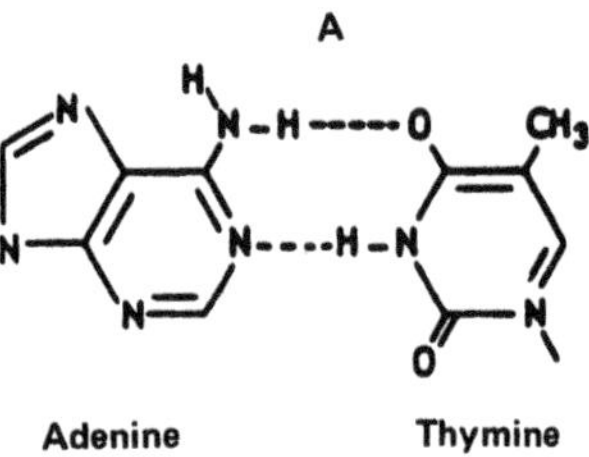

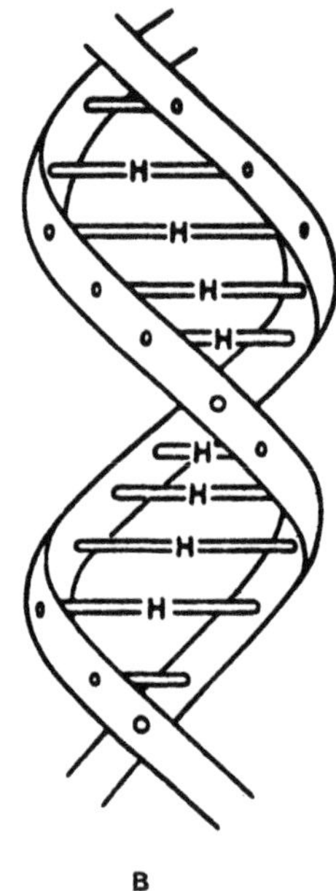

Figure 3. A) Hydrogen bonding between adenine and thymine.
B) Diagram of DNA.

which contain amino acid molecules attached on a different site from the one that recognizes the message. This process is often called translation. Finally, the amino acids are zipped up through peptide linkages to form proteins. All of these processes require the presence of metal ions and many of them involve the unwinding of the DNA molecule.

UNWINDING AND REWINDING OF DNA BY METAL IONS

One need only inspect the diagram of DNA (Fig. 3B) to anticipate that metal ions binding to the surface of this molecule, i.e., to the phosphate groups, will exert quite a different effect than metals binding to the bases, because binding to the bases means competition with hydrogen bonding. In fact, metals binding to phosphate

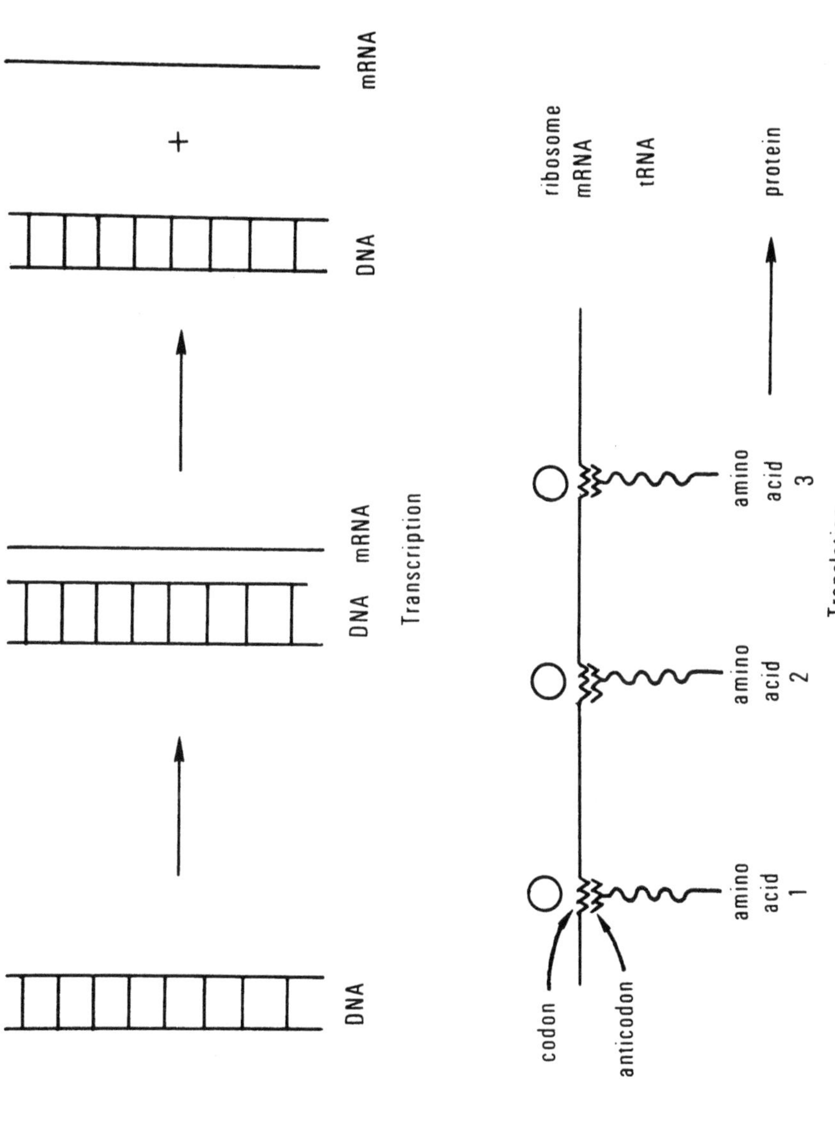

Figure 4. Scheme for some functions of nucleic acids. Transcription and translation.

stabilize the double helix, while metals binding to the bases disrupt the double helix, as is dramatically shown in Figure 5 (5). Curve A is the melting curve of DNA in the absence of any divalent metal.* In Curve B magnesium has been added to the DNA solution. The effect of the Mg(II) is to increase the temperature at which the double helix is converted to single strands, indicating that magnesium ions stabilize the double helix. They do so by binding to the phosphates on the surface of the molecule. Curve C represents what happens when copper ions are added to the DNA. The copper ions destabilize the double helix as is shown by the fact that the double helix melts out at a much lower temperature. The copper ions apparently bind to the bases and disrupt the double helix by disrupting the hydrogen bonds.

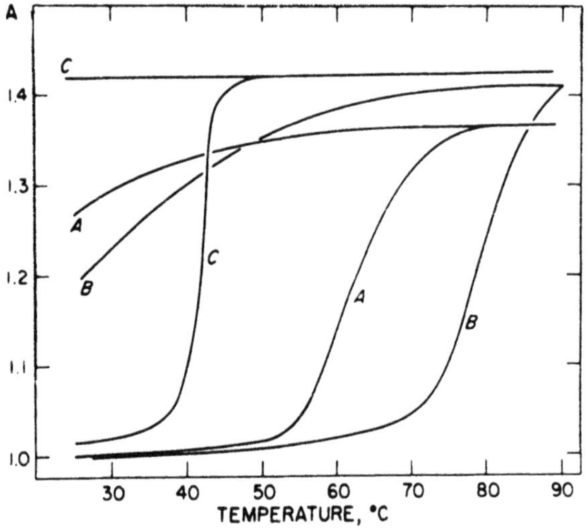

Figure 5. Heating and cooling curves of 5×10^{-5}M(P) DNA in 5×10^{-5}M $NaNO_3$ and (A) no divalent metal (B) 10^{-4}M Mg(II) (C) 10^{-4}M Cu(II) (from Ref. 5).

*The "melting curve" of DNA is a reflection of the transition from double helix to randomly coiled single strands. The absorbence of the double helix is much lower than that of the single stranded coils, because of the π-interaction of the stacked bases. The ultraviolet absorption can therefore be used to monitor the transition between these conformations. When the single stranded DNA is cooled back down, the absorbence decreases again somewhat, but this does not indicate reformation of double stranded DNA, but rather the formation of "hairpin loops" containing stacked bases within the single strands, and these hairpin loops produce the decreased absorbence in the same way as double stranded DNA. The reformation of double stranded DNA is indicated only by a decrease in absorbence all the way down to the level before the heating operation.

The ability of Cu(II) ions to unwind the DNA double helix can be curtailed at high ionic strength, as revealed in Figure 6 (6). Curve A in Figure 6 is the same as Curve C in Figure 5, i.e., the melting curve of DNA in the presence of copper ions and a low concentration of sodium nitrate. As we increase the concentration of sodium nitrate, the melting temperature of the double helical DNA is increased. This means, of course, that the double helix is stabilized by sodium ions which bind to phosphate. An excess of sodium ion binding to the phosphate overcomes the effect of the small quantity of copper (10^{-4}M) which would tend to disrupt the hydrogen bonding in the double helix. After the DNA has been melted out, cooling produces no decrease in absorbance whatever, apparently because the copper ions prevent "hairpin loops" from forming.

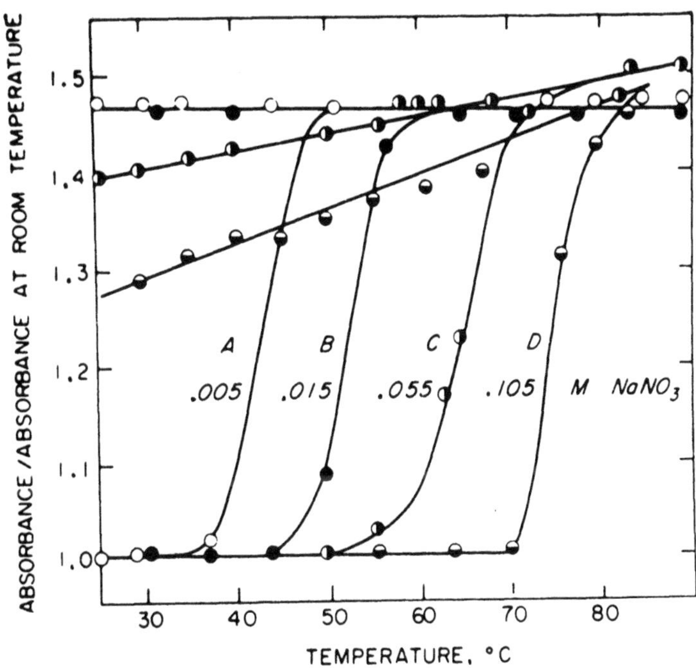

Figure 6. Effect of ionic strength on melting curve of 5×10^{-5}M(P) DNA in the presence of 10^{-4}M Cu(II) (from Ref. 6).

It occurred to us to try an experiment to determine whether the effect of sodium ions binding to phosphate could reverse the effect of copper ions binding to base. Starting with DNA in the presence of Cu(II) ions at low ionic strength, the solution is first heated to unwind the DNA and then cooled (Fig. 6, Curve A). At this point solid sodium nitrate is added to produce a concentration equivalent to that in Curve D. We then observe a gradual decrease in absorbence until the absorbence has attained its original value. The

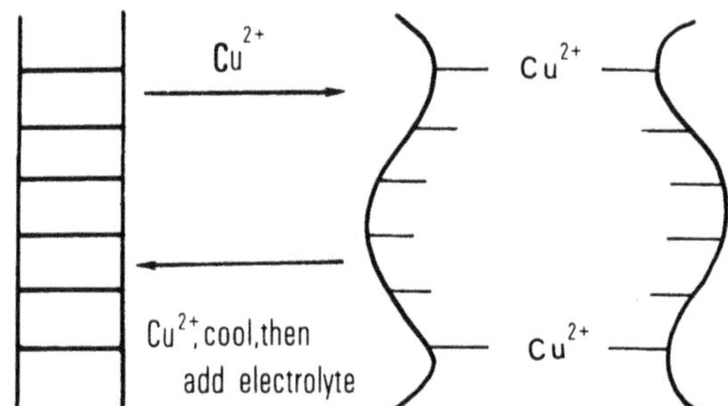

Figure 7. Scheme for unwinding and rewinding DNA in the presence of Cu(II).

explanation of this experiment may be found in Figure 7. When double helical DNA is heated in the presence of Cu(II) ion at low electrolyte concentration, the two strands unwind, but the Cu(II) ions form crosslinks between these chains. When double helix formation is favored by cooling and increasing the electrolyte concentration, the two strands are then in position for the complementary bases to come together. Consequently, rewinding is possible. When DNA is heated in the absence of copper, these two strands become so intertwined in the solution that when the thermodynamic conditions favorable for the reformation of double helix are again produced, the rewinding is prevented kinetically because of the fact that the complementary bases cannot find each other. Thus, copper ions can disrupt the double helix of DNA by binding to the bases and also make possible the reversible unwinding and rewinding of DNA by the proper manipulation of electrolyte concentration.

Having discussed one of the consequences of metals binding to bases, I would now like to consider some effects of metals binding to phosphate.

STABILIZATION OF NUCLEIC ACID STRUCTURE BY METAL IONS

It has already been noted that metal ions binding to phosphate increase the stability of the DNA double helix. There is a simple explanation for this phenomenon. The surface of the DNA double helix has the negatively charged phosphate groups arranged in close proximity, so that they tend to repel each other (Fig. 8). For this reason the solution of solid DNA in distilled water results in the unwinding of the double helix. When positive ions that bind to the phosphate groups are supplied, these negative charges are neutralized, the tendency to unwind is therefore overcome, and the double helix is stabilized.

Nevertheless, these same metal ions can have a potentially harmful effect on the propagation of the genetic information, as I shall now try to demonstrate.

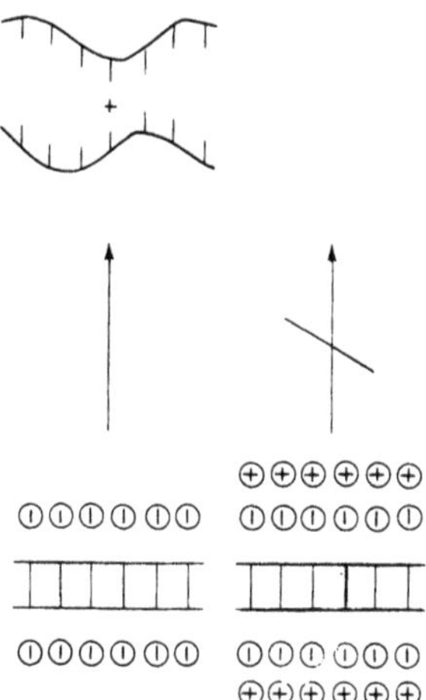

Figure 8. Scheme for the stabilization of the DNA double helix by metal ions binding to phosphate.

COMPLEMENTARY BASES AND MISPAIRING

In order for the genetic information to be propagated correctly, it is necessary for a base on one strand to be hydrogen bonded to its complementary base on the other strand. Figure 9 illustrates that complementary base pairs are not so different from those that are not complementary. In this figure the hydrogen bonding scheme for uridine and adenosine (U and A) is compared with that for inosine and adenosine (I and A). It is obvious that exactly the same hydrogen bonding scheme is involved in both cases. Differences come into play in fitting the bases into the DNA double helix; the points of attachment are different distances and different angles

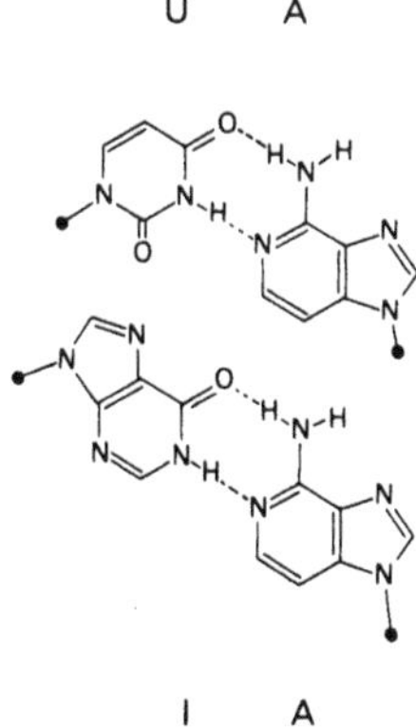

Figure 9. Comparison of complementary bases and "mispaired" bases.

apart. The most important consideration is, however, that although non-complementary bases hydrogen bond, their bonding is less table than that of complementary bases, as is evident from a comparison of the stability of poly(A)•poly(U) with that of poly(A)•poly(I) (7). Complementarity is an interaction between base pairs involving more stable hydrogen bonding that is essential for the correct propagation of the genetic information. Mispairing is used here to indicate the formation of less stable hydrogen bonded base pairs which will obfuscate the propagation of the genetic information.

Complementarity is required not only in the interaction of the two strands of the DNA molecule, but in other aspects of genetic information transfer, as for example, in translation. In this process the mRNA message is recognized by complementary base pairs on the tRNA molecule (Figs. 3, 10), which bears the amino acid that is supposed to be inserted into the protein at a particular point. The message consists of 3-component nucleotide code words or "codons," which are recognized through complementary base pairing by the "anticodons" of the tRNA. Some bases can be substituted for others without thwarting recognition. In Figure 9, U can be substituted for C, but if A or G is substituted for C a leucine tRNA is bound instead of phenylanaline tRNA. With this fact in mind, let us

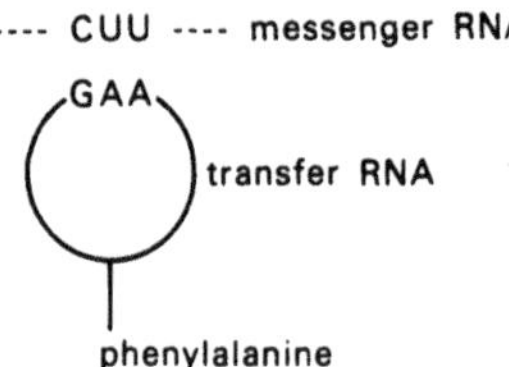

Figure 10. Scheme for recognition of codon in messenger RNA (mRNA) by anticodon in transfer RNA (tRNA).

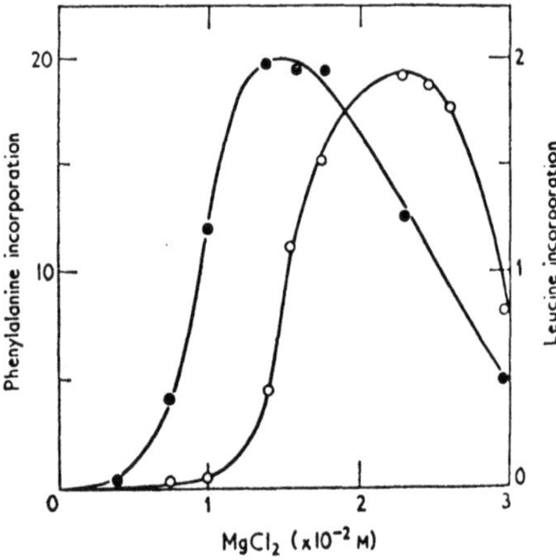

Figure 11. Amino acids incorporated into protein, using poly(U) as message, as a function of Mg(II) concentration (from Ref. 8).

consider Figure 11, which represents some work that was done some time ago by Szer and Ochoa (8). They used polyuridylic acid as the mRNA molecule for the incorporation of phenylalanine into protein and studied the amount of phenylalanine incorporation as a function of magnesium ion concentration. You will note that the amount of phenylalanine incorporated first increases with magnesium ion concentration, then passes through a maximum and finally decreases. However, as the phenylalanine incorporation decreases, a misincorporation of leucine occurs. Figure 11 exaggerates the effect, since the leucine incorporation ordinate is ten times the phenylalanine incorporation ordinate; nevertheless, a misincorporation of leucine does take place in substantial degree. How can one explain the misreading of the genetic code at high magnesium concentration?

METAL IONS AND MISPAIRING

Let us formulate a hypothesis. Consider again the recognition of the mRNA codon by the anticodon in the tRNA molecule (Figs. 3, 10). In order for this recognition to be accurate only complementary bases must interact. Only then will the mRNA codon find the particular tRNA that has phenylalanine on it and not another tRNA that has leucine on it. At low magnesium ion concentration there is relatively little tendency for strand interaction. Therefore, only the strongest base pairs, the complementary base pairs, will interact. In other words, at low metal ion concentration there is sensitivity to complementarity. On the other hand, at high metal ion concentration strand interaction becomes so strong that the sensitivity to complementarity is lost, and hydrogen bonding occurs not only between the strongly hydrogen bonded complementary bases, but also the more weakly hydrogen bonded mispairing bases, and therefore the genetic code becomes blurred.

To test this hypothesis we have carried out experiments that demonstrate that metal ions can induce mispairing (9).

Let us look again at the complementary base pair and the mispairing base pair in Figure 9. We can design an experiment in which it is possible to determine whether only complementary base pairs are formed, or whether mispairing occurs. If adenylic acid, poly(A), in which there are only A bases, is allowed to interact with a 1:1 copolymer of the two bases that bind to A in Figure 9, i.e., U and I, [copoly(I,U)], there are two ways in which these polymers can interact. Either there is complementarity, i.e., the A's recognize only the U's and the I's are looped out (10) of the helix. Or else, the A's react not only with the U's, as expected, but mispairing occurs and the A's react also with the I's (Fig. 12).

One method to study this interaction is by the method of continuous variation for determining the stoichiometry. To achieve

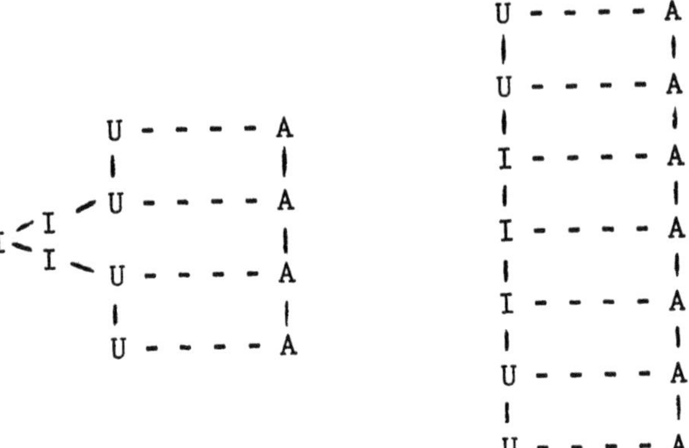

Figure 12. Complementary interaction (left) and mispairing (right) between poly(A) and poly(I,U).

complementarity, two molecules of the copolymer should react with one of poly(A); for mispairing, a 1:1 interaction of the two polymers should be observed. Figure 13 contains mixing curves obtained in the presence of sodium. A minimum occurs at 33 mole% poly(A), indicating the formation of a 2:1 complex, as expected for complementarity. Figure 14 shows what happens in the presence of magnesium ions. In this case a maximum can be seen at 50 mole% poly(A), i.e., the two polymers interact at the 1:1 ratio anticipated for mispairing. Thus, magnesium ions induce mispairing where complementarity occurs in the presence of sodium ions.

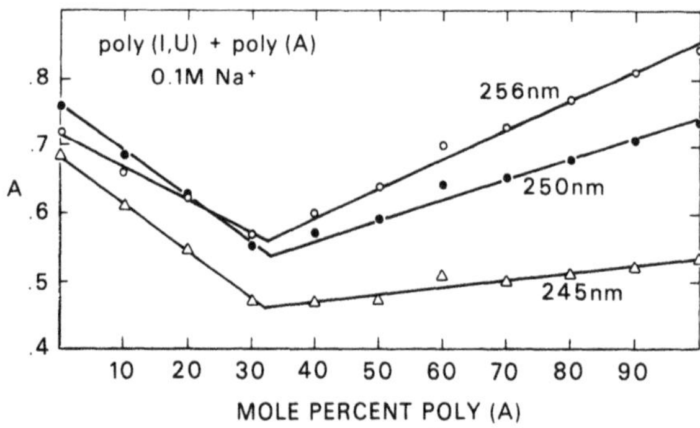

Figure 13. Mixing curves for 9×10^{-5}M poly(I,U) and poly(A) in the presence of 0.1M Na^{+}.

The question then arises whether it is possible to find a concentration of magnesium ions below which there is no mispairing, but above which mispairing does indeed take place. Figure 15 shows that, as the concentration of magnesium ions is raised, there is a transition from the 2:1 interaction characteristic of complementarity to the 1:1 interaction characteristic of mispairing. The concentration of magnesium ions in the transition range is about 1 to

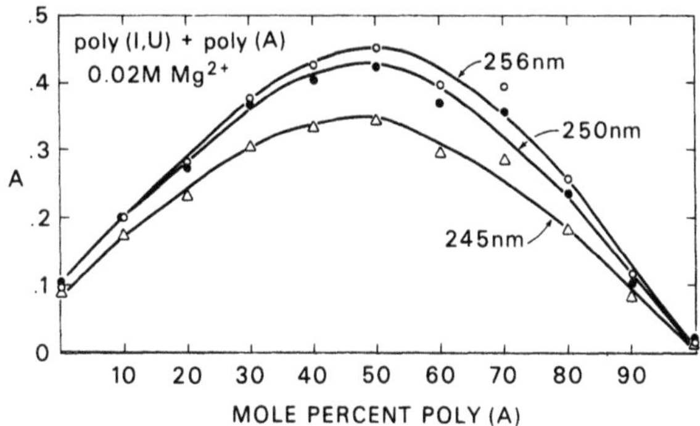

Figure 14. Mixing curves for 9 × 10^{-5}M poly(I,U) and poly(A) in the presence of 0.02M Mg(II). Solutions were centrifuged, and absorbence measured on supernatant.

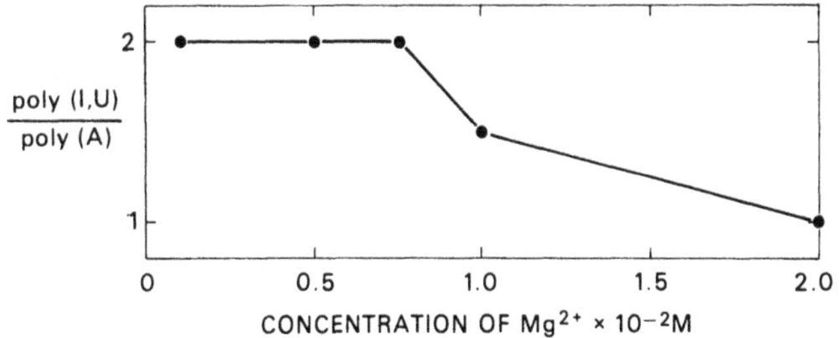

Figure 15. Stoichiometry of reaction of 9 × 10^{-5}M poly(I,U) and poly(A) as a function of Mg(II) concentration.

2 × 10^{-2}M. In the experiment of Szer and Ochoa (8) (Fig. 11) the change from correct incorporation of amino acid to incorrect incorporation also occurs between 1 and 2 × 10^{-2}M magnesium. Perhaps this experiment is circumstantial evidence that the mechanism whereby the wrong amino acid is incorporated at the high metal ion concentration involves mispairing of bases.

Although the continuous variation studies constitute good evidence for the formation of mispaired bases, it would be inexpedient to rely solely on this kind of experiment. A more direct way of looking at specific hydrogen bonding is by proton NMR on a 1:1 mixture of the poly(I,U) and poly(A). In the NMR spectra (Fig. 15) the U peaks are obscured but one can see peaks due to I protons and A protons. Hydrogen bonding interactions broaden the proton resonances so that it is possible to determine, from the extent of the broadening, which bases are hydrogen bonded. An NMR experiment on the reaction of these polymers can lead to three possible results. One is that there is no interaction at all between these two polymers. In that case, the I peaks and the A peaks should remain intact. The second is that there is complementarity; then the I peaks should remain intact but the A peaks should exhibit partial broadening. The third possibility is mispairing, which would involve A bonding to I as well as to U. Then all the peaks should be broadened. With that in mind, let us look at the actual NMR curves (Fig. 16). At 40° we can see all proton resonances, those due to I and those due to A. Therefore, at 40°, there is no interaction. If the temperature is lowered to 20°, interaction does occur between these polymers; it does not involve the I peaks at all, but broadens only the A peaks, and these only partly. This is exactly what is predicted for complementary base interaction. Finally, when magnesium ions are added at 20°, we see the extensive broadening of all peaks, demonstrating that magnesium ions do indeed induce mispairing reactions under conditions which otherwise produce complementarity. Thus, the NMR studies, like the continuous variation studies, demonstrate that divalent metal ions can cause mispairing of bases.

These experiments reveal that metal ions are indeed capable of inducing mispairing of bases, and it is reasonable to expect that excessive metal concentrations could affect replication, transcription, or translation phenomena in this way. Our results are, of course, only suggestive and do not prove that mispairing occurs in this way in a biological system.

CLEAVAGE OF PHOSPHODIESTER BONDS

We have seen that one effect of metal ions binding to phosphate is their stabilizing influence upon strand interaction, which can, in turn, lead to the mispairing of bases. I want now to discuss another effect of metal ions binding to phosphate, the cleavage of phosphodiester bonds, bringing about the depolymerization of polynucleotides. Figure 17 demonstrates the depolymerization of poly(A) by a variety of metals (11). The ordinate is a function of the extent of degradation of the molecules, the abscissa is the reaction time. Zinc is by far the most effective of these metals in depolymerizing RNA. (Lead, which is not in this series, is even more effective (12).)

It is of some interest to determine whether zinc ions will cleave every phosphodiester bond to the same extent. Table I reveals that the cleavage is selective. If the base composition at the cleavage sites is compared with the base composition in the total RNA, it can be seen that the amount of G at the cleavage sites has decreased dramatically while the amount of C and U has increased correspondingly. The significance of this result is that when the zinc ions bind to phosphate, they prefer to cleave bonds that are adjacent to uracil or cytosine bases and they prefer to leave intact those bonds that are next to a guanine. Thus, the zinc ions exhibit specificity. Such specificity could find application in the

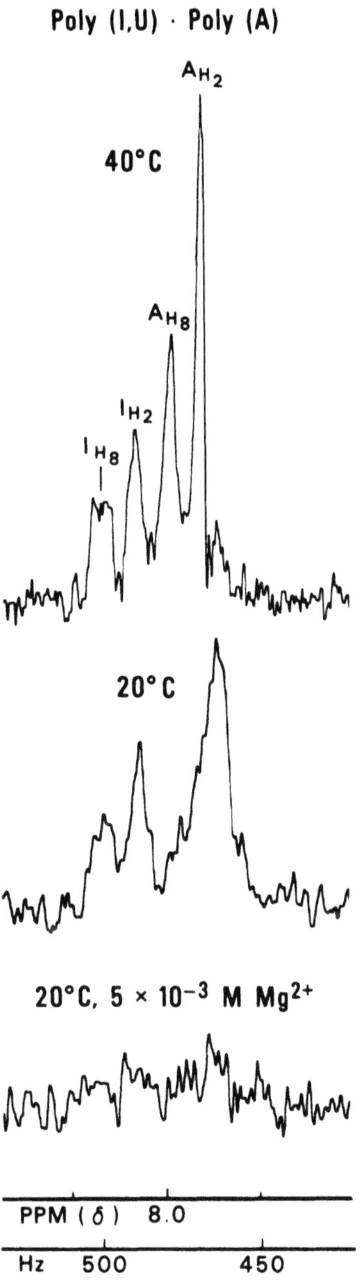

Figure 16. Proton magnetic resonance spectra (60 MHz) of a mixture of 10^{-2}M poly(I,U) and poly(A) at 40° and 20° in the absence of Mg(II), and at 20° in the presence of 5 × 10^{-3}M Mg(II).

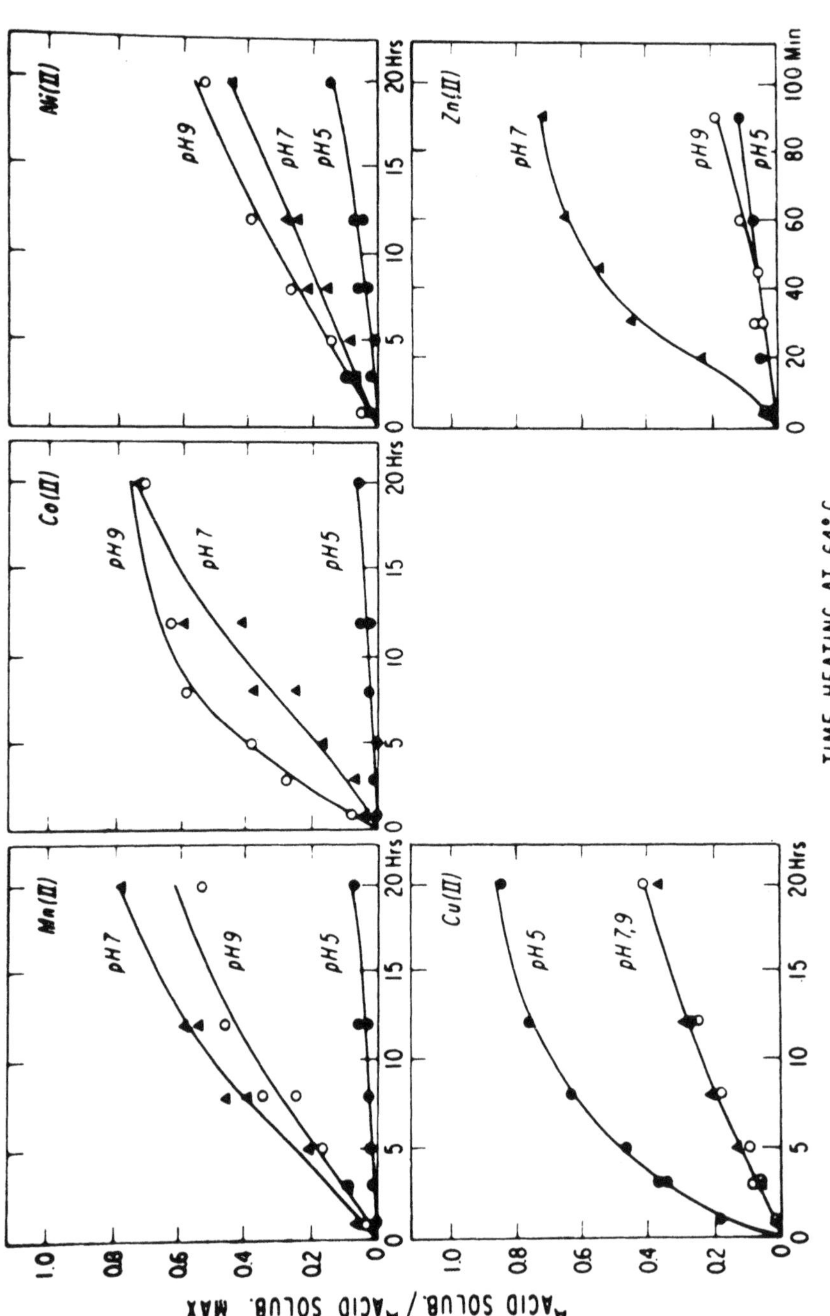

Figure 17. Depolymerization of 1×10^{-4}M(P) poly(A) by 2×10^{-4}M concentrations of various metal ions. Depolymerization measured by solubility in uranyl acetate-trichloroacetate. Note the difference in the abscissa for zinc, as compared with the other metals (from Ref. 11).

Table I

	Cleavage Sites (%)	RNA Composition (%)
G	7	28
A	20	25
C	30	19
U	42	28

determination of nucleotide sequence. The reason for the specificity is not obvious. A possible explanation is that, since the stacking tendency of guanine is the highest of all the bases, whenever guanine is one of the two stacking bases the phosphodiester bond between these two bases is more stable than other phosphodiester bonds. It so happens that uracil and cytosine stack relatively little and these bonds are perhaps for that reason more susceptible to cleavage.

Let us now consider the mechanism of this cleavage. It occurs only with RNA, and not with DNA; the 2'hydroxyl group is therefore essential (13). It appears possible that the metal forms a chelate with the 2'OH group and the phosphate group (13, 14). To settle the matter, experiments were carried out on the degradation of trimers (15). Figure 18 shows the results for the degradation of inosine trinucleotide. It can be seen that a cyclic phosphate intermediate (I > P) is formed much more rapidly than the inosine-3'phosphate end product, and that the concentration of this cyclic phosphate passes through a maximum and then returns to 0 concentration, while the inosine-3'-phosphate concentration increases until a plateau has been reached. Inosine-2'-phosphate is also produced. Similar experiments have been conducted with many trimers with similar results, and it is concluded that the mechanism of this reaction does not involve the formation of a metal chelate with phosphate and the 2'hydroxyl group, as had seemed reasonable. Instead, we picture (Fig. 19) the formation of a bond between the metal and the phosphate, the consequent formation of a positive dipole on the phosphate, followed by the formation of a cyclic phosphate and cleavage of the phosphodiester bond. Thus, the metal catalyzed cleavage of these molecules is very similar to their cleavage by enzymes or acid.

A STEREOSELECTIVE COMPLEXING REACTION FOR RIBONUCLEOSIDES

I have now discussed effects of metals binding to the bases and effects of metals binding to phosphate. I should like to consider briefly the third possible binding site of a metal: the ribose hydroxyl group. Figure 20 demonstrates the structural difference

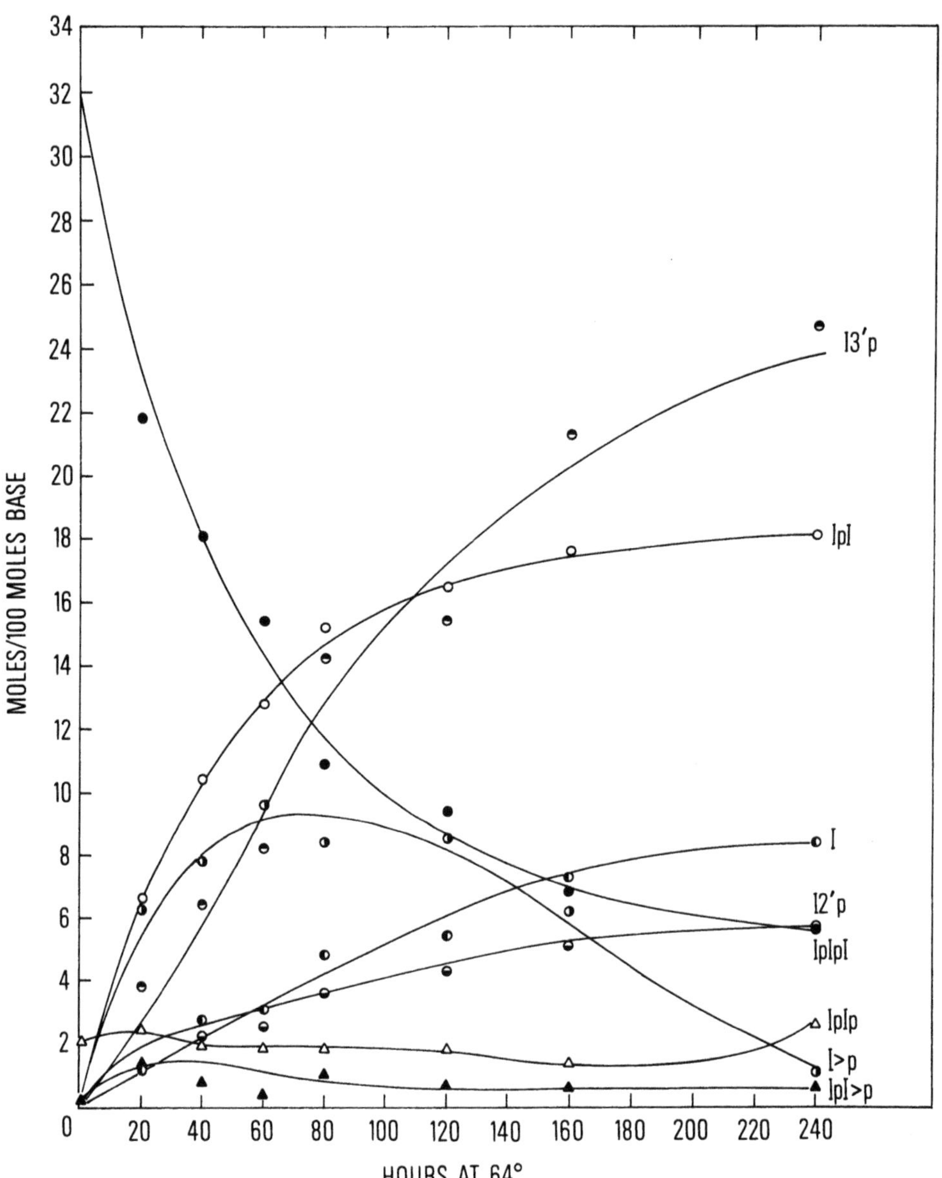

Figure 18. Kinetics of the degradation of 1×10^{-4}M (as nucleoside base) inosine trinucleotide by 2×10^{-4}M Zn(II) (from Ref. 15).

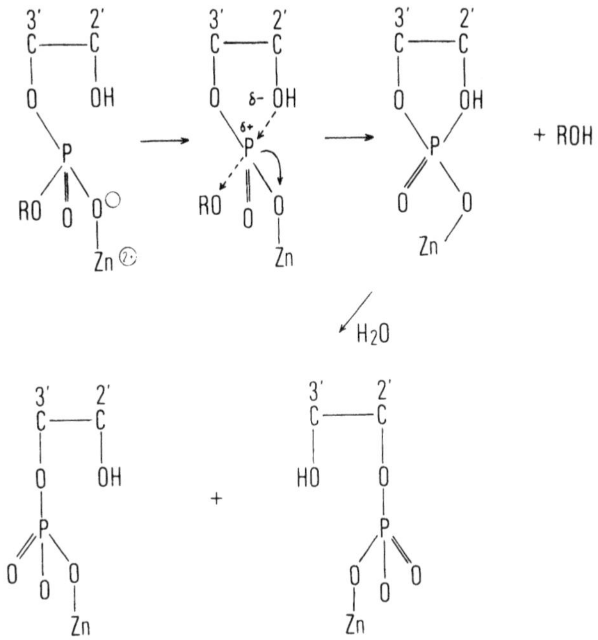

Figure 19. Scheme for mechanism of phosphodiester bond cleavage by zinc ions.

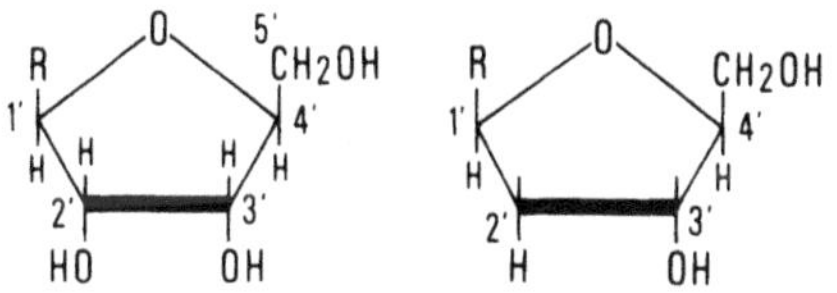

Figure 20. Comparison of structure of ribonucleoside and deoxynucleoside.

between a ribonucleoside that is the monomeric component of an RNA molecule, and a deoxynucleoside, the monomeric constituent of a DNA molecule. RNA and DNA are produced from ribonucleoside triphosphate and deoxynucleoside triphosphate monomers in the presence of RNA polymerase, or DNA polymerase, respectively. These enzymes are metalloenzymes, and other metals in addition to the intrinsic ones are also required by the active enzymes. The substitution of one metal for another causes the error incorporation of deoxynucleotides into RNA (16) and ribonucleotides into DNA (17). It therefore appears that metal ions are involved in the differentiation between ribonucleoside and deoxynucleoside structure. It is of interest to understand how metal ions can help to recognize this difference.

We do not know at this time how metal ions can aid RNA polymerase to recognize this difference, but we have found a rather simple inorganic molecule which can tell the monomers apart (18). This molecule is the copper acetate dimer (Fig. 21) which is stable in the solid state and also in non-aqueous solvents. The copper-copper distance in the Cu(II) acetate dimer is 2.64Å. This is also approximately the distance between the two hydroxyl groups in the ribonucleoside, and so it seems not unreasonable that the copper atoms in this complex could attach themselves to the two hydroxyl groups. Figure 22 provides NMR evidence that the hydroxyl peaks of uridine, in DMSO, are broadened by copper acetate. Uridine does not have much tendency to bind to the bases, and in DMSO there is almost no such tendency, as revealed by the lack of broadening of the base peaks.

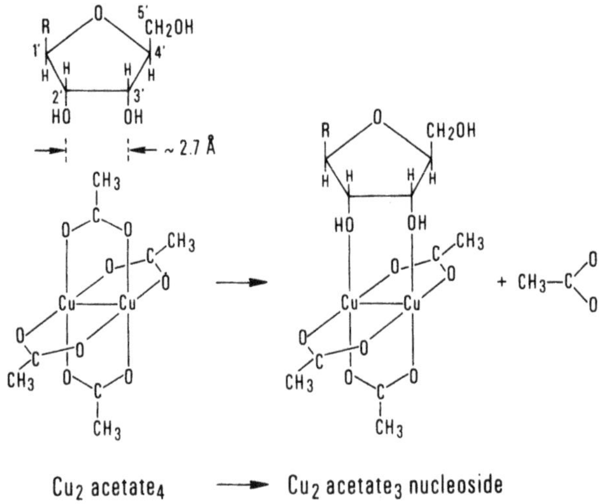

Cu_2 acetate$_4$ → Cu_2 acetate$_3$ nucleoside

Figure 21. Scheme for reaction of Cu(II) acetate dimer with ribonucleoside.

There are two peaks in the electronic spectrum of Cu(II) acetate in DMSO. One of them, in the visible, is due to the individual copper ions and the other, in the near ultraviolet, is due to the copper-copper interaction. As seen in Figure 23, the addition of deoxyuridine to a solution of Cu(II) acetate in DMSO causes no change in the spectrum. On the other hand, the addition of ribouridine causes a very dramatic change, a decrease in intensity of the visible peak and an enhancement of intensity of the near ultraviolet peak. The Cu(II) acetate dimer structure is essential to give this differentiation between ribonucleoside and deoxynucleoside; ribouridine does not react with copper nitrate at all. Figure 24 shows that, whereas the deoxynucleosides generally exhibit little or no decrease in absorbence upon addition of copper acetate, all of the ribonucleosides have a dramatic effect on the spectrum. Thus, the reaction is general for ribonucleosides, and does not occur with

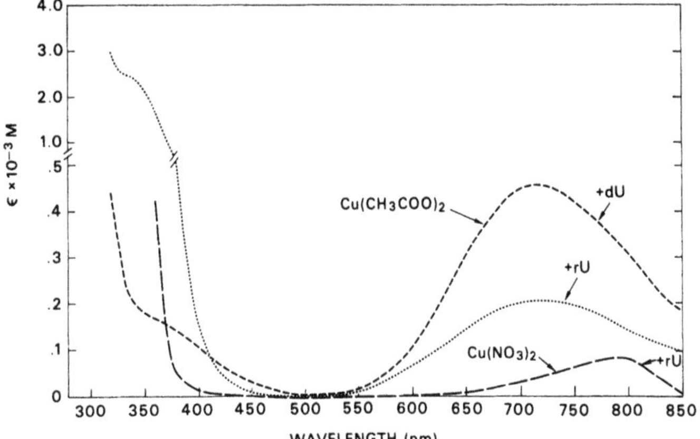

Figure 22. Proton magnetic resonance spectrum (60 MHz) of 0.1M uridine in $DMSOd_6$, without Cu(II) acetate, and with Cu(II) acetate in the concentrations indicated (from Ref. 18).

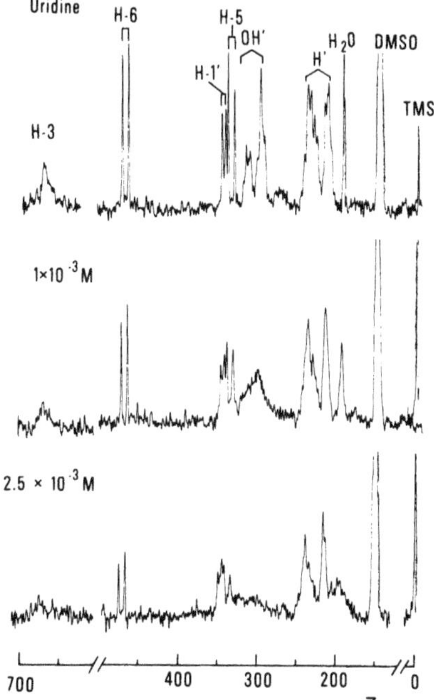

Figure 23. Electronic spectrum of 5×10^{-3}M Cu(II) acetate in DMSO, as affected by 5×10^{-3}M ribouridine and deoxyuridine. Comparison with effect of ribouridine on spectrum of Cu(II) nitrate (from Ref. 18).

deoxynucleosides. If the hydroxyl groups are blocked by acetylation or if the two hydroxyl groups are *trans* to each other instead of *cis*, the reaction also does not occur. The reaction thus requires two hydroxyl groups and they must be cis to each other. Continuous variation studies (Fig. 25) reveal a 2:1 copper/nucleoside stoichiometry (19). From all of these studies we conclude that the Cu(II) acetate dimer reacts with the 2' and 3' hydroxyl groups that are the same distance apart as the copper atoms (Fig. 21). We do not assume that such a scheme applies to RNA polymerase, but we have illustrated how a stereoselective reaction can account for the specific recognition of ribonucleosides and the lack of recognition of deoxynucleosi

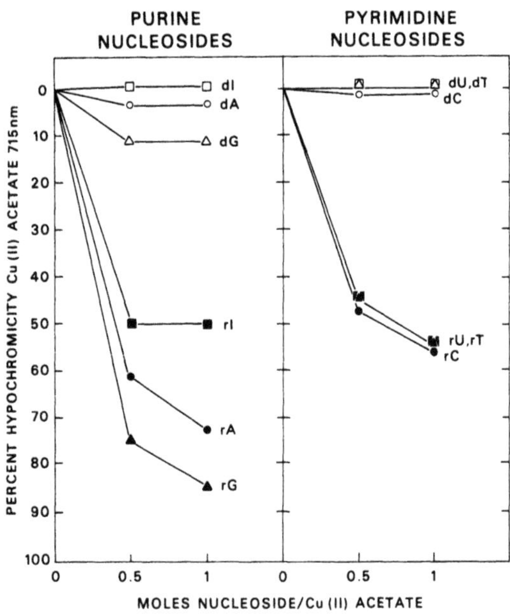

Figure 24. Decrease in absorbence of 5×10^{-3}M Cu(II) acetate in DMSO at 715 nm upon addition of nucleosides (from Ref. 18).

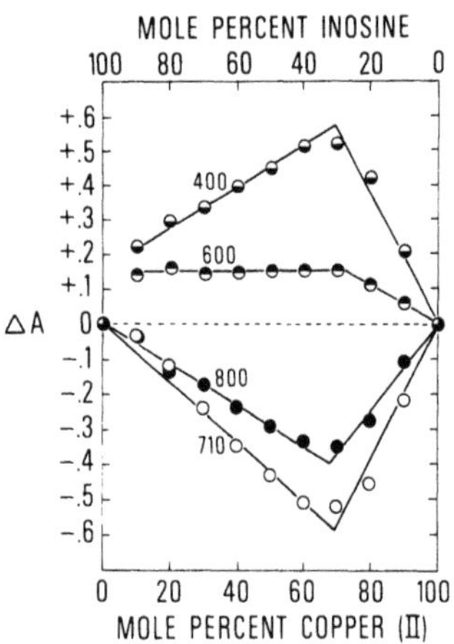

Figure 25. Continuous variation study of interaction between inosine and Cu(II) acetate in DMSO (from Ref. 19).

CONCLUSION

In summary, metal ions can bind to every conceivable electron donor site on nucleic acid molecules. They can bind to phosphate groups, base groups, and ribose hydroxyl groups. Different metal ions bind to these various sites under different conditions. Many of the reactions in which metal ions bind to nucleic acids have dramatic effects on the structure and function of these bearers of genetic information. Some of these effects are essential for the proper functioning of these molecules, while others will cause impairment in function. The same metal ions can be beneficial or deleterious under different conditions.

REFERENCES

1. Eichhorn, G. L. (1973), Inorganic Biochemistry, Amsterdam: Elsevier Publishing Co., p. 1191, 1210.

2. Izatt, R. M., Christensen, J. J., and Rytting, J. H. (1971), Chem. Rev., 71, 439.

3. Weser, U. (1968), Structure and Bonding, 5, 41.

4. Eichhorn, G. L. (1971), Adv. Chem., 100, 135.

5. Eichhorn, G. L. (1962), Nature, 194, 474.

6. Eichhorn, G. L. and Clark, P. (1965), Proc. Natl. Acad. Sci. U. S., 53, 586.

7. Felsenfeld, G. and Miles, H. T. (1967), Ann. Rev. Biochem., 36, 407.

8. Szer, W. and Ochoa, S. (1964), J. Mol. Biol., 8, 823.

9. a) Eichhorn, G. L., Richardson, C., Pitha, J. (1971), Abst. Biol. 17, 162nd National Meeting, American Chemical Society, Washington, D. C., Sept.

   b) Eichhorn, G. L., Pitha, J., Tarien, E., Richardson, C. (1973), Abstracts, Ninth International Congress of Biochemistry, p. 201.

10. Fresco, J. R. and Alberts, B. M. (1960), Proc. Natl. Acad. Sci. U. S., 46, 311.

11. Butzow, J. J. and Eichhorn, G. L. (1965), Biopolymers, 3, 95.

12. Farkas, W. R. (1968), Biochim. Biophys. Acta, 155, 401.

13. Eichhorn, G. L. and Butzow, J. J. (1965), Biopolymers, 3, 79.

14. Bamann, E., Trapmann, H., and Fischler, F. (1954), Biochem. Z., 328, 89.

15. Butzow, J. J. and Eichhorn, G. L. (1971), Biochemistry, 10, 2019.

16. Steck, T. L., Caicuts, M. J., and Wilson, R. G. (1968), J. Biol. Chem., 243, 2769.

17. Berg, P., Fancher, H., and Chamberlain, M. (1963), Symposium on Informational Macromolecules, H. Vogel, ed., New York: Academic Press, p. 467.

18. Berger, N. A., Tarien, E., and Eichhorn, G. L. (1972), Nature New Biology, 239, 237.

19. Berger, N. A. and Eichhorn, G. L. (1971), J. Am. Chem. Soc., 93, 7062.

A SURVEY OF NATURALLY OCCURRING CHELATING LIGANDS*

ARTHUR LINDENBAUM

Division of Biological and Medical Research
Argonne National Laboratory
Argonne, Illinois 60439

INTRODUCTION

It can be no great exaggeration to describe chelation as a seminal concept in modern biochemical theory and practice. Beginning with the work of Werner (A. Werner (1893), Anorg. u. Allgem. Chem., 3, 267; (1901), Ber., 34, 2584), and extending through the illuminating organic, physical-chemical, and pharmacological studies of such pioneers as Schwarzenbach, Martell, and Albert, the idea of molecular rearrangements, alterations in charge, conformational changes, etc., undergone by both metal and organic ligands as a result of their interaction to form complexes and chelate structures could not help but evoke visions of biological control, especially with respect to the action of enzymes and drugs. By now, several thousand papers dealing with chelation phenomena in living systems have appeared in the literature, and no modern textbook of biochemistry or pharmacology is without numerous examples of chelation reactions between metals such as calcium, magnesium, iron, zinc, copper, manganese, molybdenum, cobalt, or chromium, and organic metabolites ranging in complexity from glycine to coenzyme-mediated enzymes.

Unfortunately, the literature dealing with chelation of natural substances is quite scattered and presents formidable problems in assembling the highly diversified and fragmented information to cover a particular viewpoint with any sense of completeness. (Indeed, at this stage probably no single text could cover the scattered literature adequately.) Setting a more modest goal, therefore, I shall concentrate this discussion mainly on a number

*Work supported by the U. S. Atomic Energy Commission.

of naturally occurring organic compounds for which there is clear or reasonably good evidence of chelation. My objectives in this survey will be: (1) To review the chelation process briefly, suggesting some physiological implications; (2) To list some of the important metabolites known to form chelates in living systems; (3) To point out the portions of these molecules which act as ligands in reacting with metal ions; and (4) To describe several biochemical reactions in which chelation plays a part.

This review is intended to be provocative. On the basis of the knowledge that has been accumulated so far it is clear that further systematic studies of chelation reactions in living systems must yield fruitful physiological information, just as the results of physical-organic research dealing with chelation reactions have led to useful applications in pharmacology and in industry.

THE CHELATION PROCESS IN BIOLOGICAL SYSTEMS

Although most of us are familiar with complex formation in general, a brief summary of the basic chemistry of chelation may be useful. It is a common error, especially among clinically oriented scientists, to refer to most metal-organic complexes as chelates. Actually, the terms "chelate" and "chelation" should be applied only to a special form of complex formed between a metal ion and a ligand.* The chemical factors that need to be considered in chelation reactions are (1) ring formation, (2) dentation, (3) resonance, (4) pH, and (5) specificity.

Ring Formation. What distinguishes chelation as a special form of complexation is that the ligand molecule contains at least two electron-donating atoms; these atoms, usually oxygen, sulfur or nitrogen, are spaced along the ligand molecule in such a way that binding to a metal results in the formation of a heterocyclic ring containing at least one covalent bond. Covalency generally limits the metal member of the ring to members of the alkaline earth, transition, and rare earth series. Due to the limitations imposed by bond angles the most stable chelate rings contain 5 to 7 atoms, although some stable 4-membered rings are known, usually containing sulfur.

Dentation. The term bidentate refers to a ligand with two donor atoms. Other ligands may be multidentate (e.g., ethylenediaminetetraacetic acid, EDTA), although all donor atoms need not be involved. As we shall note, multivalency of the metal ions allows the formation of more than one chelate ring.

*Frequently the entire organic molecule is referred to as the ligand. Preferably, the term ligand should apply only to the portion of the molecule involved in chelation.

Resonance. Unsaturated chelate rings are stabilized by resonance; these rings are planar, as compared to saturated rings, which are puckered.

pH. In most organic ligands some of the oxygen, nitrogen or sulfur atoms are already sharing electrons with hydrogen atoms. Thus, for a metal chelate to be formed, the pH of the medium must be sufficiently high to ensure that mass action will allow the metal to compete with protons bound to these ligand atoms. But it is important to note that in some cases a pH high enough to release protons from the ligand may result in the formation of insoluble metal hydroxides, thus lowering the concentration of metal ions available for chelation.

Specificity. From a manipulative point of view, it often would be desirable if a specific ligand were to form a chelate only with a specific cation. In practice, only relative specificity can be expected. Thus, it can be arranged, taking the above factors into account, that of two metal cations in solution with a ligand, a greater fraction of the cation in lower concentration is bound to the ligand, despite competition from the more abundant cation. It might be pointed out also that organic ligands containing sulfur tend preferentially to bind Cu^{+} as well as Hg^{++}, As^{+++}, and Sb^{+++}.

PHYSIOLOGICAL EFFECTS OF CHELATION

Although the following physiological effects are, in part, speculative, there is sufficient supporting evidence of the homeostatic role of chelation reactions, both *in vivo* and *in vitro*, to warrant brief discussion in this review:

1) Regulation of metal ion concentrations. This function is probably of most importance when the concentration of a metal ion is very low and its presence in chelated or ionic form is critical.

2) Transport regulation. The oxidation-reduction potential of a metal is always altered by chelation. A valence change resulting from chelation could alter the subsequent affinity of the metal for the ligand, or could shift the location of the chelate as the result of charge reduction, increased liposolubility and, consequently, increased membrane permeability. On the other hand, if chelation leads to polymerization, the mobility of a metabolite could be restricted, or its biological function severely altered.

3) Stereochemical rearrangement of an organic metabolite through chelation may mask or unmask active centers, thus controlling participation of the molecule in competing

reactions. In the metal-activated enzymes, chelation may result in either activation or inhibition; it is pertinent to note, in this connection, that about 10% of the known enzymes require a metal cofactor (Co, Cr, Cu, Mn, Mo, Zn) for activation.

4) Macromolecule formation. There is evidence that metal ions can act as bridges binding two or more chelated organic molecules together. Some examples are given below.

5) Electron transport. It is particularly appropriate in the case of enzymes requiring metals to suggest that, in addition to conformational changes induced by chelation, chelates could serve as conducting pathways for the addition or removal of electrons.

NATURALLY OCCURRING CHELATING MOLECULES

The following list describes several prominent categories of naturally occurring organic compounds possessing chelatable ligands. Where a particular metal is known to be involved, its symbol is given in parentheses. Some examples of specific molecules containing these ligands are also mentioned. It will be noted that rather than listing these compounds alphabetically, a rough attempt has been made to group them according to biological function or chemical structure. Many of these compounds also represent important functional moieties of protein molecules.

A. Mammalian Metabolites. Many of the compounds in this list are also widely distributed in non-mammalian organisms.

Krebs cycle acids	(especially citrate, whose concentration is $\sim 10^{-3}$M in extracellular fluids and ∼1% of dry weight of the skeleton)
Hormones	Thyroxine Histamine Cortisone (?)
Vitamins	Pteridines: folic acid Riboflavine: flavin enzymes; FMN; FAD; etc. Cobalamines: vitamin B_{12} (Co)
Porphyrins	Hemoglobin (Fe) Catalase (Fe) Cytochrome c oxidase (Fe, Cu)

	Purine and pyrimidine derivatives: nucleic acids, nucleoproteins
Proteins	Metalloenzymes: carbonic anhydrase (Zn) Metal-activated enzymes: phosphoglucomutase (Cr); leucine aminopeptidase (Mn)
Amino acids	(especially histidine, cysteine, serine)
Polyamines	Spermine Spermidine "Catecholamines": epinephrin, norepinephrin; adrenaline; DOPA and analogs

B. Chelating Metabolites From Other Animals, Plants, or Micro-organisms.

Chlorophyll (Mg, Cu)
Tetracyclines: Tetracycline, terramycin, etc.; origin: Streptomyces
Penicillin and Penicillamine (Cu); origin: Penicillium
Kojic acid; origin: Aspergillus
Mycobactin T; origin: Mycobacterium
Itoic acid (binds Fe^{+++}); origin: B. subtilis
Ferrioxamine B (binds Fe^{+++}); origin: Streptomycin pileosis
Pyrimine (binds Fe^{++}); origin: Pseudomonas

C. Synthetic Chelating Compounds Used Therapeutically.

EDTA
DTPA (diethylenetriaminepentaacetic acid)
Isoniazid
Aspirin (after hydrolysis)

Some examples of organic molecules containing ligands capable of chelate formation are shown in the illustrations below. Most of these occur normally in a variety of simple and complex organisms, and nearly all possess rather simple chemical structures. There are undoubtedly many other more complex naturally occurring chelating molecules (e.g., proteins) in which steric factors controlling the number and kind of ligand groups exposed to the external milieu determine the kind and amount of metal ions bound.

Charge alterations resulting from ring formation are illustrated in the interactions of Cu(II) with three simple organic ligands. Note that no covalency is involved in the formation of the compound cupric oxalate (Albert, Selective Toxicity, 4th ed., 1968, p. 312).

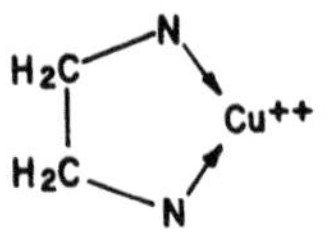

Cu(II)-ethylenediamine 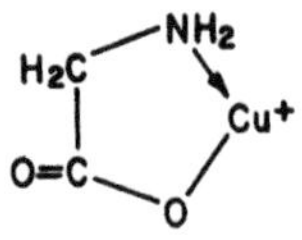 Cu(II)-glycine 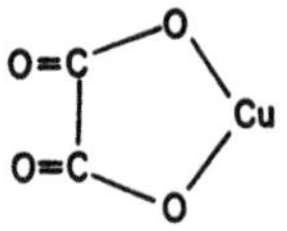 Cu(II)-oxalate

Chelate ring stability is conferred by resonance. Thus, acetylacetone forms a stronger chelate with Cu(II) than does salicylaldehyde (*ibid.*, p. 317).

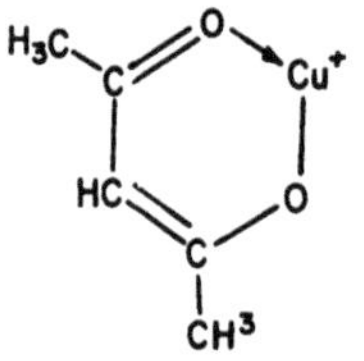

Cu(II)-acetylacetone (2 double bonds) Cu(II)-salicylaldehyde (1.5 double bonds)

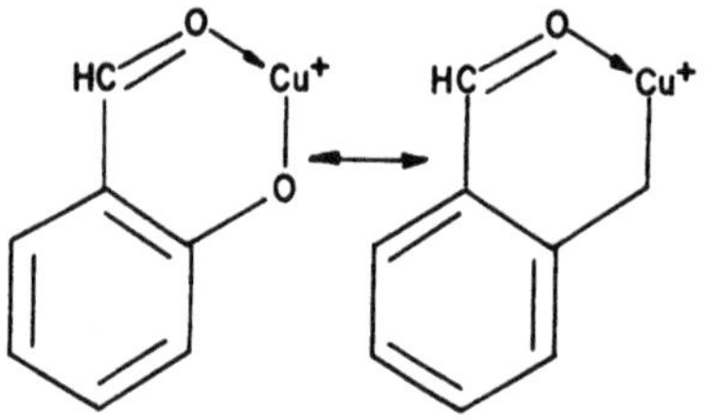

Polymer formation as a result of chelation is enhanced by the presence of more than one ligand on the same molecule. Here the Cu(II)-DOPA system is used to illustrate several possible chelate species, including formation of a monomer as well as open and closed versions of a head-to-tail dimer (J. E. Gorton and R. J. Jameson (1972), J. Chem. Soc., Dalton Trans. No. 3, 304). Tetramer formation (cis- or trans-configuration) is also possible, as indicated in the probable Cu(II)-3,4-dihydroxyphenylglycine chelate shown in head-to-head, tail-to-tail configuration (*ibid.*, p. 307). The possibilities for formation of even longer or more complex chelate polymers are obvious.

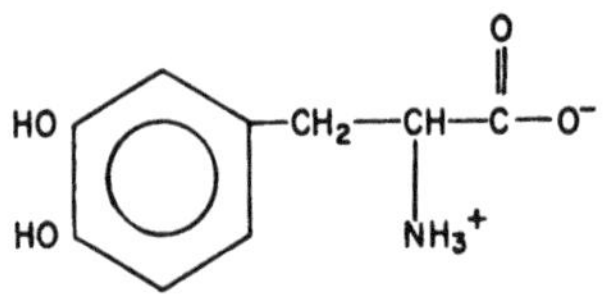

DOPA ZWITTERION 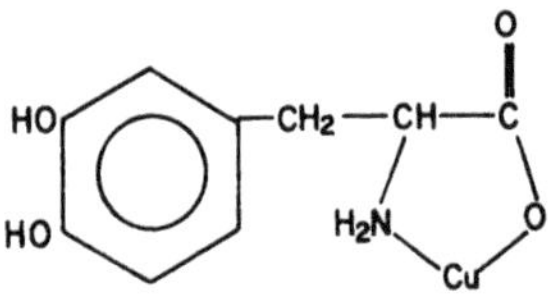 COPPER CHELATE

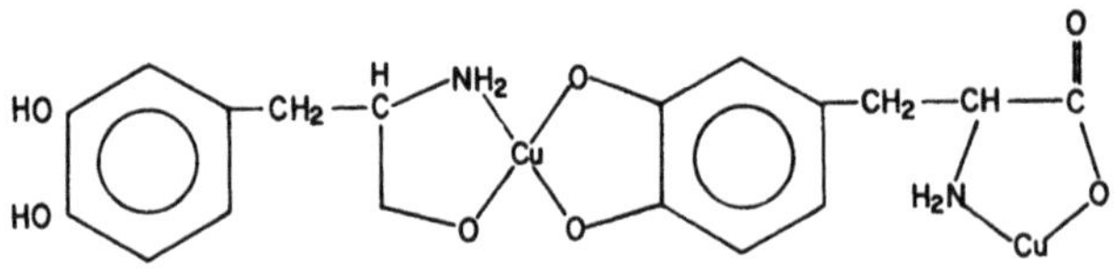

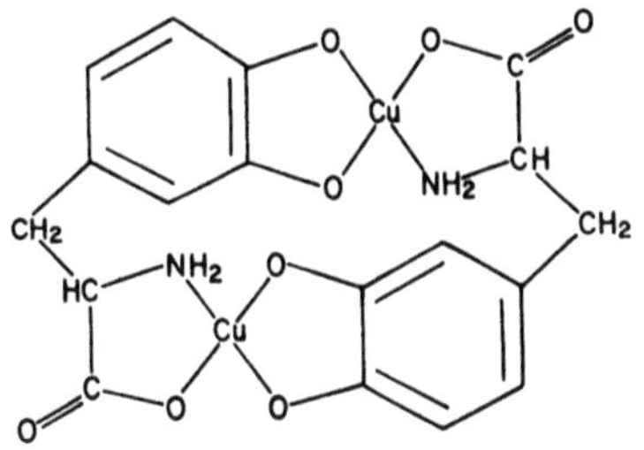

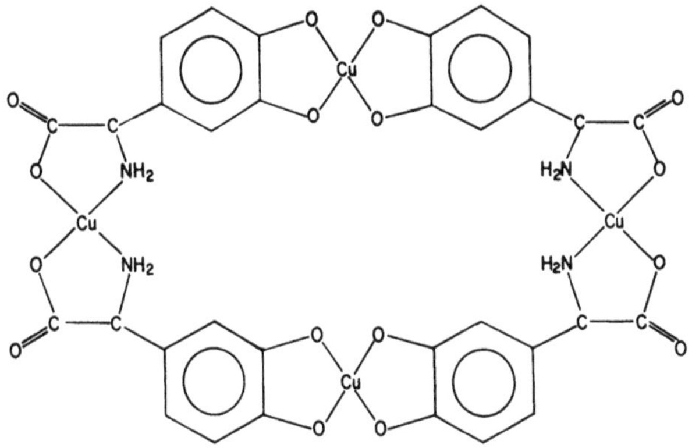

Following are several examples of natural chelating compounds found in all mammalian organisms as well as a few others which have been extracted from microorganisms for therapeutic usage:

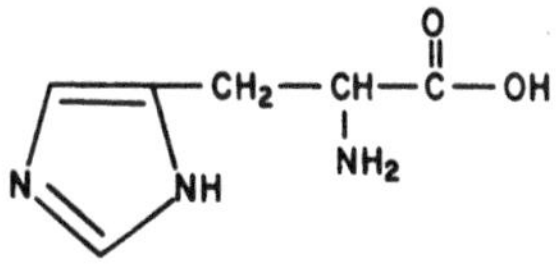

HISTIDINE

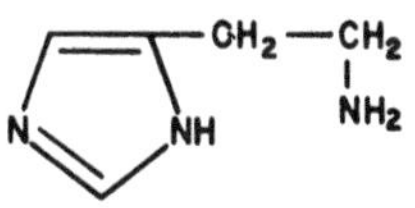

HISTAMINE

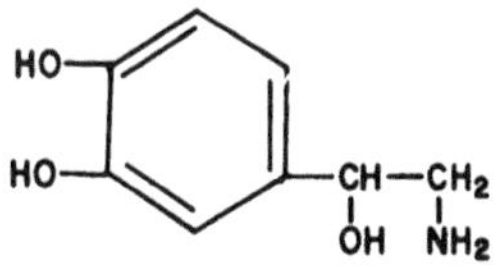

NORADENALINE

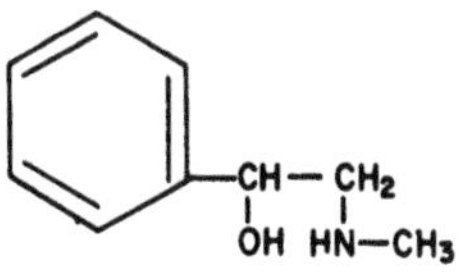

EPINEPHRINE

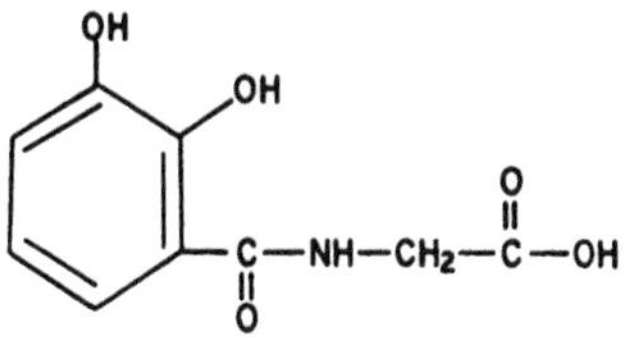

ITOIC ACID
(B.subtilis)

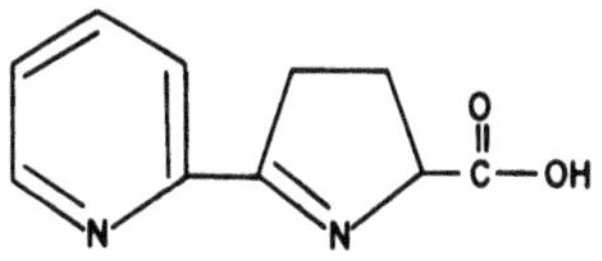

PYRIMINE
(Pseudomonas)

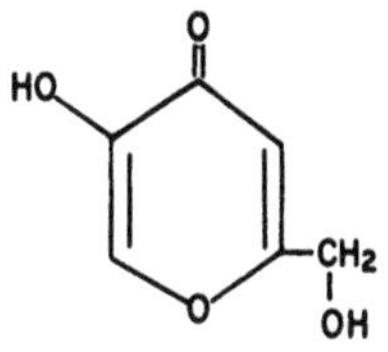

KOJIC ACID
(Aspergillus oryzae)

More complex naturally occurring compounds capable of undergoing chelation include molecules incorporating riboflavine (FMN, FAD, etc.) and molecules with catechol functional groups. A few examples are shown. The functional groups of FAD capable of participation in chelate ring formation are indicated by asterisks. Note also that tautomerization of the riboflavine moiety, as first shown by Albert ((1950), Biochem. J., 47, xxvii; (1952), ibid., 54, 646), is the basis for its participation in the chelation process. The order of binding affinities recently shown for riboflavine alone is: H > Cr(III) > UO_2(II) > Al(III) > Be(II) > Pb(II) (R. Nayan and A. K. Dey (1972), Ind. J. Chem., 10, 109).

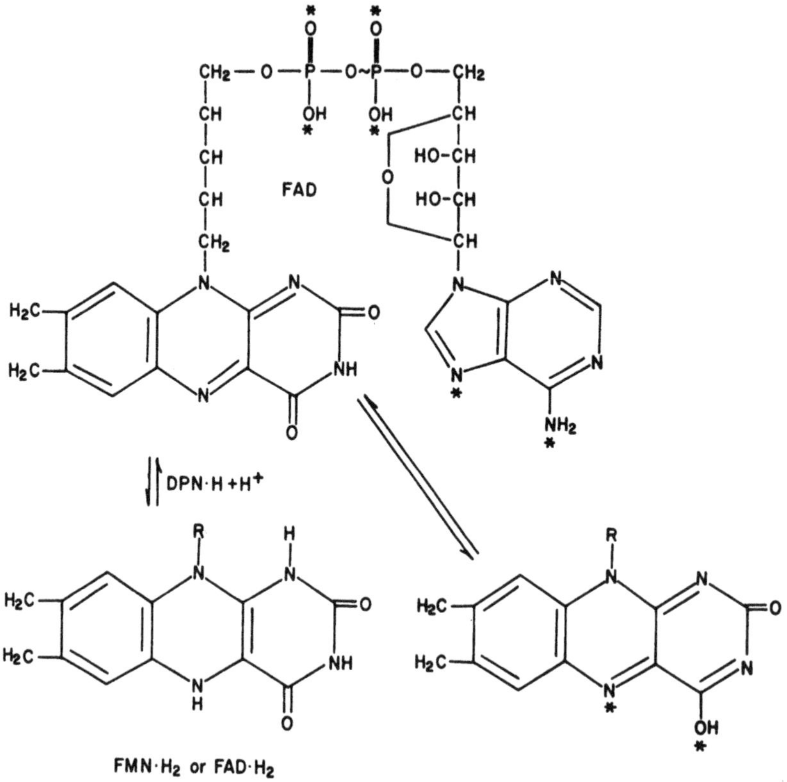

FAD

DPN·H + H^+

FMN·H_2 or FAD·H_2

The biochemical activities of a large number of enzymes are undoubtedly based on chelation reactions. For example, the methyltransferase enzymes found in many animal tissues require a cofactor, S-adenosylmethionine (SAM), to donate methyl groups to oxygen or nitrogen atoms of the substrate. In one such enzyme, catechol-O-methyltransferase, present in rat liver, the catechol groups of a large array of substrates can react with a variety of divalent metal cations and SAM to form a ligand-metal-ligand chelate. Methyl-proton transfer is visualized by Axelrod ((1965), Transmethylation and Methionine Biosynthesis, Chapter 5, S. K. Shapiro and F. Schlenk, eds., University of Chicago Press) as follows:

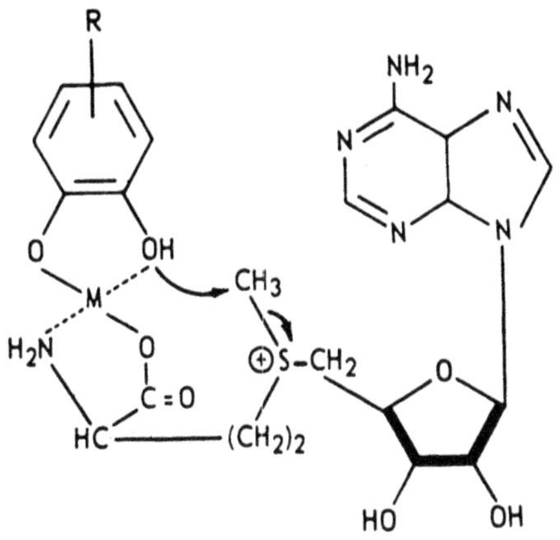

(Reproduced by permission of the University of Chicago Press)

Axelrod has proposed a "bridge complex" to link the enzyme, substrate, metal, and cofactor. More specific details of such a possible protein attachment are given by Zappia, et al. ((1969), J. Biol. Chem., 244, 4499).

In another enzymatic reaction involving SAM in E. coli it is the 3-carbon residue of the decarboxylated methionine moiety which is transferred (perhaps by a similar chelation scheme, although the questionable need for a divalent metal suggests caution). Putrescine, $H_2N\text{-}(CH_2)_4\text{-}NH_2$, is thereby converted to spermidine, $H_2N\text{-}(CH_2)_3\text{-}NH\text{-}(CH_2)_4\text{-}NH_2$, and presumably also to spermine, $H_2N\text{-}(CH_2)_3\text{-}NH\text{-}(CH_2)_4\text{-}NH\text{-}(CH_2)_3\text{-}NH_2$ (see A. White, P. Handler, and E. L. Smith (1968), Principles of Biochemistry, 4th ed., New York: McGraw-Hill, pp. 591-593).

CONCLUDING COMMENTS

In view of the rapid enlargement in our understanding of the wide variety of biochemical reactions occurring in living tissues, discussion of only one particular type of reaction would appear to be rather one-sided. Nevertheless, despite this caveat, a valid purpose is served in drawing attention to the pervasive array of reactions designated by the term chelation. There can be little doubt that chelation reactions have widespread importance in biological systems because of their effects in changing the charge, size, shape, transport, and reactivity of a large number of organic molecules; in regulating the concentration of critical polyvalent metal cations with which they can react; and, probably of utmost importance, in controlling the biological activity of many enzyme systems.

The examples of naturally occurring chelating ligands given here barely scratch the surface. The chelates formed with iron, for example, were purposely ignored because of their coverage in the article by J. B. Neilands. Discussion of the therapeutic application of synthetic chelating agents in living systems, a fascinating and expanding research area in which this writer and his colleagues have been working for many years, is of necessity out of the purview of this article.

As already pointed out, the literature on chelation reactions in living systems is widely scattered, and appending a bibliography, however long, would become something of an exercise in futility. In addition to the references already cited, however, the reader wishing to acquaint himself with this field may find the following books and articles useful, both from an historic point of view and as orientation for a further search of the literature dealing with specific aspects of chelation reactions in living systems.

Finally, it may be worth noting that a new journal, Bioinorganic Chemistry, published by American Elsevier, should now be included among the growing list of specialized journals dealing with reactions of naturally occurring organic and inorganic molecules.

SUGGESTED READING

Albert, A. (1968), Selective Toxicity, 4th ed., London: Methuen. Note particularly Chapter 9, pp. 297-352.

Brown, D. H. and MacPherson, J. (1972), J. Inorg. Nucl. Chem., 34, 1705.

Chaberek, S. and Martell, A. E. (1959), Organic Sequestering Agents, New York: John Wiley.

Davies, I. J. T. (1972), The Clinical Significance of the Essential Biological Metals, London: Heinemann.

DiStefano, V. and Neuman, W. J. (1953), J. Biol. Chem., 200, 759.

Dwyer, F. P. and Mellor, D. P. (1964), Chelating Agents and Metal Chelates, New York: Academic Press.

Korolkovas, A. (1970), Essentials of Molecular Pharmacology, New York: Wiley-Interscience.

Martell, A. E. and Calvin, M. (1952), Chemistry of the Metal Chelate Compounds, New York: Prentice-Hall.

Morgan, G. T. and Drew, H. D. K. (1920), J. Chem. Soc. 117, 1456.

Schubert, J. (1966), Scientific American, 214, 40.

Schwarzenbach, G. (1952), Helv. Chimica Acta, 35, 2344.

Seven, M. J. and Johnson, L. A. (1960), Metal Binding in Medicine, Philadelphia: Lippincott.

Sillen, L. G. and Martell, A. E. (1964), Stability Constants of Metal-Ion Complexes, 2nd ed., London: The Chemical Society.

APPLICATION OF PHYSICAL METHODS TO THE STUDY OF ENZYMES CONTAINING BOUND MANGANESE: PROBLEMS AND PROSPECTS*

MICHAEL C. SCRUTTON

Department of Biochemistry
Temple University School of Medicine
Philadelphia, Pennsylvania 19140

G. H. REED

Department of Biophysics and Physical Biochemistry
Johnson Research Foundation
University of Pennsylvania School of Medicine
Philadelphia, Pennsylvania 19140

A. S. MILDVAN

The Institute for Cancer Research
Fox Chase, Philadelphia, Pennsylvania 19111

INTRODUCTION. THE DISCOVERY OF ENZYMES CONTAINING BOUND MANGANESE

Although manganese has been long recognized as an essential micronutrient for a variety of organisms (1), the physiological basis for this metal requirement remained obscure for many years. Thus, despite the demonstration that numerous enzymes which exhibit a requirement for activation by Me^{2+}, e.g., kinases, synthetases, some dehydrogenases, can utilize Mn^{2+} at lower concentrations than Mg^{2+}, the relative concentrations of these two metal ions present in most tissues and species (2) indicate that Mg^{2+} is likely to serve as the activating metal ion under *in vivo* conditions. The recent discovery of bound manganese as a component of several proteins has

*Acquisition of the unpublished data to which reference is made in this article as well as the preparation of the article were aided by National Science Foundation Grant No. GB-30223 and by USPHS Grant No. RR0542. G.H.R. is a recipient of a Career Development Award (1K04-AM70134) from the National Institute of Arthritis, Metabolic and Digestive Diseases.

provided a rationale for the unique biological role of this metal ion. Nuclear magnetic resonance (NMR) studies were responsible for providing the first indication of the existence of protein-bound manganese. In these studies pyruvate carboxylase from chicken liver was found to cause a marked increase in the longitudinal nuclear magnetic relaxation rate ($1/T_1$) of water protons indicating the presence of a bound paramagnetic component in this enzyme. Subsequent analytical studies identified the paramagnetic component as manganese and demonstrated that this bound metal ion was present in equimolar ratio with the biotin residues which are an integral part of the active site of pyruvate carboxylase (3,4). Although the presence of manganese appears characteristic of the enzyme obtained from avian liver (3,5), pyruvate carboxylases purified from other sources contain zinc (*Saccharomyces cerevisiae*) (6) or, in part, magnesium (calf liver) (5) as the bound metal ion. Subsequently, several other manganese-proteins have been discovered including concanavalin A (7); manganin, a phytohaemagglutinin obtained from peanut seeds (8); superoxide dismutases purified from *Escherichia coli* (9); and *Streptococcus mutans* (10); and avimanganin, a protein of unknown function which is present in chicken and rat liver mitochondria (11).

It is of interest that when chickens are raised on a diet essentially devoid of manganese, a catalytically active carboxylase is elaborated which contains bound magnesium in place of manganese (5). Since, in contrast to the bacterial enzyme, the superoxide dismutase of vertebrate tissues contains bound copper and zinc (12, 13), the perotic syndrome, which is characteristic of manganese deficiency in this species (14), may indicate the existence of a further biological role for manganese which has not yet been identified.

Proteins containing bound manganese provide ideal systems for application of physical methods such as NMR and EPR to the elucidation of the role of the bound metal ion (cf. 15). In the remainder of this article we hope to illustrate the use of these, and some other, physical methods using, for the most part, data obtained in studies on pyruvate carboxylase and avimanganin. In addition, we will review the original interpretation of some of the earlier data in the light of more recent studies on these, and other, systems which have focused our attention on certain problems which are peculiar to macromolecular systems.

ANALYSIS OF THE EFFECT OF Mn(II) ON THE LONGITUDINAL ($1/T_1$) AND TRANSVERSE ($1/T_2$) NUCLEAR MAGNETIC RELAXATION RATES OF WATER AND LIGAND MAGNETIC NUCLEI. THE ENHANCEMENT PHENOMENON.

Before discussing studies performed with manganese metalloproteins, it is necessary to summarize briefly some theoretical aspects. The paramagnetic contribution to the longitudinal relaxation rates ($1/T_{1p}$) and transverse ($1/T_{2p}$) of a magnetic nucleus, e.g., a water proton, in an Mn(II)-complex, e.g., $Mn(H_2O)_6^{2+}$, is given by equation (1) (16,17):

$$\frac{1}{T_{1p}(T_{2p})} = \frac{pq}{T_{1M}(T_{2M}) + \tau_M} + \frac{1}{T_1(T_2)_{os}} \qquad (1)$$

In equation (1) p is the ratio of the concentrations of Mn(II) to that of the magnetic nucleus; q, the number of coordinated magnetic nuclei; τ_M, the mean residence time of the magnetic nucleus in the complex; and T_{1M}, T_{2M}, the longitudinal and transverse relaxation rates of the coordinated nucleus. The value of $1/T_{1p}(T_{2p})$ is usually obtained from the observed relaxation rate ($1/T_1$ or $1/T_2$) by subtracting the relaxation rate of the buffer system ($1/T_1^o$ or $1/T_2^o$).

If the scalar term, which arises from contact interaction between Mn(II) and the magnetic nucleus (and contributes little to $1/T_{1p}$) is neglected, $1/T_{1M}$ reflects the extent of dipolar interaction between Mn(II) and the magnetic nucleus and is given by equation (2) (18,19):

$$\frac{1}{T_{1M}} = B\cdot\frac{f(\tau_c)}{r^6} \qquad (2)$$

where B is a combination of constants having a value of 819 for Mn(II)-proton interaction; r is the distance in Å between Mn(II) and the magnetic nucleus; and $f(\tau_c)$ is defined by equation (3):

$$f(\tau_c) = \frac{3\tau_{c1}}{1 + \omega_I^2\tau_{c1}^2} + \frac{7\tau_{c2}}{1 + \omega_s^2\tau_{c2}^2} \qquad (3)$$

In equation (3) ω_I, ω_s are, respectively, the nuclear and electronic frequencies; and τ_{c1}, τ_{c2} are defined by equations (3a) and (3b) (20):

$$\frac{1}{\tau_{c1}} = \frac{1}{\tau_r} + \frac{1}{T_{1e}} + \frac{1}{\tau_M} \qquad (3a)$$

$$\frac{1}{\tau_{c2}} = \frac{1}{\tau_r} + \frac{1}{T_{2e}} + \frac{1}{\tau_M} \qquad (3b)$$

In equations (3a) and (3b) τ_r is the rotational correlation time which in many instances is identified as the tumbling time of the complex; and T_{1e}, T_{2e} are the longitudinal and transverse electron spin relaxation times. T_{1e} is often designated as τ_s in the biochemical literature.

In $Mn(H_2O)_6^{2+}$ the tumbling time of the complex (τ_r)(3×10^{-11} sec) (20) is several orders of magnitude shorter than T_{1e} (1×10^{-8} sec) (21) or τ_M (2.7×10^{-8} sec) (16) and, therefore, dominates τ_c for Mn(II)-water proton dipolar interaction in the aquocation. Hence, when another small ligand, e.g., oxalate, interacts with $Mn(H_2O)_6^{2+}$ displacing water molecules, $1/T_{1p}$ of the water protons decreases since this ligand substitution decreases q but has little effect on τ_r. If, however, a larger ligand such as ATP^{4-} is introduced, a slight increase in $1/T_{1p}$ of water protons is observed on complex formation since the decrease in q is more than offset by a decrease in τ_r which reflects the influence of the larger ligand on the tumbling time of the complex (22). Since the size and shape of the component ligands in a given complex determine the tumbling time, it is clear that interaction of Mn(II) with a macromolecule is likely to lead to a significant increase in $1/T_{1p}$ of water protons. This effect, which is known as the enhancement phenomenon, was first described by Cohn and Leigh for proteins (23), and by Eisinger, Shulman, and Szymanski for nucleic acids (24). In such macromolecular complexes τ_c may, in fact, no longer be dominated by τ_r since (i) the tumbling time of the complex, which is in the range of 10^{-8} - 10^{-7} secs for most Mn(II)-proteins, is considerably slower than for the manganese aquocation (3×10^{-11} sec); and (ii) significant shortening of T_{1e}, and possibly also of τ_M, may occur in a macromolecular complex if the metal ion is located in an asymmetric ligand field (21,25). Although processes having a correlation time shorter than the tumbling time of the entire complex _may_ determine τ_r (and, hence, possibly τ_c) in some instances, this appears not to be the case in most Mn(II)-enzyme complexes which have been examined. For these complexes the temperature and frequency dependence of $1/T_{1p}$ indicates that electron spin relaxation is the process which modulates the dipolar interaction, i.e., τ_c in these complexes is identified as T_{1e} (26,27).

If the process which modulates the dipolar interaction between Mn(II) and the magnetic nucleus can be identified from the temperature and frequency dependence observed for $1/T_{1p}$ (cf. 15,28) and a value assigned to τ_c, much valuable information is accessible utilizing equations (1)-(3). Two specific examples are considered here:

1. _Determination of a coordination scheme by estimation of q for water protons in a series of complexes._ From equation (1) calculation of q for water protons requires that values be obtained for $1/T_{1p}$ and either $1/T_{1M}$ and/or $1/\tau_M$. In the case where $1/T_{1M}$ determines the Mn(II)-water proton dipolar interaction, i.e., where $1/T_{1p}$ is not exchange limited, such a calculation of q requires the assignment of values for τ_c, which can be obtained from examination of the temperature and frequency dependence of $1/T_{1p}$; for r, the Mn(II)-water proton distance and for $1/T_{1(os)}$ (equation (3)). A value of 2.86Å is usually assigned for r in such calculations since

this is the average Mn(II)-water proton distance obtained from x-ray studies of Mn(II) in various crystal lattices (2) and $1/T_{1(os)}$ is typically assumed to be negligible compared to $1/T_{1p}$. The predominant contribution to $1/T_{1p}$ is thus assumed to be provided by the protons of water molecules which are directly coordinated to the Mn(II) and the contribution from water molecules which are not directly coordinated is taken to be not significantly greater than that characteristic of $Mn(H_2O)_6^{2+}$. Such an assumption is likely to be justified for complexes of Mn(II) with small ligands, but is much less secure in the case of macromolecular complexes. Thus, water molecules, which are not directly coordinated, may be immobilized on the macromolecule in the vicinity of the bound Mn(II) and, hence, $1/T_{1(os)}$ may no longer be insignificant as compared to $1/T_{1p}$. The "outer sphere" contribution to $1/T_{1p}$ may be of special importance for macromolecular complexes in which no, or very few, water molecules are directly coordinated to the bound Mn(II). Since no satisfactory method is available at present for estimation of the relative values of $1/T_{1(os)}$ (outer sphere) and $1/T_{1p}$ (inner sphere), care must be exercised if the relative values of q for a series of complexes calculated from equation (3), but assuming $1/T_{1p} >> 1/T_{1(os)}$ form the basis for a postulated coordination scheme. Such values are designated here as q^*_{obs}. It is also apparent that coordination schemes derived from the ratios of the enhancements observed for a series of macromolecular complexes are of very questionable validity, especially when $1/T_{1p}$ is small.

2. Detection of enzyme-Mn(II)-ligand bridge complexes by calculation of Mn(II)-ligand nuclei distances. When $1/T_{1p}$ for a magnetic nucleus present in a ligand, e.g., a substrate proton, is measured, the Mn(II)-magnetic nucleus distance (r) may be determined using equation (2) if (i) $1/T_{1p}$ is not limited by the rate of ligand exchange into the coordination sphere of the bound Mn(II) (τ_M); (ii) q is known or assumed; and (iii) a value is obtained for τ_c, the correlation time for the process(es) which modulates the dipolar interaction. The first two criteria generally cause few problems since exchange limitation is excluded if $1/T_{2p}$ is significantly greater than $1/T_{1p}$ (15), and q for the ligand is usually taken as 1 unless there is good reason to suspect multiple coordination. However, despite the relative insensitivity of r to the precise value of τ_c ($\tau_c \propto r^{1/6}$) (equation (3)), serious errors in interpretation have occurred when τ_c is assigned rather than calculated directly from either the frequency dependence of $1/T_{1p}$ (28) or, if $\tau_c < 10^{-9}$ sec., from the T_{2p}/T_{1p} ratio (30). Use of the latter method requires the explicit assumption that the scalar contribution to $1/T_{2p}$ is negligible. Hence, measurement of the frequency dependence of $1/T_{1p}$ is the preferred method for determination of τ_c.

Two procedures which have been employed for assignment of a value for τ_c give some cause for concern. In the first of these τ_c for the Mn(II)-enzyme-ligand complex is calculated from τ_c for the analogous Mn(II)-ligand complex and the enhancement factor obtained by comparison of $1/T_{1p}$ for the ligand nucleus in these two complexes (31). This calculation is, however, valid only if neither the structure of the complex nor the dipolar relaxation mechanism are altered by the presence of the macromolecule (vide supra). Alternatively, the correlation function, $f(\tau_c^*)$, for the Mn(II)-enzyme-ligand complex has been calculated from equation (4) (32,33):

$$\varepsilon_T = \frac{f(\tau_c^*)\cdot q^*}{f(\tau_c)\cdot q} \qquad (4)$$

In equation (4) ε_T is the enhancement factor calculated for $1/T_{1p}$ of water protons and q^*, the coordination number for water molecules in the Mn(II)-enzyme-ligand complex while $f(\tau_c)$ and q are, respectively, the correlation function and coordination number of $Mn(H_2O)_6^{2+}$. In using this latter procedure one must <u>determine</u> a value for q^* and must assume that τ_c for Mn(II)-ligand nucleus dipolar interaction is identical with that for Mn(II)-water proton dipolar interaction in the Mn(II)-enzyme-ligand complex. This latter assumption is likely to be justified if τ_c is dominated by T_{1e}, but may be questionable if other mechanisms contribute to the dipolar relaxation. Determination or assignment of a value for q^* may, however, present more severe problems due to uncertainty of the extent of the outer sphere contribution to ε_T. Although r is relatively insensitive to the value of q^* if this is finite, use of equation (4) is clearly not appropriate if $q^* = 0$, i.e., no rapidly exchanging water molecules remain in the first coordination sphere of the bound Mn(II).

However, calculation of the Mn(II)-ligand nucleus distance (r) from $1/T_{1p}$ provides <u>conclusive</u> evidence for formation of an enzyme-Mn(II)-ligand bridge complex only if the relaxing nucleus is <u>directly</u> coordinated to the metal ion. For other nuclei in the ligand calculation of r can exclude that a metal bridge complex is formed but does not provide unequivocal evidence for the existence of such a species. Detection of scalar interaction between Mn(II) and the ligand nucleus provides further evidence supporting direct coordination of the ligand since this relaxation mechanism operates only through chemical bonds (34). If τ_c is less than ω_I^{-1} (i.e., 2.6×10^{-9} sec at 60 MHz), the extent of scalar interaction can be calculated from the difference between $1/T_{2p}$ and $1/T_{1p}$ using equation (5) and is expressed as a coupling constant (A/h) (20):

$$\frac{1}{T_{2M}} - \frac{7}{6T_{1M}} = 114\ \tau_e (A/h)^2 \qquad (5)$$

In equation (5) τ_e is the correlation time for scalar interaction and is defined by equation (5a):

$$\frac{1}{\tau_e} = \frac{1}{T_{1e}} + \frac{1}{\tau_M} \tag{5a}$$

This approach generally cannot be used to detect scalar interaction in macromolecular complexes of Mn(II) since T_{1e}, which typically modulates the dipolar interaction, may be greater than ω_I^{-1} (27,28). Thus, long correlation times also contribute to the inequality of $1/T_{2p}$ and $1/T_{1p}$ and, hence, the scalar contribution to this inequality cannot be resolved.

Some of the problems which arise in the interpretation of relaxation rate studies are well illustrated using data obtained for the interaction of substrates, cofactors, and inhibitors with the bound manganese of pyruvate carboxylase.

NMR STUDIES ON THE ROLE OF THE BOUND Mn(II) IN PYRUVATE CARBOXYLASE

Since at 25° the bound Mn(II) of pyruvate carboxylase has a fourfold greater effect on $1/T_1$ of water protons than an equimolar concentration of $Mn(H_2O)_6^{2+}$ (3), it appeared possible to use the PRR method to probe the role of the bound metal ion. Substrates and inhibitors of the transcarboxylation partial reaction (Reaction 2) (35) cause marked reductions in this enhanced effect of the bound manganese whereas substrates and cofactors of the carboxylation partial reaction (Reaction 1) when added singly or in combination had no significant effect (Table 1) (36):

$$\text{E-biotin} + MeATP^{2-} + HCO_3^- \xrightleftharpoons{Me^{2+},\ Me^{+},\ \text{Acetyl-CoA}} \text{E-biotin} \sim CO_2 + MeADP^- + P_i \tag{1}$$

$$\text{E-biotin} \sim CO_2 + \text{Pyruvate} \rightleftharpoons \text{E-biotin} + \text{Oxalacetate} \tag{2}$$

The kinetic significance of the observed effects was established by comparison of dissociation constants for the enzyme-ligand complex (as obtained from the decrease in $1/T_{1p}$ as a function of ligand concentration) with inhibitor constants determined in initial rate studies of the overall reaction and/or dissociation constants

Table I. The enhanced effect of the bound manganese of pyruvate carboxylase on $1/T_1$ of water protons in the presence and absence of substrates, alternate substrates, cofactors and inhibitors.

Ligand	$\epsilon_b(\epsilon_t)$ (at 25°)[a]	q^*_{obs}[b]
A. No Additions	4.2	0.38 (0.30)
B. Substrates and Cofactors of the Carboxylation Partial Reaction (Reaction 1)		
ATP, Mg^{2+}, acetyl-CoA, HCO_3^-, or P_i (singly or in combination)	3.9-4.2	0.35-0.38
C. Substrates and Alternate Substrates of the Transcarboxylation Partial Reaction (Reaction 2)		
Pyruvate	1.7	0.15
Oxalacetate	2.3	0.21
α-Ketobutyrate	0.5	0.05
β-Methyloxalacetate	0.6	0.05
D. Inhibitors of the Transcarboxylation Partial Reaction (Reaction 2)[c]		
Type I (oxalate, oxamate, fluoropyruvate, phenylpyruvate)	0.2-0.4	0.02-0.04

Table I. Continued.

Ligand	$\varepsilon_b(\varepsilon_t)$ (at 25°)[a]	q^*_{obs}[b]
Type II (mesoxalate, malonate, tartronate)	0.3-0.5	0.04-0.05
Type III (L-malate)	0.3	0.04

[a] Data summarized from the studies of Mildvan, Scrutton, *et al*. (4,36).

[b] Calculated using $\tau_c = 3.2 \times 10^{-9}$ sec. The figure in parenthesis is calculated using $\tau_c = 6.5 \times 10^{-9}$ sec. which is obtained from the relationship $\tau_c = 1/\omega_I$ for studies of $1/T_1$ of water protons at 24.3 MHz (4). In this calculation r (the Mn-water proton distance) is taken as 2.86Å. Thus, if the calculated value of q^*_{obs} is significantly less than 1.0, as is the case here, it is clear that the value assumed for r is too small, i.e., that no rapidly exchanging water molecules are directly coordinated to the bound Mn(II). For example, if q is taken as 1 for the Mn(II)-pyruvate carboxylase complex, r may be calculated as approximately 3.5Å. The calculation of q^*_{obs} also assumes that τ_c for the Mn(II)-water proton dipolar interaction is similar for all these complexes. Although alterations in T_{2e} and, hence, probably in T_{1e}, are suggested by the narrowing of the EPR resonances on addition of ligands (cf. Fig. 3), it is unlikely that the resultant effect on τ_c will alter the interpretation of these data.

[c] The inhibitors are divided into three types on the basis of (i) the nature of the inhibition observed with respect to pyruvate and/or oxalacetate as the variable substrate; (ii) the extent of protection observed against inactivation by avidin at infinite [inhibitor]; and (iii) the nature of the relationship between $1/\Delta T_{1p}$ (water protons) and 1/[inhibitor] (36).

obtained by measurement of the reactivity of the biotin residues with avidin as a function of ligand concentration (36).

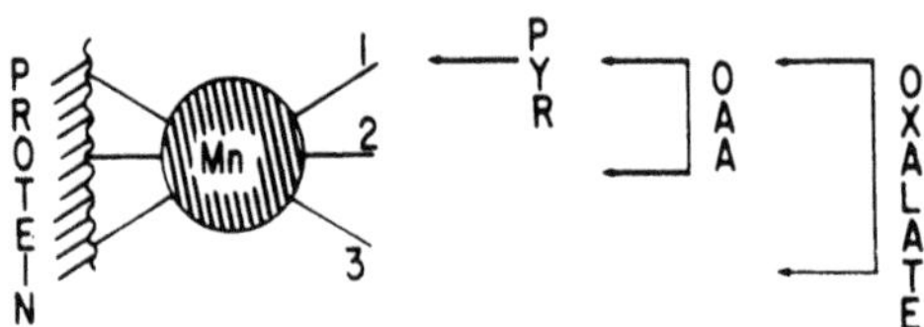

Figure 1. Coordination scheme for bound Mn(II) of pyruvate carboxylase with H_2O, pyruvate (PYR), oxalacetate (OAA), and oxalate which was proposed on the basis of the enhancement ratios for these complexes (cf. 36).

On the basis of the observed values of $\varepsilon_b(\varepsilon_T)$ (Table I) a tentative coordination scheme (Fig. 1) was proposed for the interaction of the substrates and inhibitors of the transcarboxylation partial reaction with the bound Mn(II) (36). In this scheme it was proposed that the inhibitors occupied two, and the substrates one, of three cis ligand positions which were not preempted by the protein and were, therefore, accessible to the solvent. Several additional premises were used in derivation of the tentative scheme (36) and may be summarized as follows: (i) the observation of $\varepsilon_b > 1.0$ indicates that the bound Mn(II) is accessible to the solvent; (ii) observation of reduction in ε^* to a value less than 1.0 on addition of a ligand, e.g., oxalate ($\varepsilon_T = 0.3$), indicates that this ligand is directly coordinated to the bound Mn(II); (iii) the similarity of ε_T for the pyruvate carboxylase-oxalate and Mn(II)-EDTA complexes suggests that one water molecule remains in the first coordination sphere of the bound Mn(II) in the former complex since x-ray crystallographic studies have shown the presence of a single coordinated water molecule in Mn(II)-EDTA (37). Data obtained in studies of the effect of pyruvate carboxylase on $1/T_1$ and $1/T_2$ of the methyl protons of pyruvate and the methylene protons of oxalacetate appeared to support the proposed coordination scheme (Fig. 1) since, using assigned values for τ_c in the range 10^{-10} to 10^{-12} sec., Mn(II)-proton distances consistent with direct coordination were calculated for both of these enzyme-substrate complexes (31,39). However, the distance estimates obtained were too imprecise to define the structure of these complexes. In particular, neither these nor other data have provided convincing evidence in support of the bidentate structure for the pyruvate carboxylase-oxalacetate complex (Fig. 1) which was originally proposed on the basis of considerations of mechanism; and despite the observed similarity in ε_T to the pyruvate carboxylase complex (Table I) (32).

However, more recent studies (40,41) have cast doubt on the earlier interpretation of the relaxation rate data. The first indication that all might not be well with the proposed coordination scheme (Fig. 1) was obtained from EPR studies of the various pyruvate carboxylase complexes. These studies are discussed below. However, the extent of the problem is perhaps best illustrated by calculation of q^*_{obs} for the various complexes using a mean value for τ_c (3.2×10^{-9} sec) obtained from the frequency dependence of $1/T_{1p}$ for the methyl protons (40), carbonyl carbon and carboxyl carbon (41) of pyruvate in the pyruvate carboxylase-pyruvate complex. In addition, q^*_{obs} for Mn(II)-pyruvate carboxylase is also calculated using a value of τ_c (6.5×10^{-9} sec) obtained from studies of $1/T_1$ of water protons at 24.3 MHz. in the presence of this complex. Since the variation of $1/T_{1p}$ with the reciprocal of the absolute temperature consistently exhibits a maximum in the range 20-25° (4), the correlation time which modules the Mn(II)-water proton dipolar interaction may be obtained from equation (6) (42):

$$\tau_c = \frac{1}{\omega_I} \tag{6}$$

Using these values for τ_c, q^*_{obs} for the Mn(II)-pyruvate carboxylase complex is obtained as 0.30-0.38 (Table IA). Hence, even in the absence of ligands which cause reductions in ε^* (Table I) less than one rapidly exchanging water molecule appears to be present in the first coordination sphere of the bound Mn(II). The bound metal may, therefore, be inaccessible to the solvent and the apparent paramagnetic contribution to $1/T_1$ of water protons may result from dipolar interaction between the bound Mn(II) and water molecules which are not directly coordinated to the metal ion. It is of particular interest that, as predicted by Reed, et al. (28), an enhanced effect of Mn(II) on water proton relaxation rates ($\varepsilon_b = 4.2$) (Table IA) can be observed in such a system indicating that the magnitude of the outer sphere interaction is much greater than might be anticipated from studies on complexes of Mn(II) with small ligands. The large outer sphere effect in the pyruvate carboxylase system can probably be attributed to interaction of the bound Mn(II) with water molecules which are transitorily immobilized by the structure of the protein in the vicinity of the bound metal ion.

In the presence of substrates and inhibitors of the transcarboxylation partial reaction (Reaction 2) q^*_{obs} decreases markedly to values of 0.15-0.2 (substrates) and < 0.05 (alternate substrates and inhibitors) (Table IC and D). Although the original interpretation of the $\varepsilon_T/\varepsilon_b$ ratios (cf. Fig. 1) now appears incorrect, it is apparent that binding of these ligands causes a decrease in the extent of Mn(II)-water proton dipolar interaction which is especially severe in the case of the enzyme-inhibitor complexes. This decrease may be due to either displacement of the "immobilized" water molecules as a direct or indirect result of binding of these ligands, or alternatively, to an increase in the Mn(II)-water proton distance resulting from changes in protein conformation which accompany ligand binding. Evidence for conformational changes on interaction with these ligands is provided by observation of altered rates of inactivation of pyruvate carboxylase resulting from incubation either with avidin or at 2° (36,43). In the case of the enzyme-inhibitor complexes the processes which accompany binding of these ligands appear to cause the bound Mn(II) to become essentially inaccessible to even outer sphere interaction with the solvent (vide) infra).

This reinterpretation of the data obtained for the effect of the bound Mn(II) on $1/T_1$ of water protons in the various pyruvate carboxylase-ligand complexes is in accord with data obtained in two further studies. Measurement of $1/T_{1p}$ and $1/T_{2p}$ at two frequencies for the methyl protons (40), and the ^{13}C-carbonyl and-carboxyl carbons (41), of pyruvate has permitted determination of τ_c for the dipolar interaction. All the values obtained fall in the same range with a mean of 3.2 (± 1.2) × 10^{-9} sec. In the case of the pyruvate methyl protons $1/T_{1p}$ is markedly frequency-dependent but at any given frequency shows little variation with temperature over the range 0-35° (31,40). Such behavior suggests that electron spin relaxation (T_{1e}) may modulate this dipolar interaction (15). Using the τ_c values calculated from the original data, Mn - ^1H and Mn - ^{13}C (carbonyl and carboxyl carbons) distances have been calculated for the pyruvate carboxylase-pyruvate complex (40,41), and a summary of these data is presented in Figure 2. From crystallographic studies and

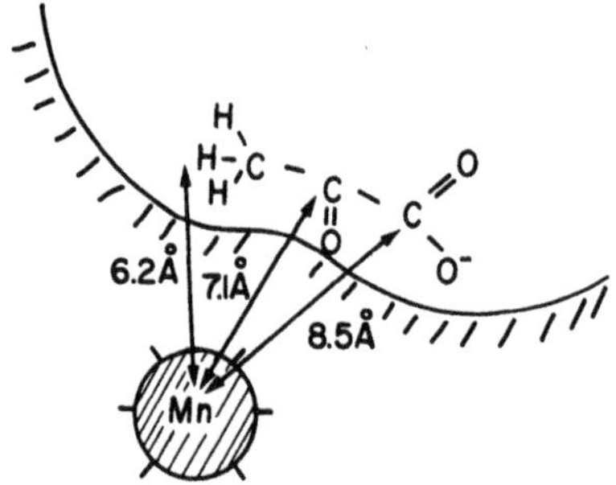

Figure 2. Manganese-proton and manganese-carbon distances in the pyruvate carboxylase-pyruvate complex obtained in ^1H and ^{13}C nuclear magnetic relaxation studies. These distances were calculated using the value for τ_c determined from the frequency dependence of $1/T_{1p}$.

molecular model studies (cf. 41) of an Mn(II)-pyruvate complex, the maximal Mn(II)-carbon distances consistent with direct coordination are 2.9Å (Mn(II)-carbonyl carbon:carbonyl coordination) and 3.5Å (Mn(II)-carboxyl carbon:carboxyl coordination). In all of these complexes the Mn(II)-methyl proton distance would be in the range 4.6-5.4Å (31). Comparison of these distances with those calculated from $1/T_{1p}$ (Fig. 2) clearly indicates that the pyruvate molecule which is observed in the NMR studies is not coordinated to the bound Mn(II) in the enzyme-pyruvate complex although the distance between the binding site for this substrate and the bound metal is such that strong dipolar interaction is observed. Although similar experiments have not been performed in the case of oxalacetate, the other substrate of the transcarboxylation partial reaction (Reaction 2), recalculation of the Mn(II)-methylene proton distance from the published value of $1/pT_{1p}$ (39) using $\tau_c = 3.2 \times 10^{-9}$ sec. gives a value (11.5Å) which also appears inconsistent with direct coordination of this substrate by the bound Mn(II) in the pyruvate carboxylase-oxalacetate complex.

However, complexes for which the residence time (τ_M) of the ligand on the metal ion exceeds 10^{-3} sec. would not have been detected in these studies. Hence, at present, we cannot exclude the existence of pyruvate carboxylase-Mn(II)-substrate bridge complexes in which the substrates exchange slowly on the bound metal ($1/\tau_M < 10^{-3}$ sec). Equilibrium dialysis experiments designed to resolve this question in the case of pyruvate are presently in progress.

These conclusions are at variance with earlier reports in which Mn(II)-proton distances consistent with formation of enzyme-Mn(II)-substrate bridge complexes were calculated from $1/pT_{1p}$ using assigned values to τ_c in the range 10^{-10} - 10^{-13} sec. (3,39). The discrepancy with the earlier reports lies in the value of τ_c since the recent determination of this parameter has given a value an order of magnitude longer than the longest assigned in the earlier studies. This situation emphasizes the importance of obtaining an experimental value for τ_c in calculations of metal-ligand distances from nuclear relaxation data. Recent data obtained for complexes of Mn(II) with macromolecules (27,28,40,41) suggests that T_{1e} for $Mn(H_2O)_6^{2+}$ (10^{-8} sec) would provide a plausible upper limit for calculation of the distance between Mn(II) and the ligand nucleus in macromolecular complexes.

The characteristics of the EPR spectra observed for the bound Mn(II) in the complexes of pyruvate carboxylase with various ligands also appear consistent with the absence of direct interaction between the bound metal and the substrates and inhibitors of the transcarboxylation partial reaction. Some representative EPR spectra are shown in Figure 3. Before analyzing the data of Figure 3, it is necessary to examine two of the parameters which determine the characteristics of EPR spectra of Mn(II) complexes. These parameters are (i) the zero-field splitting (ZFS) and (ii) the transverse electron spin relaxation time (T_{2e}). The zero-field splitting which determines the resonance positions of the fine structure transitions[3] in the EPR spectrum reflects the extent to which the coordination sphere of the bound Mn(II) deviates from perfect cubic symmetry (44). Substitution of ligands in the first coordination sphere will, in most instances, change the symmetry of this coordination sphere. Hence, such substitution appears likely to be accompanied by a change in the ZFS except

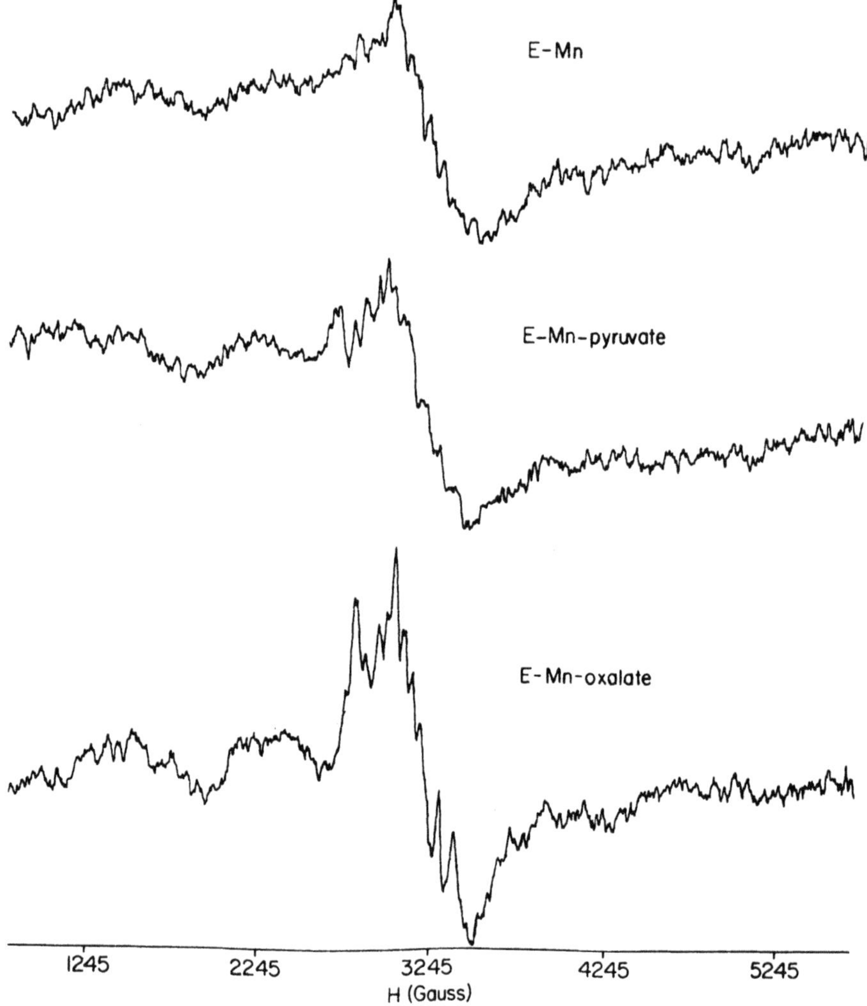

Figure 3. X-band (9 G.Hz) EPR spectra of pyruvate carboxylase, pyruvate carboxylase-pyruvate and pyruvate carboxylase-oxalate.

for the unusual case for which insertion of the new ligand has little effect on coordination sphere symmetry. In macromolecular systems changes in the ZFS cannot, however, be used as specific indicators of ligand substitution since interaction of a ligand at another site on the macromolecule may cause conformational changes which are responsible for the observed alteration of the symmetry of the coordination sphere. The width of the resonances in the EPR spectrum is determined, in part, by the transverse electron spin relaxation time (T_{2e}) and may be used to set an upper limit on this parameter in a given complex. Since electron spin relaxation results from transient perturbations of the coordination sphere caused by collision with outer sphere solvent molecules, the width of the EPR resonances provides an indication of the accessibility of the Mn(II) site to outer sphere interaction with the solvent (41A). Decreased accessibility of the site is reflected in a narrowing of the resonance lines, i.e., a longer relaxation time. A more complete discussion of factors which influence the EPR spectra of macromolecular complexes containing Mn(II) may be found in earlier publications (44,45).

In the case of pyruvate carboxylase examination of Figure 3 reveals that the positions of the EPR resonance (and, hence, the ZFS) are identical in the free enzyme (E-Mn) and in the complexes with pyruvate and oxalate. However, complex formation is accompanied by narrowing of the resonances which approximately parallels the decrease in q^*_{obs} in these complexes (Table I). This narrowing is especially striking for the pyruvate carboxylase-oxalate complex (q^*_{obs} = 0.02-0.05) (Table ID). Both the water proton relaxation and EPR data, therefore, indicate that even outer sphere interaction of the solvent with the bound Mn(II) is minimal in this complex.

Hence, all the nuclear magnetic relaxation and EPR data suggest that the bound Mn(II) of pyruvate carboxylase is buried in the enzyme protein in the vicinity of the site at which the transcarboxylation partial reaction occurs, and may not interact directly either with the solvent or with the substrates or inhibitors of the transcarboxylation partial reaction. Certain other observations also appear to be more readily understood in the context of this revised view of the relationship between the bound Mn(II) and the catalytic site of pyruvate carboxylase. First, Mn(II) is extremely tightly bound to pyruvate carboxylase. Reversible dissociation of the tetrameric enzyme ($\overline{M}_w \approx 500{,}000$) to the protomers ($\overline{M}_w \approx 125{,}000$) does not cause release of the bound Mn(II). Such release occurs only when severely denaturing conditions are employed, e.g., incubation with 1% sodium doecyl sulfate (3). Second, no evidence has been obtained for interaction of the bound Mn(II) with, or inhibition of the enzyme by, chelating agents other than oxalate (36). Restricted access to the site could explain the lack of interaction with large chelating agents, e.g., EDTA, 1,10-phenanthroline, but this postulate provides no rationale for the absence of interaction with CN^- despite the high affinity of this ion for $Mn(H_2O)_6^{2+}$ (47a). In the revised postulate inhibition by oxalate may be attributed to interaction as a transition state analog (45a), since it is isoelectronic with the enol form of pyruvate. It is of interest that several other enzymes which utilize pyruvate as substrate are also subject to potent inhibition by this dicarboxylic acid (46,47). For at least one of these enzymes (lactate dehydrogenase) inhibition by oxalate cannot be attributed to interaction with bound metal ion (48,49), indicating the absence of a correlation between the inhibitory and chelating properties of this molecule.

The mechanism proposed previously to describe the role of the bound metal ion is, therefore, not established insofar as this mechanism requires the formation of enzyme-Mn(II)-pyruvate and enzyme-Mn(II)-oxalacetate bridge complexes (36). The NMR and EPR studies have detected complexes in which the substrates are bound in the vicinity of, but not directly coordinated to the bound Mn(II). However, these studies have failed to provide any insight regarding the ligands which are present in the first coordination sphere of the bound metal. The observation of a primary isotope rate effect, and of carboxylation with retention of configuration (50), while consistent with the original proposal, do not, however, require the participation of the bound Mn(II) in either the nucleophilic activation of the methyl group of pyruvate or the geometry of carboxyl transfer. Hence, the relationship, if any, of the bound Mn(II) to catalysis of the transcarboxylation partial reaction is unclear, although the nature of the bound metal present in the protein does appear to influence the conformation of this catalytic site. For example, substitution of Mg(II) for Mn(II) alters both the apparent K_m for pyruvate and the concentration of oxalate required to cause 50% inhibition of oxalacetate synthesis in the presence of a fixed concentration of pyruvate (5). Furthermore, all pyruvate carboxylases examined in detail thus far have been found to contain bound metal in a stoichiometry equivalent to the biotin concentration (51). This correlation, which extends to the related enzyme methylmalonyl-CoA-oxalacetate transcarboxylase (52) also appears suggestive of a role either in catalysis or in the maintenance of the conformation of the catalytic site.

SPATIAL RELATIONSHIPS BETWEEN THE CATALYTIC SITES OF PYRUVATE CARBOXYLASE

In addition to defining the relationship between the bound Mn(II) and the transcarboxylation site, NMR and EPR studies can also provide insight into the spatial relationship between the carboxylation and transcarboxylation catalytic sites which, together with the biotinyl residue, constitute the active site of pyruvate carboxylase (cf. Fig. 4). Steady state kinetic characterization of the pyruvate carboxylase reaction has indicated that the mechanism of oxalacetate synthesis by this enzyme is best described by a hybrid Ping Pong mechanism (53) in which the carboxylation (Reaction 1) and transcarboxylation (Reaction 2) partial reactions occur at kinetically independent sites on the enzyme (54). Communication between the two sites which are presumably separated in space is provided by the biotinyl residue which is mounted on a long flexible arm with a maximal extension of approximately 14Å (4). Indirect evidence supporting the concept of spatial separation between the sites for the two partial reactions is provided by (i) the absence of significant changes in ε_b when components of the carboxylation partial reaction are added singly or in various combinations (Table 1) (36); (ii) the absence of any effect of acetyl-CoA on the EPR spectrum of the bound Mn(II) (since this cofactor is specifically required for the

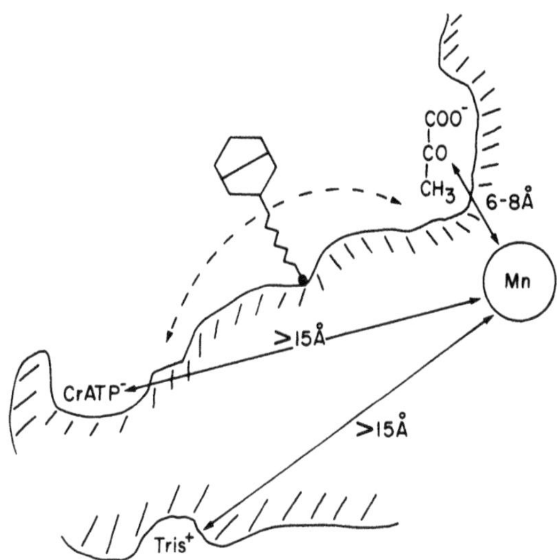

Figure 4. Spatial relationships in the active site of pyruvate carboxylase.

carboxylation partial reaction) (35); and (iii) the observed differences in the effect of substrates of the carboxylation and transcarboxylation partial reactions on the rate of inactivation of pyruvate carboxylase by avidin which reflects the environment of the biotinyl residues (36,55). However, none of these approaches provides an estimate of the distance between the two catalytic subsites. This distance can be estimated if a paramagnetic probe bound at the carboxylation sub-site increases the line width of the resonances in the EPR spectrum of the bound Mn(II) as a result of spin-spin interaction between the two paramagnetic species (cf. 56, 57). If no broadening of the resonances is observed under conditions where binding of both species is established, the distance between the sites is greater than that at which spin-spin interaction can be observed, i.e., approximately 15Å (56).

A suitable carboxylation site probe is provided by the chromium complex of ATP ($CrATP^-$) which is non-dissociable at pH's below 7.0 and is a potent competitive inhibitor of pyruvate carboxylase with respect to $MgATP^{2-}$ ($K_i \approx 10\mu M$) (58). Furthermore, since binding of $CrATP^-$ and added divalent metal ion, e.g., Mg^{2+}, Mn^{2+}, to pyruvate carboxylase is mutually interdependent (as is also the case for $MgATP^{2-}$ and Me^{2+}) (58,59), the decrease in amplitude of the EPR spectrum of added Mn^{2+} in the presence of the enzyme provided a convenient indicator of the interaction with $CrATP^-$ (60).

When such experiments were conducted, addition of 100μM $CrATP^-$ in the presence or absence of 0.5 mM Mg^{2+} as the added divalent metal ion had no significant effect on the EPR spectrum of the bound

Mn(II) in the pyruvate carboxylase-oxalate complex. Binding of $CrATP^-$ to the enzyme under these conditions was indicated by a decrease in the amplitude of the EPR spectrum of Mn^{2+} added as the divalent metal ion in place of Mg^{2+}. Hence, these data indicate that the binding site for $CrATP^-$, which is tentatively identified as the $MeATP^{2-}$ region of the carboxylation site, is at least 15Å from the bound Mn(II) which itself lies very close to the transcarboxylation site. The oxalate complex was employed in these studies in order to increase the sensitivity of the method by sharpening the resonance lines (cf. Fig. 3). Oxalate has previously been shown to be a specific inhibitor of the transcarboxylation partial reaction which has no effect on the rate of the carboxylation partial reaction at the concentrations employed in this study (0.1 mM) (36). Hence, these data provide direct evidence for spatial separation of the catalytic sites for the two partial reactions which has been inferred for this, and also for other, biotin enzymes as a result of kinetic and structural studies (cf. 61).

Additionally, insight has also been obtained into the spatial relationship between the bound Mn(II) and the binding site for $Tris^+$ which is an activator of the carboxylation partial reaction catalyzed by pyruvate carboxylase (58). When the effects of Mn(II)- and Mg(II)-pyruvate carboxylase on the nuclear magnetic relaxation rates of the Tris methylene protons are compared, it is found that both enzymes cause similar increases in $1/T_1$ and $1/T_2$, indicating the absence of a paramegnetic contribution to these increased relaxation rates. Hence, no spin-spin interaction occurs between the bound Mn(II) and the $Tris^+$ protons. Since addition of Rb^+ abolishes the effect of pyruvate carboxylase on the $Tris^+$ methylene resonance over a concentration range consistent with the activator constants determined for these monovalent cations in initial rate studies (58), the diamagnetic increases in $1/T_1$ and $1/T_2$ appear to result from kinetically significant binding to the Me^+ activator site. The absence of a detectable paramagnetic contribution then indicates that the Me^+ site is at least 15Å from the bound Mn(II) (56). Although, on the basis of these data and the specific involvement of Me^+ in the carboxylation partial reaction, it is tempting to suggest that the Me^+ site may be in the vicinity of the carboxylation sub-site, the spatial relationship between these sites has not yet been defined.

ANALYSIS OF THE VISIBLE ABSORPTION SPECTRA OF MACROMOLECULAR COMPLEXES CONTAINING HIGH SPIN Mn(III)

The discussion thus far has focused on two techniques (NMR and EPR) which when used in combination with steady state kinetic and ligand binding analysis provide the most incisive approach to the study of macromolecular complexes containing divalent manganese. However, more recently, several proteins have been described which appear to contain bound manganese in the form of high spin Mn(III), e.g., superoxide dismutase purified from E. coli and S. mutans (9, 10), avimanganin (11). Although high spin Mn(III) has limited potential as a paramagnetic probe due primarily to a short electron spin relaxation time ($T_{1e} \simeq 10^{-12}$ sec) (62), some insight regarding the environment and possible role of the bound metal in catalysis may be obtained by analysis of the visible and near infra-red absorption spectrum. In contrast to Mn(II) which exhibits very weak visible absorption bands in either octahedral (ε = 0.01-0.04 M^{-1} cm^{-1}) or tetrahedral (ε = 1.0-4.0 $M^{-1}cm^{-1}$) coordination, the visible absorption bands characteristic of octahedral complexes of high spin Mn(III) are relatively intense with extinction coefficients in the range 100-500 $M^{-1}cm^{-1}$ (63) and, hence, are readily observed in macromolecular complexes. Furthermore, the location and intensity of the visible absorption band(s) are sensitive functions of the nature of the ligands and of the coordination geometry since high spin d^4 ions such as Mn(III) are subject to Jahn-Teller distortions (63). Such effects are seen in the visible absorption spectra of model high spin Mn(III) complexes (64). Insight into the environment of bound Mn(III) in avimanganin and the bacterial superoxide dismutases may, therefore, be obtained by comparison of the visible absorption spectra observed for these proteins with these reported for the model complexes. In the case of avimanganin this comparison suggests that the manganese may be buried in the structure of the protein since the spectrum bears a close resemblance to model tris-bidentate complexes in which oxygen and nitrogen atoms are liganded to the metal ion (11,64). The effect of avimanganin on $1/T_1$ and $1/T_2$ of water protons under various conditions appears consistent with the proposed absence of direct interaction between the bound Mn(II) and the solvent, although other possibilities cannot be excluded (11).

The visible absorption spectrum of superoxide dismutase from S. mutans appears similar to that observed for avimanganin except that the major absorption band is shifted to shorter wavelength (10). Consideration of the spectra of the model complexes (64) suggests that such a shift is expected if the bound Mn(III) of superoxide dismutase is accessible to the solvent.

CONCLUSION

The studies described above provide two examples of the extent of information which can be derived when a variety of physical techniques are applied to the study of a biochemical system containing a paramagnetic transition metal ion. In addition, the discussion outlines the evolution of our insight into the relationship between the bound Mn(II) and the active site of pyruvate carboxylase and emphasizes the problems which may arise in the interpretation of magnetic resonance data obtained in studies on macromolecular complexes. Our experience suggests that such problems are most readily avoided if multiple approaches are employed in characterization of the system. For example, the initial misinterpretation of the relaxation rate data obtained from pyruvate carboxylase can be traced to a failure to recognize that in macromolecular complexes processes which exhibit time constants in the range 10^{-8} - 10^{-9} sec. could modulate the dipolar interaction between the ligand nuclei and the bound Mn(II). This possibility was first suggested when EPR analysis failed to show the anticipated changes in the ZFS on addition of ligands and was confirmed by a direct calculation of τ_c from the frequency dependence of $1/T_{1p}$. However, despite such problems in interpretation of the data, it is clear from these and many other (cf. 15) studies that magnetic resonance analysis provided unique insights into the relationship between bound Mn(II) and the structure and/or function of the macromolecule which contains this metal.

APPENDIX

1. In an external magnetic field nuclei with a finite magnetic moment align their magnetic vectors in the magnetic field in a discrete manner. For nuclei with a spin value of 1/2, e.g., protons, the moments are aligned either with or against the direction of the external magnetic field. At equilibrium the distribution of nuclei in each direction is given by the Boltzman factor:

$$\frac{N_{higher}}{N_{lower}} = e^{-\Delta E/kT}$$

where N_{higher} is the population in the higher energy orientation (against the magnetic field for protons), and ΔE is the energy separation of the two levels or orientations. The spin lattice, or longitudinal, relaxation time (T_1) is the time constant for return of the spins to this equilibrium distribution after the distribution is perturbed from equilibrium by, for example, a radio frequency field of energy ΔE.

The transverse relaxation time is a phase memory time constant for the ensemble of nuclei. This relaxation mechanism does not require an exchange of energy with the lattice or surroundings. Every T_1 relaxation event is also a T_2 event; thus, $T_2 \leq T_1$. In addition, there are several other phenomena which create a spread in resonance frequencies and, hence, cause a loss of phase memory.

2. The correlation time τ_c is the time constant for periodic interruption of the magnetic field between two magnetic dipoles, e.g., a nuclear spin and an electron spin. The resultant fluctuating fields produced by this interruption have components at the nuclear resonance frequency. The intensity of the fluctuating field at the resonance frequency depends on τ_c and is at a maximum $\omega_I\tau_c = 1$.

Hence, the efficiency of the dipolar relaxation mechanism hinges on the time constant τ_c.

3. The fine structure splitting referred to here should be distinguished from the nuclear hyperfine splitting which results from the presence of ^{55}Mn (44).

REFERENCES

1. Cotzias, G. C. (1958), Physiol. Revs., 38, 503.

2. Thiers, R. E. and Vallee, B. L. (1957), J. Biol. Chem., 226, 911.

3. Scrutton, M. C., Utter, M. F., and Mildvan, A. S. (1966), J. Biol. Chem., 241, 3480.

4. Scrutton, M. C. and Mildvan. A. S. (1968), Biochemistry, 7, 1490.

5. Scrutton, M. C., Griminger, P., and Wallace, J. C. (1972), J. Biol. Chem., 247, 3305.

6. Scrutton, M. C., Young, M. R., and Utter, M. F. (1970), J. Biol. Chem., 245, 6620.

7. Agrawal, B. B. L. and Goldstein, I. J. (1968), Arch. Biochem. Biophys., 124, 218.

8. Dieckert, J. W. and Rozacky, E. (1969), Arch. Biochem. Biophys., 134, 373.

9. Keele, B. B., McCord, J. M., and Fridovich, I. (1970), J. Biol. Chem., 245, 6176.

10. Vance, P. G., Keele, B. B., and Rajagopalan, K. V. (1972), J. Biol. Chem., 247, 4782.

11. Scrutton, M. C. (1971), Biochemistry, 10, 3897.

12. Carrico, R. J. and Deutsch, H. F. (1970), J. Biol. Chem., 245, 723.

13. McCord, J. M. and Fridóvich, I. (1969), J. Biol. Chem., 244, 6049.

14. Wilgus, H. S., Norris, L. C., and Heuser, G. L. (1936), Science, 84, 252.

15. Mildvan, A. S. and Cohn, M. (1970), Adv. in Enzymol., 33, 1.

16. Swift, T. J. and Connick, R. E. (1962), J. Chem. Phys., 37, 307.

17. Luz, Z. and Meiboom, S. (1964), J. Chem. Phys., 40, 2686.

18. Solomon, I. (1955), Phys. Rev., 99, 559.

19. Bloembergen, N. (1957), J. Chem. Phys., 27, 572.

20. Eisinger, J., Shulman, R. G., and Szymanski, B. M. (1962), J. Chem. Phys., 36, 1721.

21. Luz, Z. and Shulman, R. G. (1965), J. Chem. Phys., 43, 3750.

22. Mildvan, A. S. and Cohn, M. (1966), J. Biol. Chem., 241, 1178.

23. Cohn, M. and Leigh, J. S. (1962), Nature, 193, 1037.

24. Eisinger, J., Shulman, R. G., and Blumberg, W. E. (1961), Nature, 192, 963.

25. Letter, M. S., Grant, M. W., Wood, E. J., Dodgen, H. W., and Hunt, J. P. (1972), Inorg. Chem., 11, 2701.

26. Reuben, J. and Cohn, M. (1970), J. Biol. Chem., 245, 6539.

27. Reed, G. H., Diefenbach, H., and Cohn, M. (1972), J. Biol. Chem., 247, 3066.

28. Peacocke, A. R., Richards, R. E., and Sheard, B. (1969), Mol. Physics, 16, 177.

29. Montgomery, H., Chastain, R. V., and Lingalfelter, E. C. (1966), Acta Crystallogr., 20, 731.

30. Leigh, J. S. (1971), Ph.D. Thesis, Univ. of Pennsylvania.

31. Mildvan, A. S. and Scrutton, M. C. (1967), Biochemistry, 6, 2978.

32. Nowak, T. and Mildvan, A. S. (1972), Biochemistry, 11, 2819.

33. Jones, R., Dwek, R. A., and Walker, I. O. (1972), Europ. J. Biochem., 28, 74.

34. Barfield, M. and Karplus, M. (1969), J. Am. Chem. Soc., 91, 1.

35. Scrutton, M. C., Keech, D. B., and Utter, M. F. (1965), J. Biol. Chem., 240, 574.

36. Mildvan, A. S., Scrutton, M. C., and Utter, M. F. (1966), J. Biol. Chem., 241, 3488.

37. King, J. and Davidson, N. (1958), J. Chem. Phys., 29, 787.

38. Hoard, J. L., Pedersen, B., Richards, S., and Silverton, J. V. (1961), J. Am. Chem. Soc., 83, 3533.

39. Scrutton, M. C. and Mildvan, A. S. (1970), Arch. Biochem. Biophys., 140, 131.

40. Reed, G. H. and Scrutton, M. C. Unpublished data.

41. Fung, C. H., Mildvan, A. S., Allerhand, A., Komoroski, R., and Scrutton, M. C. (1973), Biochemistry, 13, in press.

41A. Reed, G. H., Leigh, J. S., and Pearson, J. E. (1971), J. Chem. Phys., 55, 3311.

42. Cohn, M. and Reuben, J. (1971), Accounts. Chem. Res., 4, 214.

43. Irias, J. J., Olmsted, M. R., and Utter, M. F. (1969), Biochemistry, 8, 5136.

44. Reed, G. H. and Ray, W. J. (1971), Biochemistry, 10, 3190.

45. Reed, G. H. and Cohn, M. (1972), J. Biol. Chem., 247, 3073.

45A. Wolfenden, R. (1972), Accounts. Chem. Res., 5, 10.

46. Reed, G. H. Unpublished data.

47. Novoa, W. B., Winer, A. D., Glaid, A. J., and Schwert, G. W. (1959), J. Biol. Chem., 234, 1143.

48. Terayama, M. and Vestling, C. (1956), Biochem. Biophys. Acta, 20, 586.

49. Pfleiderer, G., Jackel, D., and Wieland, T. (1958), Biochem. Z., 330, 296.

50. Rose, I. A. (1970), J. Biol. Chem., 245, 6025.

51. Scrutton, M. C. and Young, M. R. (1972), vol. 6, p. 1, The Enzymes, 3rd ed., P. D. Boyer, ed., New York: Academic Press.

52. Northrop, D. B. and Wood, H. G. (1969), J. Biol. Chem., 244, 5801.

53. Northrop, D. B. (1969), J. Biol. Chem., 244, 5808.

54. Barden, R. E., Fung, C. H., Utter, M. F., and Scrutton, M. C. (1972), J. Biol. Chem., 247, 1723.

55. Scrutton, M. C. and Utter, M. F. (1965), J. Biol. Chem., 240, 3714.

56. Leigh, J. S. (1970), J. Chem. Phys., 52, 2608.

57. Taylor, J. S., Leigh, J. S., and Cohn, M. (1969), Proc. Natl. Acad. Sci. U. S., 64, 219.

58. Scrutton, M. C. Unpublished data.

59. Bias, R. and Keech, D. B. (1972), J. Biol. Chem., 247, 3255. McClure, W. R., Lardy, H. A., and Kneifel, H. P. (1971), J. Biol. Chem., 246, 3569.

60. Cohn, M. and Townsend, J. (1954), Nature, 173, 1090.

61. Moss, J. and Lane, M. D. (1971), Adv. Enzymol., 35, 321.

62. Schwartz, R. W. and Carlin, R. L. (1970), J. Am. Chem. Soc., 92, 6763.

63. Cotton, F. A. and Wilkinson, D. (1966), Advanced Inorganic Chemistry, 2nd ed., New York: Interscience.

64. Dingle, R. (1966), Acta Chem. Scand., 20, 33.

CRYSTALLOGRAPHY OF A METAL-CONTAINING PROTEIN, CONCANAVALIN A

KARL D. HARDMAN

Division of Biological and Medical Research
Argonne National Laboratory
Argonne, Illinois 60439

INTRODUCTION

Concanavalin A* is a protein of the class called lectins. These proteins agglutinate or bind to the surfaces of various types of cells, specifically by interaction with carbohydrates on the surface membranes, and are of current interest in various types of cancer research. Many proteins of this class have been referred to as phytohemagglutinins or phytoagglutinins (Boyd and Shapleigh, 1954), but the term <u>lectin</u> has come into prominance more recently because not all of these substances are found in plants. Some have been isolated from invertebrates and lower vertebrates such as fish (Pardoe and Uhlenbruck, 1970). A short review on lectins by Sharon and Lis (1972) has recently appeared and is an excellent source for additional references.

Most lectins are metalloproteins, with the metal ions being a prerequisite for carbohydrate binding, and Con A requires divalent transition metal ions (Kalb and Levitzki, 1968). Carbohydrates with the minimum configuration for Con A binding contain residues with the <u>arabino</u> configuration for carbons 3, 4, and 5 (Poretz and Goldstein, 1970) and unmodified hydroxyl groups at carbons 3, 4, and 6. αMeMan<u>p</u> has been found to be the most specific monosaccharide and αMeGlc<u>p</u> which differs only by the configuration of C-2 binds 1/4 as strongly (Goldstein, <u>et al</u>., 1965).

---

*Abbreviations used are: Con A, concanavalin A;
αMeMan<u>p</u>, α-methyl-D-mannopyranoside;
αMeGlc<u>p</u>, α-methyl-D-glucopyranoside;
βIphGlc<u>p</u>, β-(<u>o</u>-iodophenyl)-D-glucopyranoside;
βIphGal<u>p</u>, β-(<u>o</u>-iodophenyl)-D-galactopyranoside.

The ability to bind carbohydrate leads to some very unusual biological effects. For example, Con A agglutinates erythrocytes from various animal species, starch granules, and some types of bacteria and yeasts, and has been shown to precipitate glycogens, dextrans, mannans, and various glycoproteins (Sumner and Howell, 1936a,b; So and Goldstein, 1968; Leon and Young, 1970). Large polysaccharides must have sugar residues with the minimum configuration on the non-reducing termini of the chains for Con A agglutinability (So and Goldstein, 1968). Con A induces transformations of lymphocytes by reversibly binding to specific sites on the cell surfaces (Powell and Leon, 1970; Novogrodsky and Katchalski, 1971). These transformed cells continue development to blast-like cells. Con A agglutinates embryonic tissue cells, whereas the binding sites in the corresponding adult cells appear to be masked (Moscona, 1971). Con A has also been shown to agglutinate leukemic cells and tissue culture cells which have been transformed by chemical carcinogens, viruses, or x-rays, whereas normal cells under the same conditions are not agglutinated (Inbar and Sachs, 1969). Sachs and coworkers (Inbar, *et al*., 1971) report that the site for Con A binding on the cell surface membrane of hamster cells has two components, one which actually binds Con A and another which is involved in the agglutination.

Nicolson (1971) has shown by electron microscopy of cells binding ferritin-conjugated Con A that the topographic distribution of the receptor sites on the cell membranes *differs* between virally-transformed fibroblasts and the normal cells. Also, he has shown (Nicolson, 1972) that a similar result is obtained after mild proteolysis of normal cell membranes and suggests that increased agglutination of transformed cells is not a result of synthesis of new sites or unmasking of sites, but migration of existing randomly-dispersed sites to clustered sites, favorably spaced for intercellular Con A cross-bridges. Other curious properties of Con A are its immunosuppressive action in mice (Markovitz, *et al*., 1969) and its interaction with some enveloped viruses, Sendai and herpes simplex, rendering them non-infectious (Okada and Kim, 1972). Con A has also been shown to inhibit the growth of various tumors *in vivo* (Shoham, *et al*., 1970).

Because of its carbohydrate binding properties, Con A has been used for the detection and purification of blood group substances (Lloyd, *et al*., 1969), glycopeptides from surface membranes of hepatoma cells (Smith, 1972) and from neuronal membranes (Gombos, *et al*., 1972), receptor sites from lymphocytes (Allan, *et al*., 1972), chorionic gonadotropin (Dufau, *et al*., 1972), and immunoglobulins (Nordin, *et al*., 1969). Moreover, it is now commercially available (Pharmacia Fine Chemicals, Inc.) covalently linked to Sepharose beads for uses in affinity chromatography.

Thus, Con A and other lectins have many properties that are

useful and unique for studies of the chemical architecture of cell surfaces and changes which these surfaces undergo during neoplastic transformations.

The features of the Con A molecule which made this protein appealing for x-ray crystallographic analysis were (1) the existence of some type of subunit structure, (2) its ability to specifically bind certain carbohydrates, and (3) a requirement for transition metal ions for the carbohydrate binding. I shall describe here the complete three-dimensional structure of Con A, with emphasis on the course of the polypeptide chain, the unique β-structure regions, the subunit structure, a site for binding non-polar molecules, a possible carbohydrate-binding site, and, of course, the binding sites for manganese and calcium. The experimental methods used to determine the structure are described elsewhere (Hardman, *et al.*, 1971a,b; Hardman and Ainsworth, 1972). Also note that an independent study of the crystal structure has been carried out in the laboratories of W. P. Lipscomb and G. M. Edelman (Quiocho, *et al.*, 1971; Edelman, *et al.*, 1972) and a tentative amino acid sequence has been reported (Edelman, *et al.*, 1972).

THE CRYSTAL STRUCTURE

Molecular Weight

The first report of crystalline Con A was by J. B. Sumner (1919) over 50 years ago. He studied some of the globular proteins from the jack bean, *Canavalia ensiformis*, and thereby isolated, crystallized, and named Con A. Note that this was about seven years before he reported the first crystalline enzyme, urease, from the same source (Sumner, 1926). Sumner and coworkers (1938) later obtained a molecular weight for Con A of 98,000 and more recent ultracentrifuge results have varied between 50,000 and 120,000 (Agrawal and Goldstein, 1967; Olson and Liener, 1967; Kalb and Lustig, 1968). Studies carried out below pH 6 indicated molecular weights near 50,000, whereas data obtained near pH 7 and above gave higher values. An explanation of these disparate results became clear on examination of low resolution electron density maps (Hardman, *et al.*, 1971a,b). In the crystalline state four *identical* subunits of 26,000 daltons each form a pseudotetrahedral cluster with a total molecular weight of 104,000. In solution below pH 6, these molecules then split into dimers. Each chemically unique subunit or protomer is a single polypeptide chain of 238 amino acids* with

*We currently have 237 amino acids built into our model. According to the tentative sequence (Edelman, *et al.*, 1972), residues 71 and 72 are Asx. This region is at a corner (see Figure 4) and the model fits the contours better by deleting one of these residues. In numbering the residues in the figures in this article, residue

approximately 1,900 non-hydrogan atoms. An atomic model of this monomer has been built on the scale of two centimeters per angstrom from an electron density map at 2.4Å resolution which was calculated by Ainsworth and Hardman from data of native crystals and 5 heavy atom derivatives.

Pleated Sheet Regions

The most striking feature of the general course of the polypeptide chain is the large amount of β-structure (or β-pleated sheets). In Con A there are 3 such regions, which consist entirely of anti-parallel strands. The atoms in these regions can easily be fitted into the electron density map while maintaining oxygen to nitrogen distances between strands very close to the hydrogen bond distances found by Pauling and Corey (1951) for silk fibers. The first of these regions is shown in Figure 1, which is an α-carbon

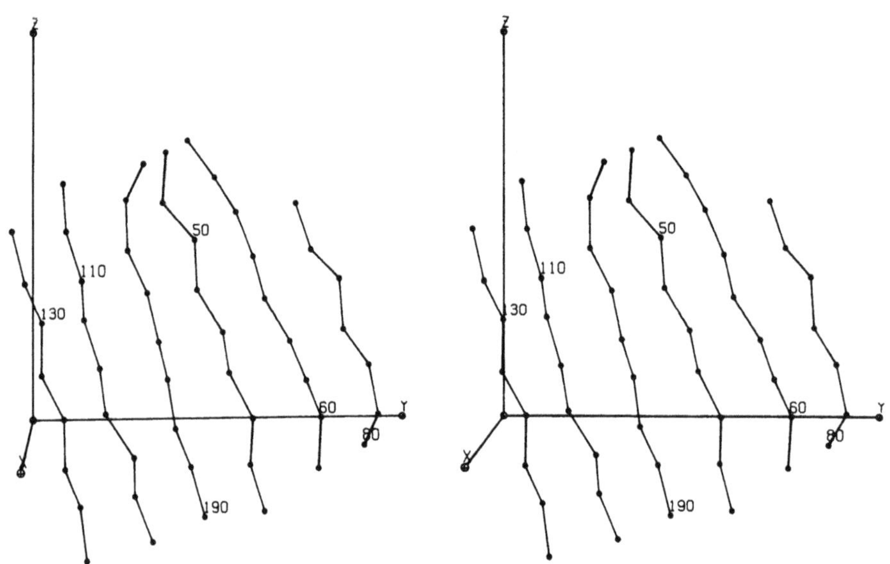

Figure 1. Alpha-carbon stereogram of the Sheet I β-structure region, which forms the back or "base" of each subunit.

stereogram produced by the ORTEP computer program written by Dr. C. K. Johnson (1965). This region is referred to as Sheet I and, as will be seen later, forms the "base" of the subunit. Each point in this figure represents the location of the α-carbon of each amino acid residue for this region and the 6 antiparallel

72 has been skipped so that the numbered amino acids correspond to those published by Edelman, et al. (1972).

strands present contain about 50 of the total 238 amino acids. The Sheet I region of one subunit continues across the X-axis, a crystallographic twofold rotation axis, and will be discussed later.

The 3-D arrangement of all three β-structure regions in the subunit is shown in Figure 2. Again, each amino acid residue is

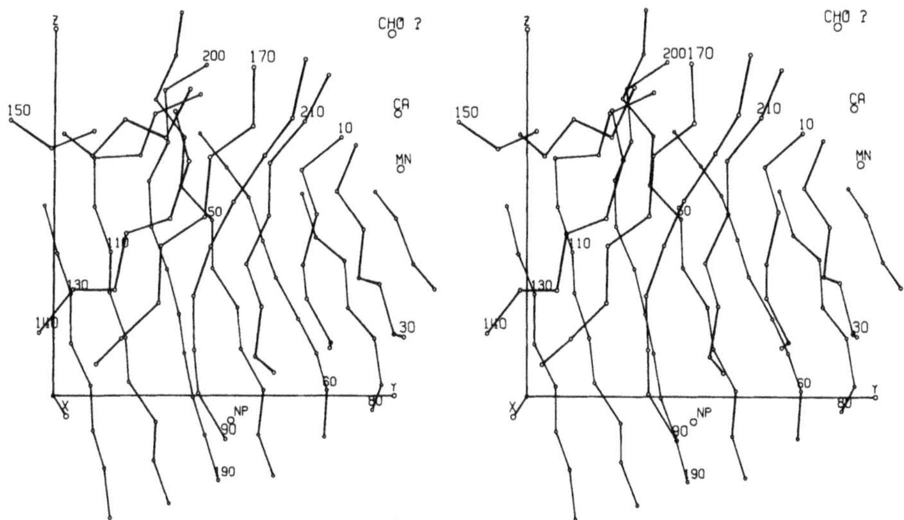

Figure 2. Alpha-carbon stereogram of the three β-structure regions.

represented by the α-carbon position only. Sheet I lies to the back, Sheet II through the center, and Sheet III, the smallest and least regular, is towards the top; in all, they involve 124 residues, or 52% of the total. The manganese, Mn^{2+}, and calcium sites are indicated here also. They are approximately 5Å apart and will be discussed later. The non-polar binding site is marked (NP) and lies about 22Å from the manganese ion. Figure 3 is a diagram of the hydrogen bond arrangement in the three pleated sheet regions. The numbers correspond to the amino acid sequence numbers published by Edelman, et al. (1972), and the slanted hydrogen bonds represent probable bifurcated bonds or positions where the regular bonding pattern of the pleated sheet is broken.

Completion of the Monomer

The completed polypeptide chain of the monomer is represented in Figure 4. Major sections of the backbone not found in the β-pleated sheet sections which thus complete the subunit are found

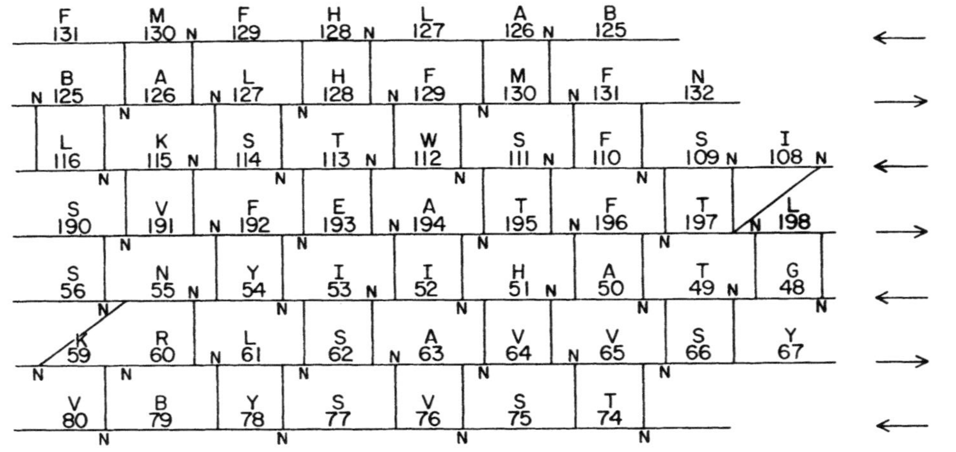

Figure 3a. β-Sheet I.

Figure 3. A diagram of the hydrogen bonding pattern and the amino acids involved in the three β-structure regions. Figures 3a, b, and c are Sheets I, II, and III, respectively. The amino acid residues are given by the standard 1-letter symbols [see Biochemistry, 5, 1445 (1972)] and the N-terminal to C-terminal directions of all strands are indicated by the arrows. In Figure 3a, the top strand is identical to the next strand but is rotated 180° around the X-axis and belongs to a second subunit.

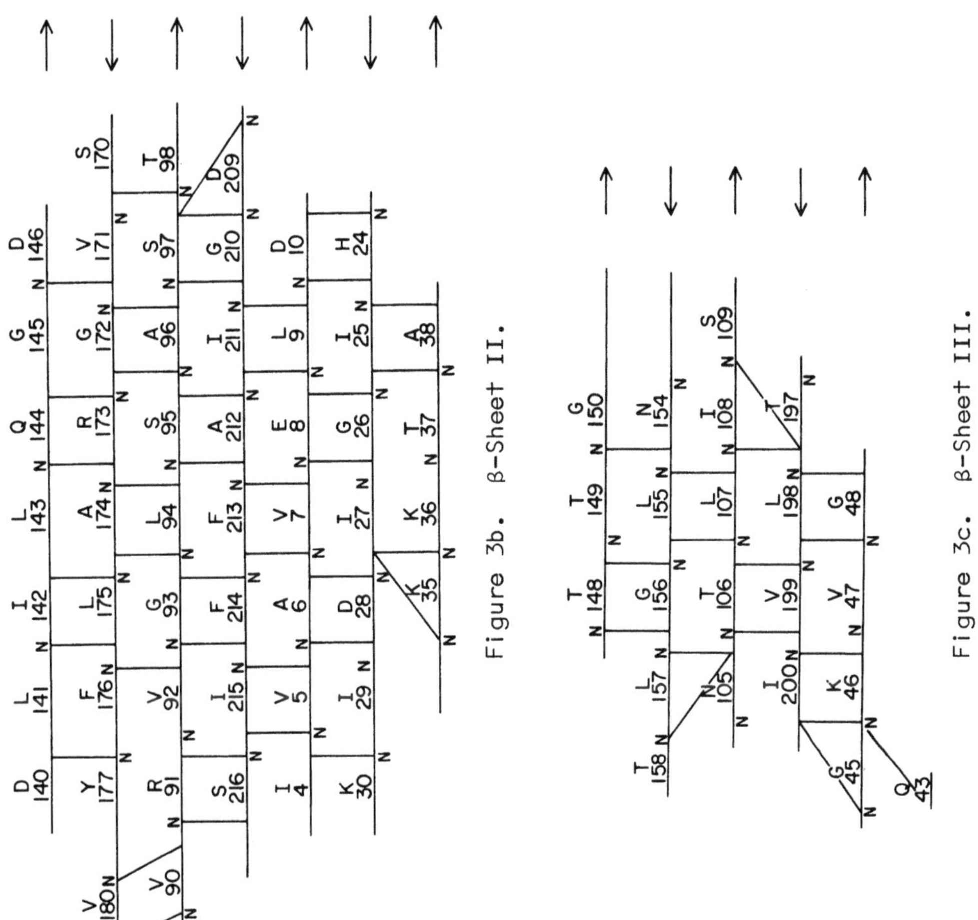

Figure 3b. β-Sheet II.

Figure 3c. β-Sheet III.

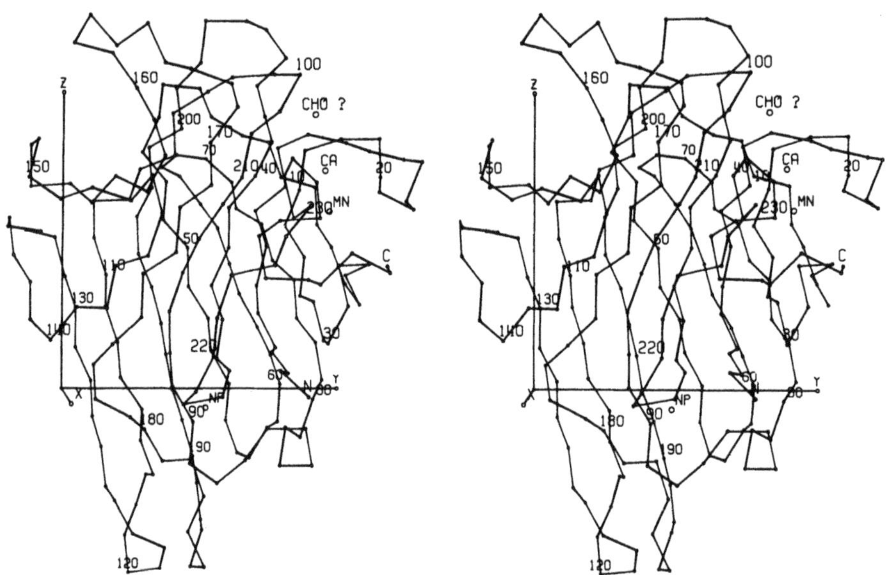

Figure 4. Alpha-carbon stereogram of the monomer. Additional positions marked are sites for the manganese (MN), calcium (CA), non-polar binding (NP), and possible carbohydrate binding (CHO?).

around the manganese and calcium sites (residues 11-23 and 30-35), near the top connecting Sheets I and II (131-140), and the upper right where a "loop" is found exposed to solvent (159-170). Another "loop" on the left side (116-124) connects the first two strands of Sheet I. Sheets I and II near the non-polar binding site are connected through residues 80-90 and 180-190. Finally, a section of 22 residues (217-238) terminates the chain.

While Sheet I is very flat and has the same type of "twist" from strand to strand as found in other proteins with large amounts of β-structure, such as carboxypeptidase A (Lipscomb, *et al.*, 1968) and carbonic anhydrase (Kannan, *et al.*, 1971), Sheet II is "cup-shaped" and appears to be unique. The lower-inside portion of the "cup" contains *polar* side chains involved in the manganese and calcium binding plus others which extend into solvent, whereas the upper-inside contains *non-polar* side chains which have many van der Waals interactions with non-polar side chains of the C-terminal region. The "outside" of this cup contains non-polar residues interacting with similar residues from Sheet I (Figs. 2 and 4), forming a non-polar "sandwich." The non-polar binding site (NP) and a likely candidate for the carbohydrate-specific site (CHO) are also marked (Fig. 4) and will be discussed later.

A portion of the 2.4Å electron density map is shown in Figure 5. This stereogram shows 12 serial sections (8Å thick) which contain

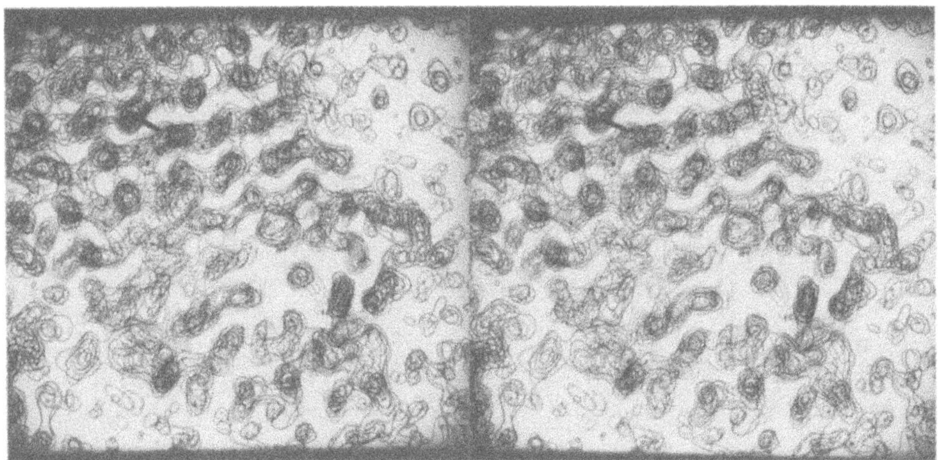

Figure 5. A portion of the 2.4Å electron density map. Note that the orientation of this stereogram (and Figure 11) is rotated 90° around the X-axis in comparison to all other figures. Phe 196 is incorrectly labeled 189. Reproduced from Hardman and Ainsworth (1972b) with permission of publishers.

most of the Sheet I region of the subunit. Four of the strands are clearly visible, beginning at the X-axis (solid column of dyad symbols near the upper left). The carbonyl oxygens appear as pronounced "bumps" on the chain and examples of their positions are marked by solid circles in the two strands nearest the X-axis. A few places can be seen where there are connecting positive electron density contours where a hydrogen bond is expected to bridge different strands. A number of large side chains are easily discernible, of which Trp-40 is most prominent. Immediately to its right is Met-42. Further examples of some of the larger side chains are Phe-196 and its neighbor to the right, Leu-198, which can be seen to be forked. On the left side of this stereogram, below the position of the non-polar binding, is Tyr-54 and nearby, Phe-192 and Trp-112. Smaller side chains also nearby are Ser-114 and -190. It is quite easy to see the subunit boundary in these sections, with solvent regions to the lower left and upper right. A few side chains from a <u>different</u> molecule appear in the lower-right corner.

The Dimer and Tetramer

Rotation of the subunit shown in Figure 5 around the X-axis by 180° produces an <u>equivalent</u> subunit to the upper left, of which only the first two strands can be seen. Figure 6 represents the resulting dimer. The six anti-parallel chains of Sheet I in the

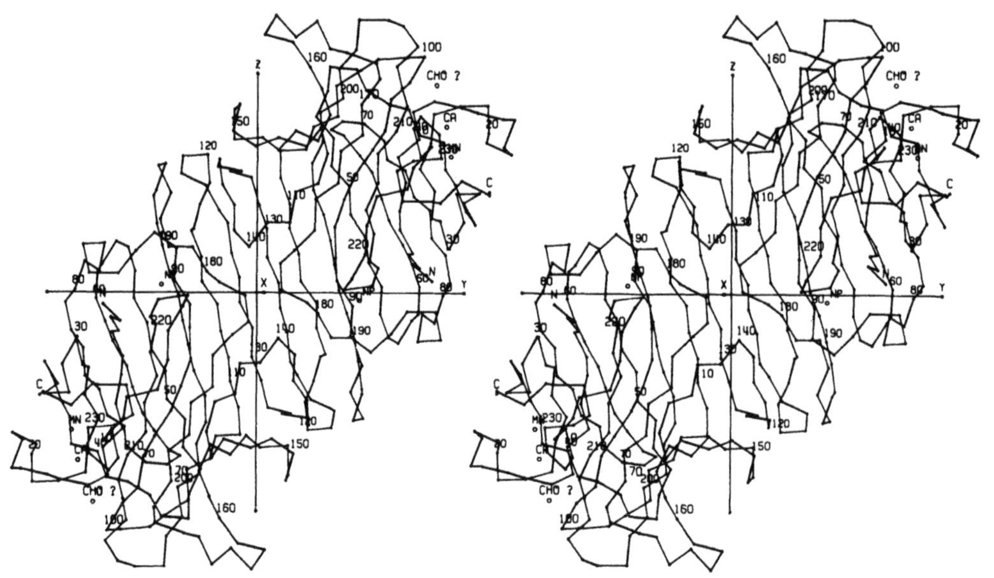

Figure 6. Alpha-carbon stereogram of the dimer.

first subunit are continued without interruption across the X-axis, forming a 12-strand pleated sheet. The distances between strands are quite uniform. The manganese and calcium sites are marked in both subunits of the dimer (Fig. 6).

Figure 7 gives a detailed look at the atoms in the two equivalent strands around the X-axis. It should be emphasized that this is a crystallographic twofold axis and not an approximate (or "local") axis. The sequence of the right-hand strand in the N to C direction (bottom to top) is Asx-Ala-Leu-His-Phe-Met-Phe and the side chain nearest the rotation axis is His-128. A structure very similar to this is found in insulin (Blundell, et al., 1972), where two strands of the B chains in the dimer form an antiparallel pleated sheet around a twofold axis, which contributes greatly to the stability of the dimer. In the case of insulin, however, the twofold rotation axis is a "local" and not a crystallographic axis, and there are no additional strands involved to form an extensive pleated-sheet region. Note that in both cases there are 2 Phe residues in each strand. It is also of interest that Con A mimics the effect of insulin on fat cells (Cuatrecasas and Tell, 1973). At first glance, the structures of these two proteins appear greatly

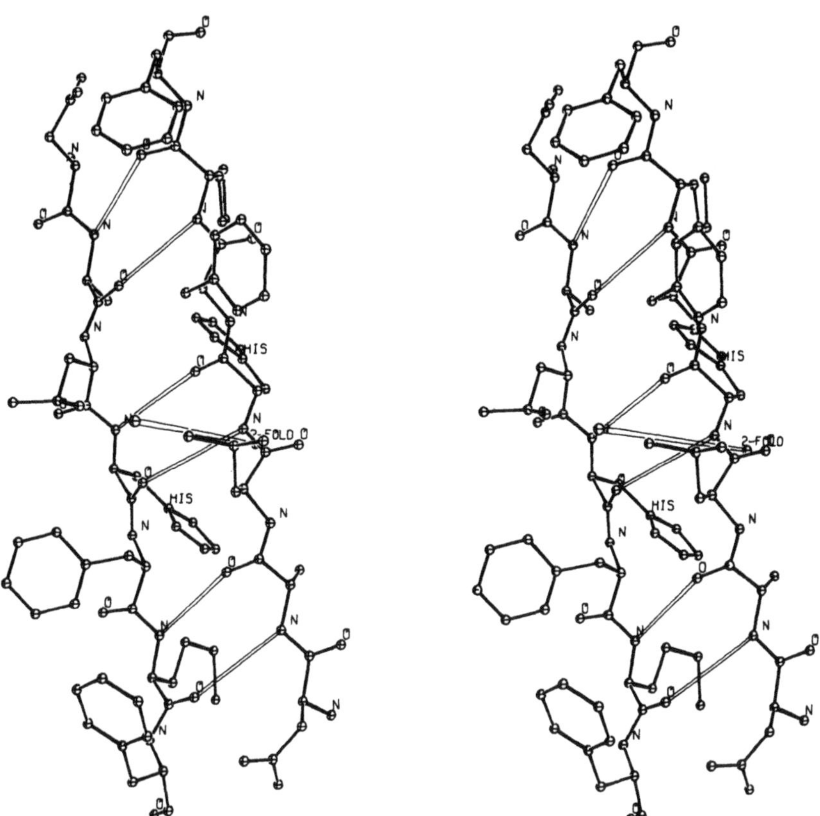

Figure 7. Stereogram of the strand of β-Sheet I nearest the X-axis and its symmetry related equivalent in the dimer.

different except for the feature just mentioned, and a close comparison of the surface regions is necessary to see whether there are, indeed, any subtle similarities which may explain this effect on fat cells.

In the crystal, four subunits cluster about a point of 222 symmetry (the intersection of three twofold rotation axes--the X, Y, Z axes) to form the tetramer (Figs. 8 and 9). This can be thought of as being generated by a 180° rotation of the dimer (Fig. 6) about either the Y or Z axes. The residue of each subunit nearest the center is His-128 (Figs. 7-9). If in solution the subunits of the tetramer are arranged similarly, this cluster of four histidines will, indeed, play an important role in the dimer-tetramer

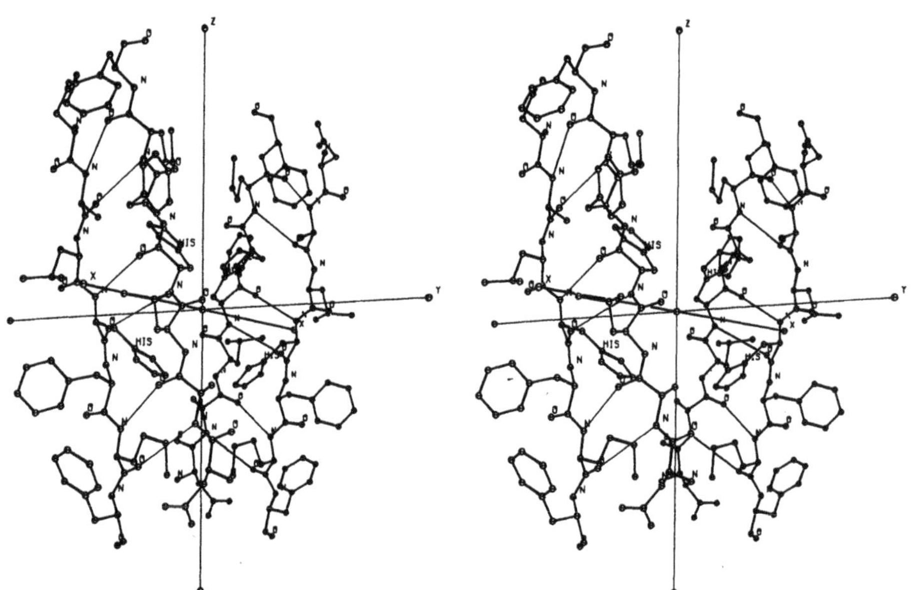

Figure 8. Stereogram of the same strand of β-Sheet I shown in Figure 7, which here is multiplied by four, showing the full 222 symmetry of the tetramer.

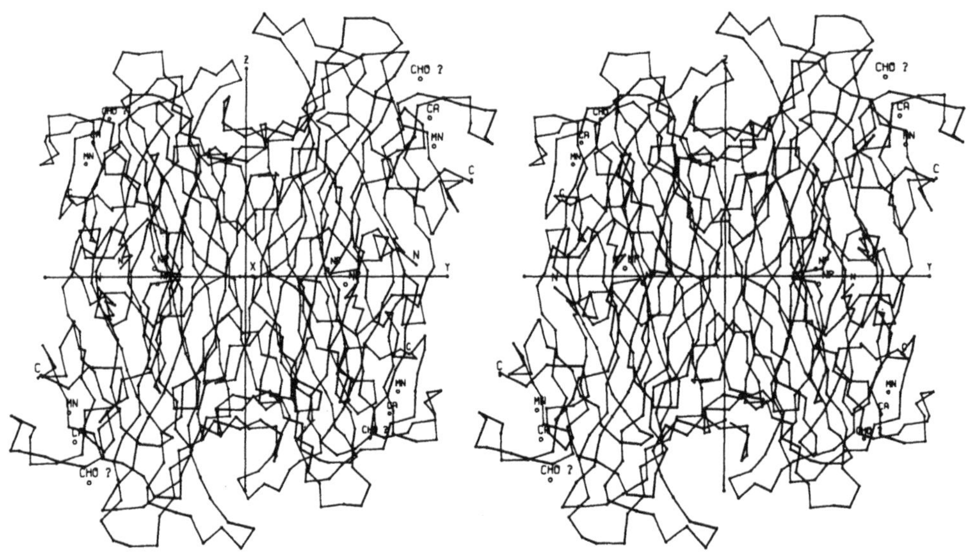

Figure 9. Alpha-carbon stereogram of the tetramer.

dissociation near neutral pH. The two sets of two histidines paired across the Z-axis (Fig. 8) are close enough togehter that if the pH were lowered from, say 7.4, to below 6 so that these groups became protonated, charge repulsion between these pairs would be expected. In addition, another histidine (51) is involved in the interface between the two dimers. Other interactions are hydrogen bonds between residues such as Ser, Thr, and Asn, which would be solvated when the pH were lowered and the tetramer split.

The overall shape of the tetramer and the relationships that the four identical subunits have to one another can be seen in Figure 9. The longest dimension of each dimer is about 82Å and forms an angle of about 40° with the Y-axis. The tetramer is approximately 60Å thick along the X-axis. Figure 10 shows the four

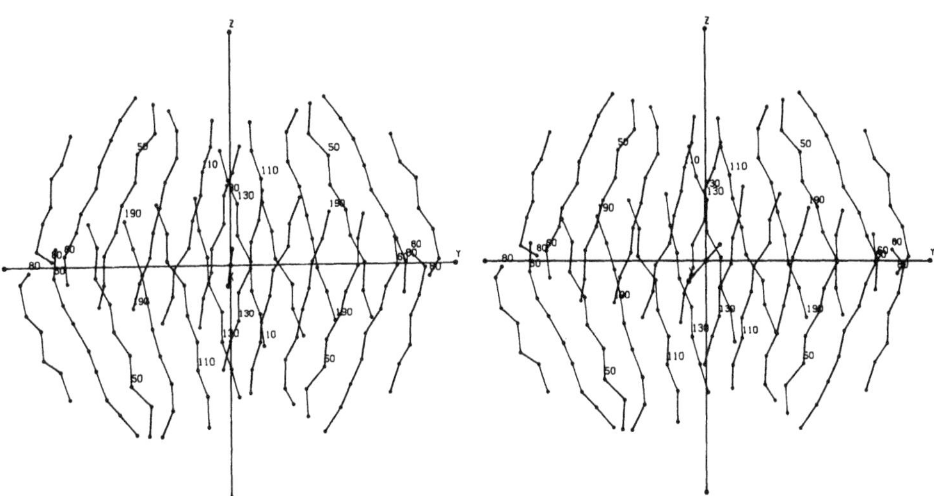

Figure 10. Alpha-carbon stereogram of the four β-Sheet I regions of the tetramer.

Sheet I regions of the tetramer from which the dimer-dimer interactions are formed. In Figure 3, amino acid residues which are aligned in vertical columns with Ala-126, His-128, and Met-130 all are pointed to the outside of each subunit and are involved in contacts either with the other dimer or with solvent. Residues aligned with Leu-127, Phe-129, and Phe-131 are non-polar and point to the interior of each subunit. This unique subunit structure found here in Con A, involving "back to back" β-structure sheets and histidine residues, leads one to speculate that similar structures may be found in oligomeric enzymes whose catalytic function is under allosteric control.

Manganese and Calcium

In keeping with the theme of this conference, the most important part of the Con A molecule is shown in Figure 11, the sites of

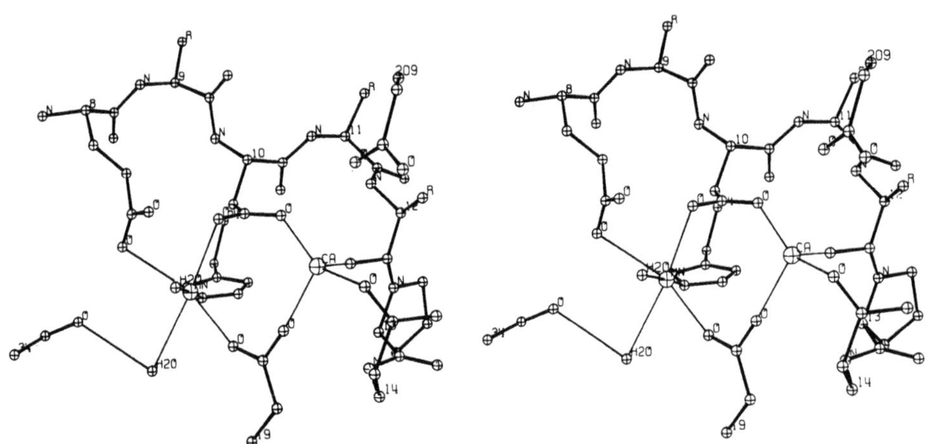

Figure 11. Stereogram of the amino acid residues which form the Mn^{2+} and Ca^{2+} double ion site.

manganese and calcium binding, which are about 5.0Å apart. The amino acid residues which directly bind the manganese and calcium lie between residues 8 and 24. Two aspartate residues, 10 and 19, each contribute both carboxyl oxygens as ligands, one to each ion. The manganese is the highest electron density peak in the map and corresponds to the site where Weinzierl and Kalb (1971) found cadmium to bind in the apoprotein. Beginning at the N-terminus, the polypeptide chain runs through Sheet II β-structure, through the center of the subunit, supplying Glu-8 to the Mn^{2+}, and the carboxyl of Asp-10 to both Mn^{2+} and Ca^{2+}. The chain then wraps around the two ions and supplies as ligands the carbonyl oxygen of Tyr-12 to Ca^{2+}, the oxygen (or perhaps the nitrogen) of the amide Asn-14 to Ca^{2+}, and the two carboxyl oxygens of Asp-19 to both Ca^{2+} and Mn^{2+}. After this loop at the surface of the molecule, the chain continues back into Sheet II. The imidazole ring of His-24 contributes N-3 to Mn^{2+} and N-1 forms a hydrogen bond with the carbonyl oxygen of residue 22. At Lys-30 the chain again turns back past the Mn^{2+} where a H_2O molecule forms a bridge between the Ser-34 side chain and the Mn^{2+}. Edelman, et al. (1972) have interpreted their electron density map as having this H_2O bridging the carbonyl oxygen of Val-32; however, in our map the density contours and atomic distances are more consistent with Ser-34. Note that before reaching residue 40, the polypeptide chain has contributed three consecutive strands of Sheet II, which construct the framework of the double ion site, and has supplied all the amino acid side chains directly binding the Mn^{2+} and Ca^{2+}. The electron density at the sixth ligand position of Mn^{2+} is relatively weak in comparison to the other ligands and is interpreted as resulting from partial occupancy of a water molecule. The fifth side chain close to the calcium is Asp-209. A continuous region of electron density bridges this side chain and the Ca^{2+}, and probably involves a water molecule (not shown in Figure 11). The symmetry of the manganese is very nearly octahedral and the calcium is penta-coordinated with four of the atoms being close to tetrahedral, and distortion arising from the water molecule between the Ca^{2+} and residue 209.

It is obvious from the course of the polypeptide chain in this region that Mn^{2+} and Ca^{2+} greatly stabilize the conformation of the subunit. In their absence, the four carboxyl groups clustered at this site would strongly repel one another, forcing the chain from about residues 10 through 22 out into the solvent. In the crystalline lattice, this region is close to a neighboring molecule, and it is thus not surprising that the apoprotein cannot be crystallized in the same space group (Jack, et al., 1971).

The apoprotein first must bind Mn^{2+}, or some other transition metal ion, e.g., Ni^{2+}, before it is able to bind Ca^{2+}. The association constants for Ni^{2+} and Ca^{2+} have been found to be 1.3×10^5 and 3.0×10^3 ℓ/mol, respectively (Kalb and Levitzki, 1968). The transition metal ion, therefore, appears to stabilize the conformation of the polypeptide chain in this region so that at least some of the groups involved in binding Ca^{2+} are properly positioned. This is easily explainable by examination of the Mn^{2+} - Ca^{2+} double site. On binding Mn^{2+}, which is closer to the interior of the subunit (Fig. 4), Asp-10 and -19 carboxyls would be properly positioned to initiate Ca^{2+} binding.

Non-Polar Binding Cavity

Con A contains a distinct cavity which binds a number of relatively non-polar molecules (Figure 4 (NP)). This cavity has been previously thought to contain the carbohydrate-specific binding site. Becker, et al (1971) have previously shown that βIphGlcp binds in this cavity in Con A crystals and, since this protein specifically binds sugars with the manno- or gluco-configuration, it was assumed to indicate the general site for carbohydrate binding. Additionally, we had previously reported that myo-inositol likewise bound in this cavity (Hardman and Ainsworth, 1972a). However, the binding of βIphGlcp appears to be bound in this site solely by the iodophenyl group, and the electron density contours previously reported to be due to myo-inositol binding were apparently produced by trace amounts of the mold inhibitor, methyl(p-hydroxy)benzoate, which evidently had not been completely removed from the protein preparation.

The galactopyranose configuration, the C-4 epimer of glucopyranose, does not bind to Con A (Goldstein, et al., 1965; Hassing and Goldstein, 1970), and yet we have shown that βIphGalp binds in this site as well as the glucopyranose analogue (Hardman and Ainsworth, manuscript submitted for publication). Therefore, the binding of these iodophenyl sugar derivatives at this site is due to the iodophenyl group and not the stereospecificity of the carbohydrate. Furthermore, βIphGlcp does bind to the carbohydrate-specific site in solution, as shown by dextran precipitation (Poretz and Goldstein, 1971), whereas βIphGalp does not (Hardman and Ainsworth, manuscript submitted for publication). It thus appears that the carbohydrate-specific binding site lies elsewhere in the subunit and is discussed later.

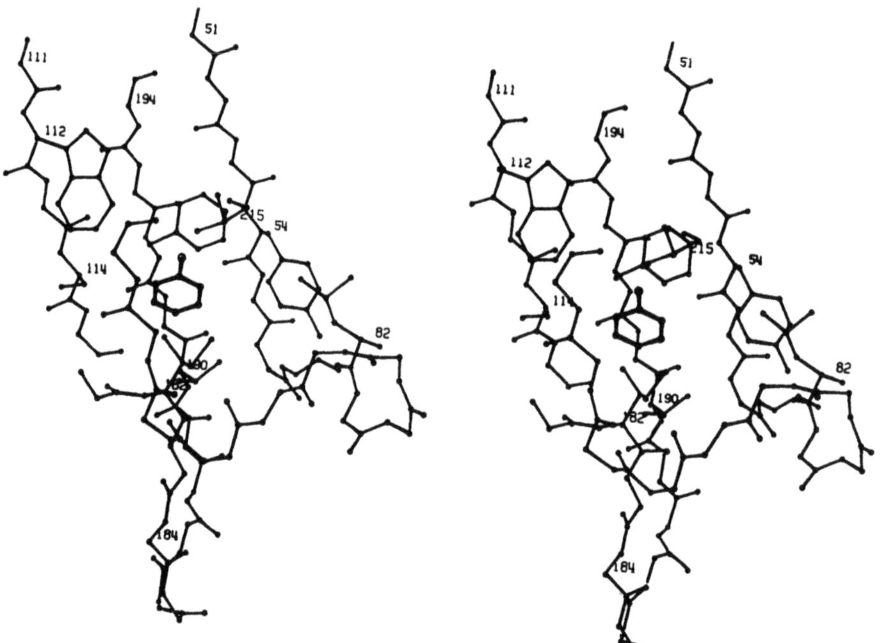

Figure 12. Stereogram of the non-polar binding cavity. The darker ring represents the position of the bound iodophenyl group of a number of compounds which complex with Con A.

Figure 12 shows the amino acid residues involved with the non-polar binding site and the position of the iodophenyl ring as determined by the βIphGalp derivative. Other compounds which we have shown to bind at this site are methyl(p-hydroxy)benzoate, o-iodobenzoic acid, o-iodoaniline, o-iodophenol, phenylphosphate, and dimethyl mercury, and they exhibit very little or no inhibition of the dextran precipitation. The basic framework for this region is formed by strands of β-structure Sheets I and II. Portions of three strands of Sheet I form the "back" of this site and two strands through the center of Sheet II form the "front" (Figs. 2 and 4). These two sheets are held together by two sections of polypeptide chain, 81-90 and 180-188, which form the opening of this cavity (Figs. 4 and 12). The bound iodophenyl group is surrounded by many non-polar side chains; Tyr-54, Leu-82 and -86, Val-92 and -180, Ile-182 and -215, Trp-112 and Phe-192. The cavity dimensions are approximately 15Å deep with a cross section of about 9 × 6Å. The only residues which appear to move to accommodate binding are the side chains of Phe-192, which rotates upward slightly (Fig. 12), and residues Leu-182 through Glu-184, which move downward about 2Å. It appears that the center of this cavity is accessible to solvent molecules since some of these benzene derivatives which bind possess ionized groups, i.e., *o*-iodobenzoic acid and phenylphosphate. No electron density contours are produced by the binding of βIphGal*p* which could represent the galactopyranose moiety (Hardman and Ainsworth, manuscript submitted for publication), so it appears that this group must occupy the center of this cavity and has enough freedom of motion to produce very little increase in average electron density.

There has apparently been no report of any natural function in the jack bean (speculated or otherwise) for binding of non-polar molecules by Con A. Also, the apparently fortuitous biological properties which Con A exhibits (Sharon and Lis, 1972) have generally been shown to result from its ability to bind specific carbohydrates. However, non-polar molecules have been extracted from Con A preparations with non-polar solvents (Jaffe, *et al.*, 1971). The possibility should not be overlooked that perhaps Con A functions as a regulator of germination or cell division through binding molecules which fall into one of the classes of plant growth factors, i.e., cytokinins.

Possible Carbohydrate-Specific Site

We have observed that crystals of native Con A are cracked and dissolved by concentrations of αMeMan*p* of 1 mM and above, whereas lower concentrations produce no significant changes in electron density difference maps (Hardman and Ainsworth, unpublished observations). Concentrations of 5 mM and above of βIphGlc*p* also dissolved native crystals, but 5 mM βIphGal*p* had no effect. Therefore, molecules binding at the carbohydrate-specific site either sterically interfere with the lattice packing of the crystals or produce conformational changes in the protein (allosteric effects) which do so. Conformational changes have been shown to occur upon binding carbohydrate by circular dichroism studies (McCubbin, *et al.*, 1971; Pflumn, *et al.*, 1971) and by NMR studies (Carver, personal communication). Studies of binding of αMeGlc*p* and αMeMan*p* to Con A by NMR techniques have indicated that the distances from the Mn^{2+} to the carbon atoms of the sugars are between 9 and 12Å (Brewer, *et al*., 1973; Villafranca, personal communication; Carver, personal communication).

It is then probable that the carbohydrate-specific site involves a region with a number of intermolecular contacts in the crystalline lattice of the native protein (space group I222). The region with the largest number of intermolecular contacts in these crystals involves side chains of Asp-16, Arg-229, Tyr-12, and Tyr-101, which cluster around a point about 12Å from the Mn^{2+} and 7Å from the Ca^{2+}. It should be noted here that the only peptide bond in the subunit that is interpreted as existing in the "cis" configuration (between residues 208 and 209) is adjacent to both this region and the Ca^{2+} site. Also adjacent is the only charged side chain, Asp-103 (other than those directly involved in the Mn^{2+} and Ca^{2+} double site), which is on the "interior" of the subunit. It will be interesting to see in the future if these two features are involved in conformational changes upon binding carbohydrate.

In order to get a preliminary indication of the carbohydrate-specific site using the I222 crystals, we have covalently cross-linked the outer surface of crystals with glutaraldehyde to prevent them from dissolving and subsequently soaked them in 20 mM βIphGlc*p*. Data were collected on one such crystal to 3.0Å resolution, and the highest electron density difference peak, indeed, appeared in this region (near Tyr-12 and -101). Smaller peaks also appeared at the non-polar binding site and elsewhere.

It appears obvious, however, that in order to study the crystal structures of Con A-carbohydrate complexes adequately, the complexes will have to be formed *first* and then crystallized, particularly for di- and trisaccharides, or larger polymers. We have, in fact, grown crystals of Con A complexed with αMeMan*p* (Hardman and Ainsworth, unpublished experiments) which have a completely different habit and space group (triclinic) from the native Con A, and the asymmetric unit contains the tetramer of 104,000 molecular weight. The three-dimensional structure of such a complex should not only describe the carbohydrate-specific site, but should also go far in explaining relationships among the Mn^{2+} and Ca^{2+} ions and conformational changes which occur on the binding of carbohydrates.

ACKNOWLEDGEMENTS

I am deeply indebted to Clinton F. Ainsworth for his many contributions throughout this project and to the U. S. Atomic Energy Commission for their support.

REFERENCES

Agrawal, B. B. L. and Goldstein, I. J. (1967), Biochim. Biophys. Acta, 133, 367.

Allan, D., Auger, J., and Crumpton, M. J. (1972), Nature (London), New Biol., 236, 23.

Becker, J. W., Reeke, G. N., Jr., and Edelman, G. M. (1971), J. Biol. Chem., 246, 6123.

Blundell, T., Dodson, G., Hodgkin, D., and Mercola, D. (1972), Advances in Protein Chemistry, 27, 279.

Boyd, W. C. and Shapleigh, E. (1954), Science, 119, 419.

Brewer, C. F., Sternlicht, H., Marcus, D. M., and Grollman, A. P. (1973), Proc. Nat. Acad. Sci. U. S., 70, 1007.

Cuatrecasas, P. and Tell, G. P. E. (1973), Proc. Nat. Acad. Sci. U. S., 70, 485.

Dufau, M. L., Tsuruhara, T., and Catt, K. J. (1972), Biochim. Biophys. Acta, 278, 281.

Edelman, G. M., Cunningham, B. A., Reeke, G. N., Jr., Becker, J. W., Waxdal, M. J., and Wang, J. L. (1972), Proc. Nat. Acad. Sci. U. S., 69, 2580.

Goldstein, I. J., Hollerman, C. E., and Smith, E. E. (1965), Biochemistry, 4, 876.

Gombos, G., Hermetet, J. C., Reeber, A., Zanetta, J. P., and Treska-Ciesielski, J. (1972), FEBS Letters, 24, 247.

Hardman, K. D. and Ainsworth, C. F. (1972a), Nature (London), New Biol., 237, 54.

Hardman, K. D. and Ainsworth, C. F. (1972b), Biochemistry, 11, 4910.

Hardman, K. D., Wood, M. K., Schiffer, M., Edmundson, A. B., and Ainsworth, C. F. (1971a), Proc. Nat. Acad. Sci. U. S., 68, 1393.

Hardman, K. D., Wood, M. K., Schiffer, M., Edmundson, A. B., and Ainsworth, C. F. (1971b), Cold Spring Harbor Symp. Quant. Biol., 36, 271.

Hassing, G. S. and Goldstein, I. J. (1970), Eur. J. Biochem., 16, 549.

Inbar, M., Ben-Bassat, H., and Sachs, L. (1971), Proc. Nat. Acad. Sci. U. S., 68, 2748.

Inbar, M. and Sachs, L. (1969), Proc. Nat. Acad. Sci. U. S., 63, 1418.

Jack, A., Weinzierl, J., and Kalb, A. J. (1971), J. Mol. Biol., 58, 389.

Jaffe, W. G. and Palozzo, A. (1971), Acta Cient. Venezolana, 22, 102.

Johnson, C. K. (1965), Oak Ridge Tech. Manual, 3794.

Kalb, A. J. and Levitzki, A. (1968), Biochem. J., 109, 669.

Kalb, A. J. and Lustig, A. (1968), Biochem. Biophys. Acta, 168, 366.

Kannan, K. K., Liljas, A., Waara, I., Bergstén, P. C., Lövgren, S., Strandberg, B., Bengtsson, U., Carlbom, U., Fridborg, K., Jarup, L., and Petef, M. (1971), Cold Spring Harbor Symp. Quant. Biol., 36, 221.

Leon, M. A. and Young, N. M. (1970), J. Immunol., 104, 1556.

Lipscomb, W. N., Hartsuck, J. A., Reeke, G. N., Jr., Quiocho, F. A., Bethge, P. H., Ludwig, M. L., Steitz, T. A., Muirhead, H., and Coppola, J. C. (1968), Brookhaven Symp. Biol., 21, 24.

Lloyd, K. O., Kabat, E. A., and Beychok, S. (1969), J. Immunol., 102, 1354.

Markowitz, H., Person, D. A., Gitnick, G. L., and Ritts, R. E., Jr. (1969), Science, 163, 476.

McCubbin, W. D., Oikawa, K., and Kay, C. M. (1971), Biochem. Biophys. Res. Com., 43, 666.

Moscona, A. A. (1971), Science, 171, 905.

Nicolson, G. L. (1971), Nature (London), New Biol., 233, 244.

Nicolson, G. L. (1972), Nature (London), New Biol., 239, 193.

Nordin, A. A., Cosenza, H., and Hopkins, W. (1969), J. Immunol., 103, 859.

Novogrodsky, A. and Katchalski, E. (1971), Biochim. Biophys. Acta, 228, 579.

Okada, Y. and Kim, J. (1972), Virology, 50, 507.

Olson, M. O. J. and Liener, I. E. (1967), Biochemistry, 6, 3801.

Pardoe, G. I. and Uhlenbruck, G. (1970), J. Med. Lab. Technol., 27, 249.

Pauling, L. and Corey, R. B. (1951), Proc. Nat. Acad. Sci. U. S., 37, 729.

Pflumn, M. N., Wang, J. L., and Edelman, G. M. (1971), J. Biol. Chem., 246, 4369.

Poretz, R. D. and Goldstein, I. J. (1970), Biochemistry, 9, 2890.

Poretz, R. D. and Goldstein, I. J. (1971), Biochem. Pharmacol., 20, 2727.

Powell, A. E. and Leon, M. A. (1970), Exp. Cell Res., 62, 315.

Quiocho, F. A., Reeke, G. N., Jr., Becker, J. W., Lipscomb, W. N., and Edelman, G. M. (1971), Proc. Nat. Acad. Sci. U. S., 68, 1853.

Sharon, N. and Lis, H. (1972), Science, 177, 949.

Shoham, J., Inbar, M., and Sachs, L. (1970), Nature, 227, 1244.

Smith, D. F. (1972), Ph.D. Dissertation, University of Texas.

So, L. L. and Goldstein. I. J. (1968), J. Biol. Chem., 243, 2003.

Sumner, J. B. (1919), J. Biol. Chem., 37, 137.

Sumner, J. B. (1926), J. Biol. Chem., 69, 435.

Sumner, J. B. and Howell, S. F. (1936a), J. Bacteriol., 32, 227.

Sumner, J. B. and Howell, S. F. (1936b), J. Biol. Chem., 115, 583.

Sumner, J. B., Gralen, N., and Eriksson-Quensel, I. B. (1938), J. Biol. Chem., 125, 45.

Weinzierl, J. and Kalb, A. J. (1971), FEBS Letters, 18, 268.

TRANSFER OF MERCURY AND CADMIUM FROM TERRESTRIAL TO AQUATIC ECOSYSTEMS*

JOHN W. HUCKABEE and B. G. BLAYLOCK

Environmental Sciences Division
Oak Ridge National Laboratory
Oak Ridge, Tennessee 37830

INTRODUCTION

The ecological problems of toxic heavy metals have generated a progression of review articles from specialty journals to semi-popular magazines to national news media (1), and inevitably the backlash has set in (2). Still, the ecological status of mercury and cadmium remains incompletely known. While many of the pollutive discharges of mercury and cadmium have been identified and assessed, the natural background concentrations of these elements in the environment due to release by weathering of rocks and volcanic events are not well understood.

Added to the natural background level of mercury and cadmium is that derived from pollution. The metals are introduced to the environment through releases to both the aquatic and terrestrial systems. Although quantitative data are scarce, it is safe to say that a significant portion of the effluents is discharged to the terrestrial environment. These accumulations are then transferred to lakes and streams; ergo, we cannot disregard the terrestrial cycling of mercury and cadmium when considering the impact of mercury and cadmium on the aquatic system.

Organisms acquire mercury and cadmium in their tissues through direct contact (air and/or water) and through the food chain. Body burdens at any time are a function of the sum of these uptake routes and the elimination rate, which is expressed as the biological half-life. Biological half-lives for the same substance in the

*Research sponsored by the National Science Foundation under Union Carbide Corporation contract with the U. S. Atomic Energy Commission.

same species may vary according to route of uptake.

Sometimes a toxic substance tends to accumulate in predatory organisms to a much higher concentration than in the prey species. This process is called biomagnification or trophic level magnification. The significance of biomagnification is that toxic levels will build up in top carnivores (including humans) and these species then may be the most affected in the ecosystem. The best known example of this situation is the decline of raptorial birds because of chlorinated hydrocarbons concentrating in their tissue.

It has been widely reported that mercury is biomagnified, but a closer look at the evidence shows that we cannot so generalize. Both mercury and cadmium accumulate with age of the animal. The distinction between biomagnification and bioaccumulation is non-trivial; in the first case--biomagnification--toxic levels may accumlate much more rapidly than in the case of bioaccumulation.

The heavy metals research strategy outlined here proceeds in three steps:

1) determination of natural background concentrations of mercury and cadmium in biota in an unpolluted area (Great Smoky Mountain National Park);

2) microcosm experiments in which radiotracers are employed to illustrate the transfer of mercury and cadmium from terrestrial to aquatic ecosystems;

3) specific laboratory and field experiments suggested by the microcosm experiments or other data.

BACKGROUND LEVELS OF MERCURY AND CADMIUM

Unlike the chlorine and phosphate insecticides and other toxic organic molecules discharged to the environment by agricultural, industrial, and municipal pollution, heavy metals discharged to the environment are usually not something entirely novel for the biota to cope with. Mercury and cadmium are almost ubiquitously distributed in unpolluted waters, rocks, and soils. Hence, plants and animals are exposed to low levels of mercury and cadmium under entirely natural circumstances. The mercury and cadmium content of the various compartments of the environment are thus the sum of two sources: natural, or background; and man-made, or pollution. Background concentrations of mercury in natural waters, soils, and plants have been reported (3,4), but the values presented could be refined considerably by taking larger samples and by more careful selection of sampling sites.

Naturally occurring mercury is widely but unevenly distributed as HgS in sandstones and other clastic sedimentary rocks (5). Magmatic and metamorphic rocks usually contain smaller concentrations of mercury than sedimentaries. Due to this inhomogeneity of mercury distribution the background mercury concentrations in biota should vary from region to region (6). An example of such regional variation of background mercury concentrations in biota is found in northwestern Wyoming (7). Mercury is found widely disseminated in two clastic sedimentary formations that outcrop around the eastern side of the large intermontane valley, Jackson Hole. Populations of rodents from the high mercury outcrop areas had higher levels of mercury in hair than did populations of the same species in the low mercury areas on the valley floor. Eighty-three per cent of seven rodent species listed in the mercury outcrop area had > 0.008 ppm (detection limit) mercury in hair (mean = 0.15 ppm) while only 7% of the animals from the low mercury area had more than 0.008 ppm. Only one species from the low mercury area had mercury detectable in its hair. In addition to the rodents, mercury concentrations were also measured in coyote (*Canis latrans*) hair samples which were obtained from live-trapped and released animals. Although 85% of the coyote hair samples contained more than 0.008 ppm mercury (mean = 0.57 ppm), there was no difference in mercury concentration by area. However, coyote's home ranges are so large that they overlap the high and low mercury areas.

Few data are available on background mercury concentration in plants. Shacklette reported that mercury concentrations of 0.5 ppm (dry weight basis) appear to be about average in plants from uncontaminated areas (8).

Cadmium is always geochemically associated with the much more abundant element zinc; one can understand the environmental dynamics of cadmium only by understanding the zinc-cadmium system. The widespread distribution of zinc has resulted in the concomitant wide distribution of cadmium. Like mercury, cadmium is not evenly distributed in nature and information on cadmium background levels is even scarcer than for mercury. Much of the pertinent information available (though not a complete literature survey) on cadmium in the environment is found in the document "Cadmium: The Dissipated Element" (9).

MERCURY AND CADMIUM IN THE TERRESTRIAL ECOSYSTEM

Analyses of Mercury and Cadmium in Plants from the Great Smoky Mountain National Park

Plant samples were collected from a relatively uncontaminated area, Cades Cove, in the Great Smoky Mountains National Park,

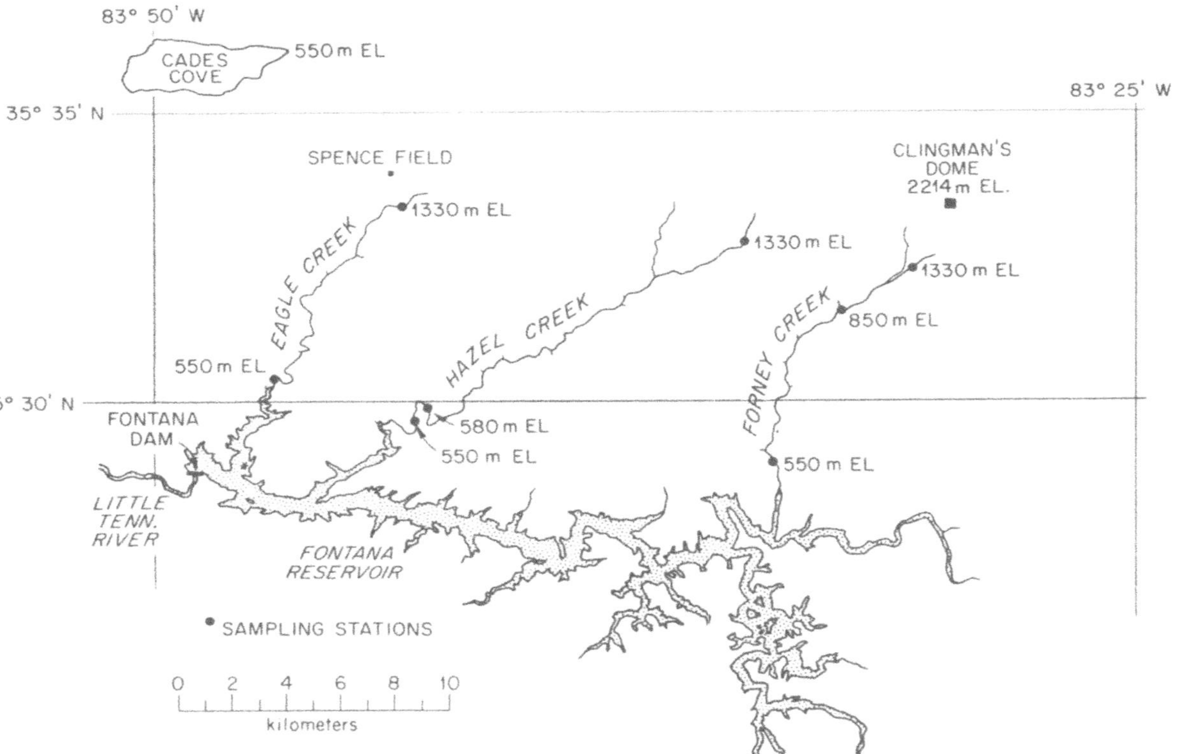

Figure 1. Location map and elevation of sample collection stations in Great Smoky Mountains National Park, Spring, 1972. Aquatic samples taken for mercury concentration analyses were brook and rainbow trout, stonerollers, rosyside dace, and banded sculpin, sediments and water. Terrestrial samples were mosses (Dicranum and Polytrichum), grasses (Festuca and Andropogon), Rhododendron, and eastern hemlock.

Tennessee, for mercury and cadmium analysis in an attempt to determine near-background levels in plants (Fig. 1).**

The mosses (*Dicranum* and *Polytrichum*) had significantly higher concentrations of cadmium and mercury than the other vegetation with the exception of mercury in the grass *Andropogon* (Table 1). Mosses have been shown to differentially accumulate mercury from the atmosphere compared to vascular plants (10). In areas subjected to fly ash pollution, mosses concentrated mercury at least 4 times higher than grasses and forbs. Since Cades Cove is relatively free from fly ash input, the mosses contained about the same mercury concentrations as grass. Background concentrations of metals in plants will vary widely because of the complexity of factors controlling metal ion uptake: soil concentration of metals, competing soil cations, species differences in uptake rate, and atmospheric input of metals.

Table 1. Mercury[†] and Cadmium[††] Content (ppm) of Vegetation in Cades Cove, Great Smoky National Park, Tennessee.

	Dicranum		*Polytrichum*		*Festuca*		*Andropogon*	
	Hg	Cd	Hg	Cd	Hg	Cd	Hg	Cd
n	14	14	12	12	6	6	6	6
$\overline{X}$	0.118	0.421	0.092	0.383	0.087	0.279	0.120	0.210
SD	0.060	0.143	0.046	0.114	0.048	0.500	0.068	0.060

	Tsuga			
	Needles		Stems	
	Hg	Cd	Hg	Cd
n	6	6	6	6
$\overline{X}$	0.027	0.083	0.028	0.326
SD	0.004	0.052	0.010	0.181

[†] Air-dried weight
[††] Oven-dried weight

**The mercury analyses [flameless atomic absorption spectrophotometry (AA)] were performed in the Analytical Chemistry Division of Oak Ridge National Laboratory (ORNL). Sample preparation was by a modification of the procedure of Smith (11). Recovery tests of both organic and inorganic mercury compounds were run to check for mercury losses during sample preparation; losses were negligible. The cadmium analyses were by spark source mass spectrometry (SSMS)

An indication of the lower limits of mercury concentrations in biota was provided by analyzing the samples from the Great Smoky Mountains National Park. Much higher levels of mercury have been reported in fish and other animals in regions thought to be free from mercury pollution, although background concentrations were not known. These reports and subsequent investigations led to identification of fossil fuel burning as a source of regional mercury contamination (13,14,15,16). Mercury in coal is converted to the vapor phase in the furnace when the coal is burned, but some of it may condense on fly ash particles when the fly ash and flue gas stream ascend the stack and the temperature drops. Mercury from coal is thus introduced into the environment in the vapor form and in the metallic form (perhaps some as mercuric oxide) adhering to particles of fly ash which ultimately settle out of suspension at points distant from the stacks. The environmental input of mercury vapor and mercury-bearing fly ash has not been adequately monitored. The mercury concentration in coal varies so much (<0.01 - >100 ppm) that it is unrealistic to cite an average value (14,16,17,18). No figures are available on the mercury concentration in coal fly ash collected in areas distant from the discharge stacks, but fly ash collected from stack precipitators has been shown to contain 0.2 ppm mercury while the coal contained 0.3 ppm mercury (15). These same authors calculated that 90% of the mercury in the coal was lost out of the stack to the environment during normal furnace operations.

Fly Ash as a Source of Mercury Pollution

Fly ash collected at ORNL over a seven-week period 2 km from the nearest stack contained a mean mercury concentration of 0.98 ppm (18). The fly ash was collected along with rain water in jars at the base of the 2-m^2 funnels. The mercury content of the fly ash may, therefore, derive in part from mercury vapor wash-out by rain. Based on the rate of dry fall and the 0.98 ppm value, the mercury input to this area is calculated to be 20-30 μg mercury $m^{-2}y^{-1}$. The availability of mercury from flue gas and fly ash to the biota is indicated by the mercury content of fish from a stream near the fly ash sampling station. The watershed where fly ash samples were collected is free from industrial or agricultural discharges (other than fly ash) and the small streams are oligotrophic. However, 38 long-nose dace (*Rhinichthys cataractae*) were found to contain a mean of 0.281 ppm mercury. Thirty fish of the same species from a stream in Cades Cove at the 550-m elevation were found

---

with isotope dilution. The AA method was compared with the SSMS method and a neutron activation analysis (NAA) method (12). The results of mercury analyses on thoroughly homogenized samples of freeze-dried tuna were (means, ppm): AA, 0.525; SSMS, 0.45; NAA, 0.45. The NAA method involved a sample preparation. Very similar analytical agreement was obtained on several samples of coal.

to contain a mean of 0.058 ppm mercury (five composite samples of 2 to 11 fish each). The significantly higher ($P \leq 0.01$) concentrations in the ORNL fish indicate that flue gas and fly ash may contribute significantly to the mercury concentration in the ORNL fish.

Fly Ash Microcosm Experiment

Since mercury derived from the atmosphere appears to be available to the aquatic biota, a microcosm experiment was conducted to determine the transfer rate of fly ash-rain water derived mercury from the terrestrial to the aquatic ecosystem and to identify functional unit of a given ecosystem, viz., one brings into the laboratory a carefully excised piece of the outdoors, supplies light, water, and the proper temperature, and the flora and fauna of interest go about as if they were still outdoors. Obviously, the size of a microcosm will be determined by the biota of interest. For the fly ash mercury investigation, duplicate tanks 0.6-m × 0.6-m were established and kept at long-day photoperiod and temperature (18-21°C) for the duration of the experiment. Sections of a stream bank were placed in one end of the tanks, and stream sediments and 60ℓ of water in the other half. Mosquito fish (*Gambusia affinis*) and snails (*Goniobasis clavaeformis*) were placed in the aquatic portion. After two weeks, radioactive ^{203}Hg-tagged fly ash was added to the terrestrial portion of each microcosm via a simulated rainfall. Each microcosm received about 110 μCi of ^{203}Hg. Water was added as simulated rainfall twice weekly in the amount of 0.5 inches per week, the average quantity for that time of year (April - October).

The fly ash-mercury tag was accomplished by placing a coil of copper wire in a solution of $^{203}Hg(NO_3)_2$ overnight, removing and placing the wire over a few grams of fly ash in a flask and heating the wire to incandescence. The ^{203}Hg that had amalgamated onto the wire vaporized off and deposited as metallic mercury on the fly ash particles. The tagged fly ash was then mixed with unlabeled fly ash to bring it to the proper activity (∿ 200 μCi/g).

Sampling and counting for ^{203}Hg activity was then conducted on a routine basis for six months. The terrestrial portion of the microcosms was sampled immediately post-tag, then once weekly for five weeks, and once monthly for the duration of the experiment. Terrestrial sampling included litter and soil. Water samples and fish were counted at 24-hour post-tag, once weekly for five weeks, and once monthly until the experiment was terminated. Snails and sediments were counted once weekly starting at one week and two weeks, respectively, until the fifth week, and then once monthly for the duration of the experiment. The distribution of ^{203}Hg activity at the termination of the experiment (139 days) is shown in Table 2. The mercury on fly ash was found throughout the soil column and in the plant roots. Soil invertebrates could have

Table 2. Distribution and concentration of 110 μCi of ^{203}Hg in microcosm experiments at 139 days after spiking with ^{203}Hg-tagged fly ash.

Component	Microcosm A			Microcosm B		
	μCi	dis/min/g	% Activity	μCi	dis/min/g	% Activity
Terrestrial						
Moss	0.14	88,000	0.12	0.03	11,000	0.03
Grass	0.25	12,000	0.22	0.008	1,000	<0.01
Forbs	0.11	14,000	0.10	0.02	4,400	0.01
Forb Roots	0.05	31,000	0.05	0.026	7,500	0.02
Litter	4.70	110,000	4.30	2.00	52,000	1.80
Soil	46.00	4,000	42.00	59.00	5,000	54.00
Aquatic						
Water	---	---	---	---	---	---
Sediment	59.00	13,000	53.00	49.00	14,000	44.00
Fish	0.005	1,100	<0.01	0.035	1,400	0.03
Snails	0.073	1,000	0.07	0.012	1,400	0.01
Lost (liner, atmosphere)	0.12	---	0.11	0.18	---	0.17

distributed the particles of fly ash down into the soil to this extent, but plants that sprouted two months after the tag contained ^{203}Hg in the vegetative parts, indicating uptake and translocation of ionic mercury.

The activity in the sediments could be due to terrestrial wash off of fly ash particles, but the activity of the aquatic biota seems more reasonably to be due to the uptake of ionic mercury rather than to a continuous accidental ingestion of fly ash particles. Figure 2 shows the transfer of activity from the terrestrial to the aquatic portion of the microcosms. Although there is wide variation in the data points, the trends are clear. The fish slowly increased in ^{203}Hg content (radioactivity/gram) while most of the ^{203}Hg input accumulated in the sediments. Much of the sediment activity is probably periphyton and fecal matter from the fish and snails. These results demonstrate the accumulation of mercury from fly ash by aquatic biota.

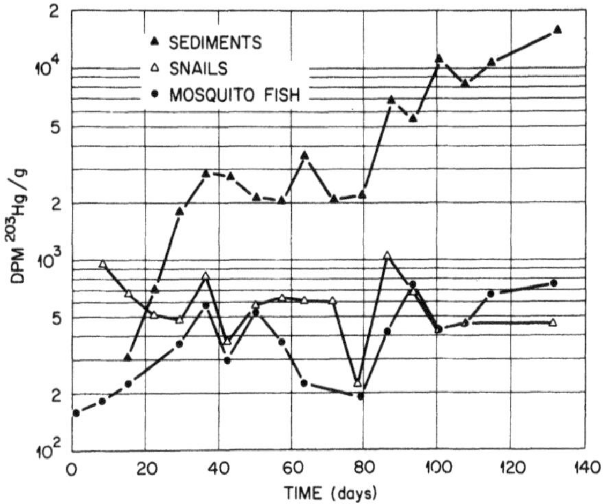

Figure 2. Transfer of ^{203}Hg from terrestrial to aquatic portion of microcosm experiment. Microcosm tagged with 110 μCi of ^{203}Hg-labeled fly ash added in simulated rainfall on terrestrial portion in April, 1972.

Natural Plot Studies of the Cycling of Mercury and Cadmium.

The transfer rate of mercury from the terrestrial to the aquatic ecosystem is strongly dependent upon the residence time of mercury in the terrestrial compartments. Matti (18) studied the cycling of mercury and cadmium in a terrestrial ecosystem by tagging 60 1-m^2 plots with $^{203}Hg(NO_3)_2$ at ORNL. Half of the plots were clipped bare of all vegetation; the other half were left natural. The plots were harvested routinely for six months and the distribution of the ^{203}Hg activity determined. The activity contained in the plants which sprouted from the bare plots represents ^{203}Hg taken up from the soil, while the activity in the plants on the undisturbed plots represents foliar uptake plus soil uptake. The per cent distribution of the ^{203}Hg activity in the vegetation and in the soil after foliar application is shown in Figure 3.

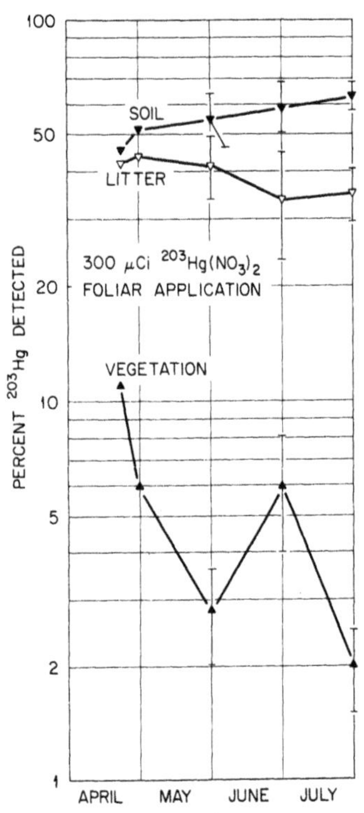

Figure 3. Mean per cent distribution of ^{203}Hg after tagging by simulated rainfall on 15 1-m^2 plots at ORNL, Tennessee, 1972, in experiment measuring the redistribution of mercury in terrestrial environments.

Foliar applications of ^{109}CdCl$_2$ through simulated rainfall showed that after 100 days, about 40% of the cadmium was retained in the top 2-cm of the soil, while about 55% of the cadmium was retained in the litter (Fig. 4). The vegetation (grasses and forbs) retained about 10% of the total cadmium applied 24 hours after application and 3% after 100 days. The same species of grasses and forbs showed soil uptake amounting to about 0.5% of the total cadmium applied after 100 days.

Witherspoon (19) studied the movement of mercury and cadmium in trees in June-August by inoculating the isotopes into the trunks and counting the resultant activity in the different parts of the tree and in the understory vegetation, litter, and soil. Table 3 shows the distribution of ^{203}Hg and ^{109}Cd eight weeks after the inoculation. Most (96.5%) of the ^{203}Hg from $^{203}Hg(NO_3)_2$ remained in the trunk. Only 3.2% of the mercury reached the crown, and 0.3% of the 3.2% was leached out of the foliage by rainfall. Of the 0.3% leached out, 82.6% was found in the soil, and 16% in the understory vegetation. These data indicate that inorganic mercury is not translocated readily in trees during the summer months.

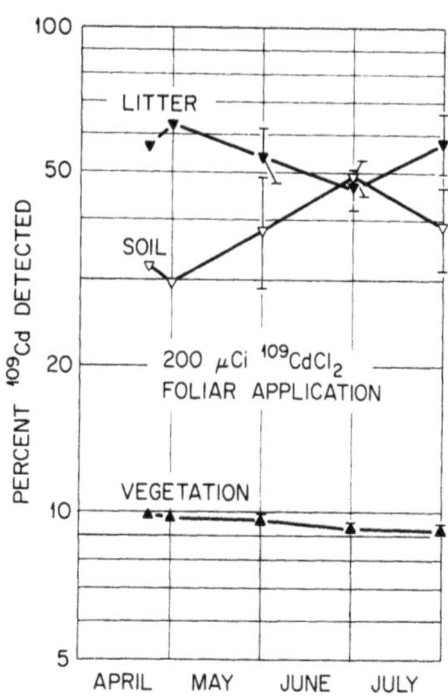

Figure 4. Mean per cent distribution of ^{109}Cd after tagging by simulated rainfall on 15 l-m^2 plots at ORNL, Tennessee, 1972, in experiment measuring the redistribution of cadmium in terrestrial environments.

Table 3. Distribution of ^{109}Cd and ^{203}Hg eight weeks after inoculation into trunks of Eastern Red Cedar Trees at ORNL, summer, 1972.

	^{109}Cd	^{203}Hg
Per cent of input		
In trunk	34.6	96.5
In branches	36.8	1.7
In foliage	28.5	1.5
Leached by rain	0.1	0.3
Per cent of activity leached by rain		
In understory	2.7	16.0
In litter	68.0	1.4
In soil	29.3	82.6

The tree inoculation study revealed that cadmium does not behave similarly to mercury (19). Unlike mercury, cadmium as $^{109}CdCl_2$ was transported to the foliage rather promptly (Table 3). About 65% of the cadmium inoculum reached the crown. Over an eight-week period, only 0.1% of the ^{109}Cd in the crown was leached out by rain (11 rains totaling 20.8 cm). About 68% of the ^{109}Cd transferred to the forest floor by rain remained in the litter under the canopy. Understory vegetation absorbed about 3%, and the remaining 29% entered the soil where most of it remained in the top 2-cm. Thus, while cadmium translocation by trees was prompt, the transfer of the element to other components of a forest floor ecosystem was low both in rate and magnitude. However, cadmium would be more available than mercury to animals whose food base is litter or herbaceous plants subjected to a chronic input of these elements in rain.

A more urgent consideration in terms of the human food chain is the uptake of methylmercury by food plants (20). Witherspoon and Gaskins (21) approached this problem at ORNL by planting snap bean (Phaseolus vulgaris) seeds and watering them with a methylmercury solution after four days. For comparative purposes an identical plot was treated with $^{203}Hg(NO_3)_2$. The plants were harvested at weekly intervals for eight weeks and activity distribution determined (Table 4). The most significant result was that ^{203}Hg from methylmercury concentrated in the bean by a factor of 33, an order of magnitude more than the inorganic form. The actual

Table 4. Mercury uptake by bean plants (Phaseolus vulgaris) (means, μCi/g dry weight) from soil at ORNL, summer, 1972.

Sample	Age (weeks)	$CH_3{}^{203}HgCl$	$^{203}Hg(NO_3)_2$
Foliage	2	3.4	0.14
	3	2.2	0.29
	5	2.4	0.41
	8	0.3	0.02
Stems	2	2.1	0.05
	3	1.1	1.1
	5	1.3	0.45
	8	0.1	0.02
Roots	2	4.0	2.4
	3	1.8	1.4
	5	2.4	1.2
	8	1.2	0.01
Pods	8	0.5	0.08
Beans	8	3.1	.3

concentration of mercury from methylmercury in the bean was 0.89 ppm and for the inorganic form, 0.08 ppm.

Cadmium Microcosm Experiments

Cadmium, like mercury, is introduced to the biosphere from industrial and agricultural (fertilizer) sources through both atmospheric and water pollution. The ecological behavior of cadmium was studied in much the same manner as mercury. Unlike mercury, almost no information was available indicating the environmental pathways, sinks, or target organisms of cadmium compounds. A microcosm experiment was conducted in order to observe and quantify the rates and pathways of cadmium ion transfer from terrestrial to aquatic systems and to analyze the sites and quantities of bioaccumulation of cadmium within terrestrial and aquatic systems.

Two microcosms were established similar to those described previously in the mercury fly ash experiment. These microcosms were maintained at natural photoperiod, rainfall, and temperature for three weeks in May, 1971. After $^{115m}CdCl_2$ (284 µCi) was applied to the terrestrial portion of each microcosm in a simulated rainfall, sampling was conducted on a routine basis in the same manner as described for the mercury fly ash microcosms.

A total cadmium budget was determined after 27 days by harvesting all components of each system. A leaching experiment with ammonium acetate was performed to determine the cadmium in the soil that was biologically available. These tests indicated that only 2-3% of the total cadmium present was leachable or available for uptake by plants or soil fauna. Table 5 shows the distribution

Table 5. Distribution of 284 µCi of ^{115m}Cd in microcosm experiments at 27 days after tagging with $^{115m}CdCl_2$ at ORNL, summer, 1971.

Component	Microcosm A		Microcosm B	
	µCi	% Activity	µCi	% Activity
Terrestrial		93.8		95.5
Moss	28.90	10.2	23.34	7.9
Higher Plants	0.38	0.1	0.78	0.3
Litter	43.40	15.3	30.20	10.6
Soil	193.50	68.1	218.00	76.8
Aquatic		3.4		4.5
Water	0.52	0.2	0.56	0.2
Sediment	8.93	3.1	8.63	3.3
Fish	<0.01	<0.1	<0.09	<0.1
Snails	0.22	0.1	0.14	<0.09
Watercress	0.02	<0.09	0.04	<0.09
Plastic Liner	7.89	2.8	3.26	1.1

of the ^{115m}Cd in the microcosms at 27 days. Most of the cadmium (93.8 and 95.5%) was found in the terrestrial portion. Only 3.4 and 4.5% were found in the aquatic portion. The data indicate that cadmium is less mobile in terrestrial systems than is mercury.

MERCURY AND CADMIUM IN THE AQUATIC ECOSYSTEM

We have shown that mercury from fly ash and/or flue gas is continuously input to streams and lakes directly from the atmosphere and from wash-off from the terrestrial environment. In addition, the industrial discharges to streams (largely curtailed) have become incorporated in stream sediments at discrete points near the outfalls. Microorganisms in the sediments convert the inorganic mercurials, whatever the source, to the more toxic dimethyl and methyl forms which are given off into the water, transported downstream or enter the aquatic food chain.

Analysis of Mercury in Fish from the Great Smoky Mountains National Park

In order to obtain data on mercury concentrations in fish at or near natural background levels, 198 fish of five species were collected from three high-altitude streams remote from pollution sources in Great Smoky Mountains National Park (83° 25'-50' west longitude by 35° 30'-35' north latitude) during May and June, 1972 (22) (Fig. 1). The section of the Park in which the study streams are located is incompletely mapped geologically, but the bedrock on which the streams flow is probably the partially metamorphosed clastic sediments of the Precambrian Ocoee Series. There are some abandoned copper mines in the study area (near Hazel Creek) in which the ore was the sulfide minerals of Cu and Fe, with some Zn and traces of Pb, Co, and As; however, no cinnabar (HgS) was detected (23). The study streams are located in remote sections of the Park and are accessible only by foot or on horseback. Multiple pH tests in all three streams always showed a value of 6.5. Historically, the area was heavily logged, but all logging activity was phased out with the establishment of the Park in 1926.

Rainbow trout (*Salmo gairdneri* Richardson) and stonerollers [*Campostoma anomalum* (Rafinesque)] were collected at 550 m in Hazel Creek and at 850 and 1225 m in Forney Creek. Eastern brook trout [*Salvelinus fontinalis* (Mitchill)] were collected at elevations between 1150 and 1225 m in all three streams. Rosyside dace (*Clinostomus funduloides* Girard) and banded sculpin [*Cottus carolinae* (Gill)] were collected at 550 m in Hazel Creek. The fish sampled represent two trophic levels, trout, dace, and sculpins being predators and stonerollers feeding on periphyton. Although the rainbow trout were introduced approximately 50 years ago, the streams are not presently stocked with hatchery fish. Sediment samples were also collected at 550 m in each of the three streams for mercury analysis.

Most of the fish collecting was by electroshocking, but the discharge and gradient of the streams at the lower reaches were too great for this method to be effective on the trout. Trout collections in the lower reaches were by angling. All fish samples were immediately frozen in Dewar flasks containing Dry Ice (CO_2) for transport to the laboratory. The gastrointestinal tracts were removed from all the stonerollers to avoid the possibility of prejudicing the analyses with mercury-containing sediment or food organisms. Comparative analyses of trout samples were made with the gastrointestinal tract removed and with it present. A comparison was also made of mercury content in trout by dissolving separately for analysis the whole fish and a strip of axial muscle tissue. Under conditions of chronic exposure to mercury, the muscle tissue of fish contains the greatest per cent of the total mercury body burden (24).

There was very little difference in the mercury content of three sediment samples taken at 550 m in all three streams. Samples from Hazel Creek had 0.004 ppm mercury; from Eagle Creek, 0.004 ppm; and from Forney Creek, 0.005 ppm. Water samples from all three streams had $<$ 0.02 ppb mercury. The brook trout, which were collected at the highest altitude and the greatest distance ($\geq$ 25 km) from the industrialized Appalachian Valley, had the lowest mean concentrations of mercury, but this could be a species difference rather than geographical difference (Table 6). Mercury concentrations have been measured in brook and rainbow trout before. Zitko, *et al.* (25) analyzed seven brook trout in Canada and found a range from 0.08 to 0.13 ppm (wet weight) methylmercury. In Idaho, Gebhards, *et al.* (26) measured mercury concentrations by neutron activation analysis in wild and hatchery-reared rainbow trout. The 16 wild fish had a mean of 0.31 ppm (wet weight) mercury with a range of $<$ 0.008 to 0.6 ppm, and the hatchery fish had a mean mercury concentration of 0.10 ppm with a range of 0.05 to 0.17 ppm. D'Itri, *et al.* (27) compared total mercury levels in stocked rainbow trout from an oligotrophic and a eutrophic lake in Michigan and found significantly higher concentrations in the oligotrophic lake samples as compared to the eutrophic lake samples. However, all of those fish were stocked, rather than native to the lakes. We conclude that the values reported here--$\sim$ 0.04 ppm--are near the lower limit of mercury concentration naturally occurring in fish.

Most reports of mercury analyses in fish and other organisms show wide variability of mercury concentration in tissue. The actual amounts--nanograms and less--are very small, so a large percentage variation actually involves only a small quantity of mercury. The known heterogeneity of mercury distribution in rocks

Table 6. Mercury concentrations (ppm) in fish from high altitude streams in Great Smoky Mountains National Park, May–June, 1972.

Species	Stream	Elevation (Meters)	n	$\overline{X}$	(Range)	±SD
Salmo gairdneri (Rainbow trout)	Hazel Creek	550	15	0.023	(0.006–0.147)	0.035
		580	12	0.060	(0.016–0.177)	0.047
	Eagle Creek	550	11	0.014	(0.008–0.020)	0.003
	Forney Creek	550	17	0.040	(0.003–0.097)	0.025
		850	12	0.042	(0.024–0.061)	0.012
		1225	8	0.042	(0.030–0.056)	0.008
	TOTAL, All Streams		75	0.036	(0.003–0.177)	0.028
Salvelinus fontinalis (Eastern brook trout)	Hazel Creek	1225	9	0.019	(0.014–0.030)	0.004
	Eagle Creek	1150	7	0.002	(0.007–0.005)	0.001
	Forney Creek	1220	3	0.053	(0.046–0.064)	0.009
	TOTAL, All Streams		19	0.018	(0.007–0.064)	0.018
Campostoma anomalum (Stoneroller)	Hazel Creek	550	18	0.031	(0.015–0.070)	0.015
	Eagle Creek	550	9	0.059	(0.031–0.076)	0.014
	Forney Creek	550	6	0.033	(0.009–0.083)	0.029
	TOTAL, All Streams		33	0.039	(0.009–0.083)	0.022
Clinostomus funduloides (Rosyside dace)	Hazel Creek	550	44	0.044	(0.026–0.129)	0.019
Cottus carolinae (Banded sculpin)	Hazel Creek	550	27	0.025	(0.006–0.105)	0.018

could easily account for the observed differences in mercury concentrations for fish from a single stream (8).

There was no significant difference in mean mercury content among species or between separate samples of one species since the error terms broadly overlapped (Table 6). There was no systematic change in mercury concentration in the fish along the elevational gradient sampled. No differences (normalized to weight) were detected between whole fish and axial musculature or between whole fish and fish with gastrointestinal tracts removed.

Mercury Stream Tag Experiments

A small stream within the AEC-ORNL Reservation was tagged with $CH_3{}^{203}HgCl$ and a similar stream with $^{197}Hg(NO_3)_2$ in an effort to clarify the fate of mercury compounds in natural stream ecosystems. Stream tag experiments, in which a radioactive isotope is spiked into the water and the distribution of the activity followed by collecting samples and determining the radioactivity can rapidly yield quantitative data on transfer rates and bioaccumulators of the element. Obviously, these parameters will change from stream to stream, but the qualitative generalizations should hold. The tagging and sampling procedures used were developed in previous studies on these same streams by Nelson, et al. (28) and Elwood and Nelson (29).

The study sections of the streams were 100 m long measured downstream from the point of isotope release with sampling stations at 10 m, 20 m, 40 m, 70 m, and 100 m. Samples taken included fish, snails, watercress, periphyton, and sediments. Fish were also collected at 200 m downstream.

The discharge of one of the streams, Walker Branch, was measured at the two weirs of the Walker Branch Watershed project located immediately upstream from this experiment (30). A Parshall Plume was installed in the other stream, White Oak Creek, to measure its discharge. The sediments of both creeks consist of subangular gravel-size rubble and poorly sorted sand. There is little organic detritus. Table 7 shows the physical parameters and tagging details of the two streams during each experiment. A greater concentration of $Hg(NO_3)_2$ than CH_3HgCl was used. If 65-hour half-life ^{197}Hg had been used, the ^{197}Hg would have decayed to levels difficult to detect after about four or five days. Indeed, poor counting statistics caused the termination of the $^{197}Hg(NO_3)_2$ experiment at nine days, even though the initial concentration of $^{197}Hg(NO_3)_2$ was 35 times greater than the concentration of $CH_3{}^{203}HgCl$. Mercury-197 was used in one stream instead of ^{203}Hg because the stream crosses a public road several hundred meters downstream from the experiment. The shorter half-life isotope further reduced any chance of detectable radioactivity at the vicinity of the road.

Table 7. Experimental parameters of stream tag experiments using $CH_3{}^{203}Hg$ and $^{197}Hg(NO_3)_2$ at ORNL, Tennessee.

Parameter	White Oak Creek $CH_3{}^{203}HgCl$ 1 Sept. 1971	Walker Branch $^{197}Hg(NO_3)_2$ 5 Oct. 1971
Isotope added (mCi)	1.65	4.48
Total Hg added (mg)	0.32	19.73
Spike duration (min)	11.8	19.1
Maximum Hg concentration at 10-m (mg/ℓ × 10^{-4})	0.832	29.7
pH	8.1	8.0
Temperature (C)	16.5	17.0
Discharge (ℓ/m)	325	347
Dissolved O_2	saturated	saturated

Fluorescein dye was added to the streams immediately before and immediately after each isotope was released. Water samples were taken at 5 minute intervals from the time the first dye appeared at a sampling station until none of the post-spike dye was visible, thus bracketing the isotope spike as it flowed past the sampling stations. In both streams, 40 minutes were required for the dye to reach the 100 m station and 90 minutes were required for all the dye to pass the 100 m point. Thus, sampling lasted 90 minutes at the lower stations.

Two different water sampling techniques were employed. A 5-ml sample of water was pipetted into a counting tube every 5 minutes and 250 ml of water was collected in a polyethylene bottle every 10 minutes. Each bottle of water was poured through a 0.45-µ Millipore filter and the gamma-ray activity of the filtered residue was counted. A portable generator was used to power a pump for the filtering apparatus, and filtrations were done as the samples were collected. This technique minimized the possibility of mercury adsorption on the bottles. Figures 5 and 6 show the water activity during isotope flow-through at each sampling station. The sampling interval (5 minutes) was too long in the CH_3HgCl experiment, and the peak activity was missed at reach I. At 4-hour post-spike, collections were made at each sampling station, although fish and emergent plants were not present at every station. Collections were then made at regular intervals for nine days in Walker Branch (^{197}Hg) and 40 days in White Oak Creek (^{203}Hg).

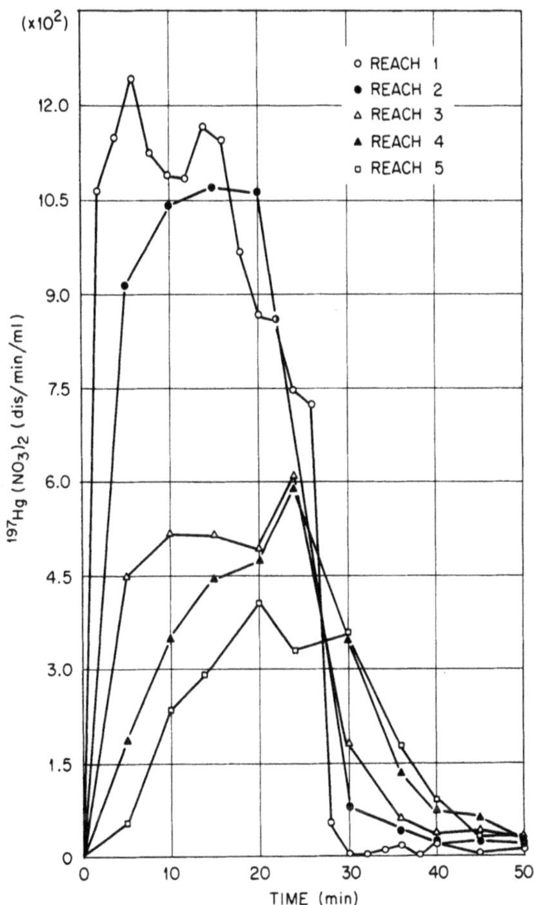

Figure 5. Activity in stream water at successive downstream sampling stations during tagging experiment with 4.5 mCi of $^{197}Hg(NO_3)_2$ at ORNL, Tennessee, 1971. Water sampling interval was 2 minutes at Reach 1, and every 5 minutes at Reaches 2-5. Activity decrease at 10 minutes in Reach 1 due to partial interruption of isotope release.

In White Oak Creek, the mean stable mercury content of fish determined by atomic absorption spectrophotometry was 0.24 ppm; sediments, 0.19 ppm; and water, < 0.02 ppb. The values for Walker Branch were: fish, 0.28 ppm; sediments, 0.09 ppm; and water, <0.02 ppb.

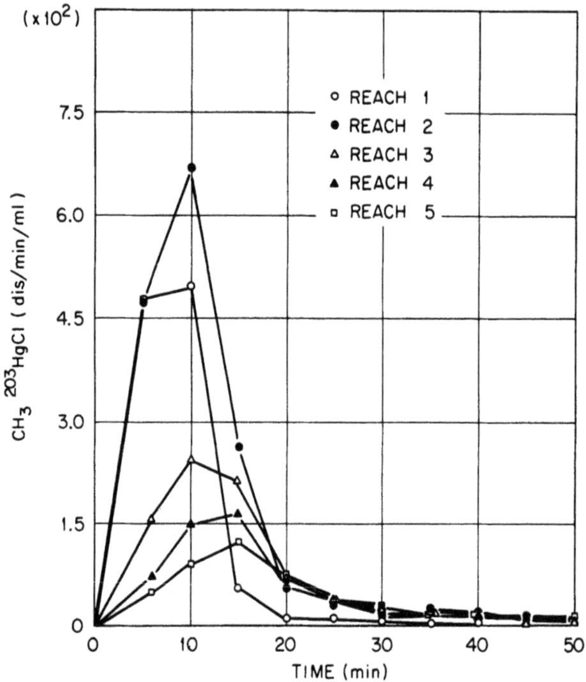

Figure 6. Activity in stream water at successive downstream sampling stations during tagging experiment with 1.65 mCi of $CH_3{}^{203}HgCl$ at ORNL, Tennessee, 1971. Water sampling interval was 5 minutes.

In the $^{197}Hg(NO_3)_2$ experiment, 25% of the isotope flowed out of the first 100 m of the stream study section, and in the $CH_3{}^{203}HgCl$ experiment, 19% of the isotope flowed out of the first 100 m of the stream study section. Twenty-five per cent of the water's activity was found on filtered particulate matter at 100 m in the $^{197}Hg(NO_3)_2$ experiment, while only 4% of the total water sample activity was found on particulate matter in the $CH_3{}^{203}HgCl$ experiment.

Tables 8 and 9 show the transport of radiomercury through the stream sediments (mean of 3 samples), periphyton (mean of 3 composite samples), and snails (composite samples) at each sampling station. The sediment activity peaks showed no systematic pattern with distance from the input. The periphyton and snails indicate peak concentrations occurring later in time with increasing distance from the isotope input. Snails of the genus Goniobasis have been shown to exhibit positive rheotaxis so we conclude that the data show a net transport of mercury, not snails, downstream (31).

The periphyton, which took all its activity directly from the water, peaked in two days [$^{203}Hg(NO_3)_2$] at 100 m, while the snails

Table 8. Transport of $^{203}Hg(NO_3)_2$ in a stream at ORNL, Tennessee. Day of maximum concentration of ^{197}Hg from $^{197}Hg(NO_3)_2$ in sediments, periphyton, and snails following tag.

Compartment	Station	Day of Maximum Concentration After Tag
Sediments	1	9
	2	7
	3	9
	4	3
	5	2
Periphyton	1	1
	2	2
	3	2
	4	2
	5	2
Snails	1	1
	2	1
	3	2
	4	3
	5	5

Table 9. Transport of $CH_3{}^{203}HgCl$ in a stream at ORNL, Tennessee. Day of maximum concentration of ^{203}Hg from $CH_3{}^{203}HgCl$ in sediments, periphyton, and snails following tag.

Compartment	Station	Day of Maximum Concentration Following Tag
Sediments	1	1
	2	8
	3	8
	4	42
	5	16
Periphyton	1	1
	2	1
	3	2
	4	4
	5	4
Snails	1	4
	2	8
	3	12
	4	28
	5	28

that took activity from the water and from the periphyton (and sediments) on which they grazed, peaked at five days. In the $CH_3{}^{203}HgCl$ experiment, the periphyton peaked at four days, and the snails at 28 days. These data show that inorganic mercury is cycled through the periphyton compartment about twice as fast as methylmercury, while the snail compartment cycles inorganic mercury about five times faster than methylmercury.

In these oligotrophic waters, $^{203}Hg^{2+}$ from methylmercury was concentrated and retained to a greater extent than mercuric ions from $^{197}Hg(NO_3)_2$ by all the animals sampled. Figures 7 and 8 show the concentrations (dis/min/g) of radiomercury in the samples of selected stations (all biota present) in each experiment. No error terms are plotted for composite samples or for less than four observations.

Fish acquire and retain methylmercury to a greater extent than other biota and inorganic sediments. The concentrations of $^{203}Hg^{2+}$ activity from methylmercury in the few crayfish and salamanders which were caught were always less than in fish from the same stream subsection. The observed difference in methylmercury concentrations in Rhinichthys cataractae (long-nose dace) and Campostoma anomalum (stoneroller) could be due to the different trophic level these species occupy. Gut content analysis showed that dace fed on a variety of small invertebrates while stonerollers fed on algae (mostly diatoms). However, dace also concentrated $^{197}Hg^{2+}$ more than Cottus carolinae (banded sculpin). In the methylmercury experiment, a few sculpins were taken, and they always had a lower mercury concentration ratio than dace from the same station. The gut contents of sculpins showed a higher percentage of macroinvertebrates than dace. In a subsequent experiment, we determined that zooplankton accumulates methylmercury faster than do insects. Thus, by feeding on the zooplankton the dace would ingest more mercury than the sculpins. Mercury concentrations in fish tissue as a function of the trophic level the species occupies has not been demonstrated. Gebhards, et al. (26) analyzed 160 fish samples representing ten species from the Snake River in Idaho for mercury residue. They reported that channel catfish, yellow perch, and suckers contained higher concentrations of mercury than did the other fish species taken from the same reach of the stream. Our measurements on the Great Smoky Mountains fish discussed earlier also reveal no trophic level magnification.

In contrast to methylmercury, the fish concentrated fewer mercuric ions from $Hg(NO_3)_2$ than any of the other samples. Both isotopes were detectable in fish at 200 m. Five days after the $Hg(NO_3)_2$ isotope release in Walker Branch, 127 dis/min $^{197}Hg/g$ (wet weight) were detected in the snail Goniobasis clavaeformis at 500 m. Sixty days after the $CH_3{}^{203}HgCl$ isotope release, 8600 dis/min $^{203}Hg/g$ (wet weight) were detected in dace and 6690

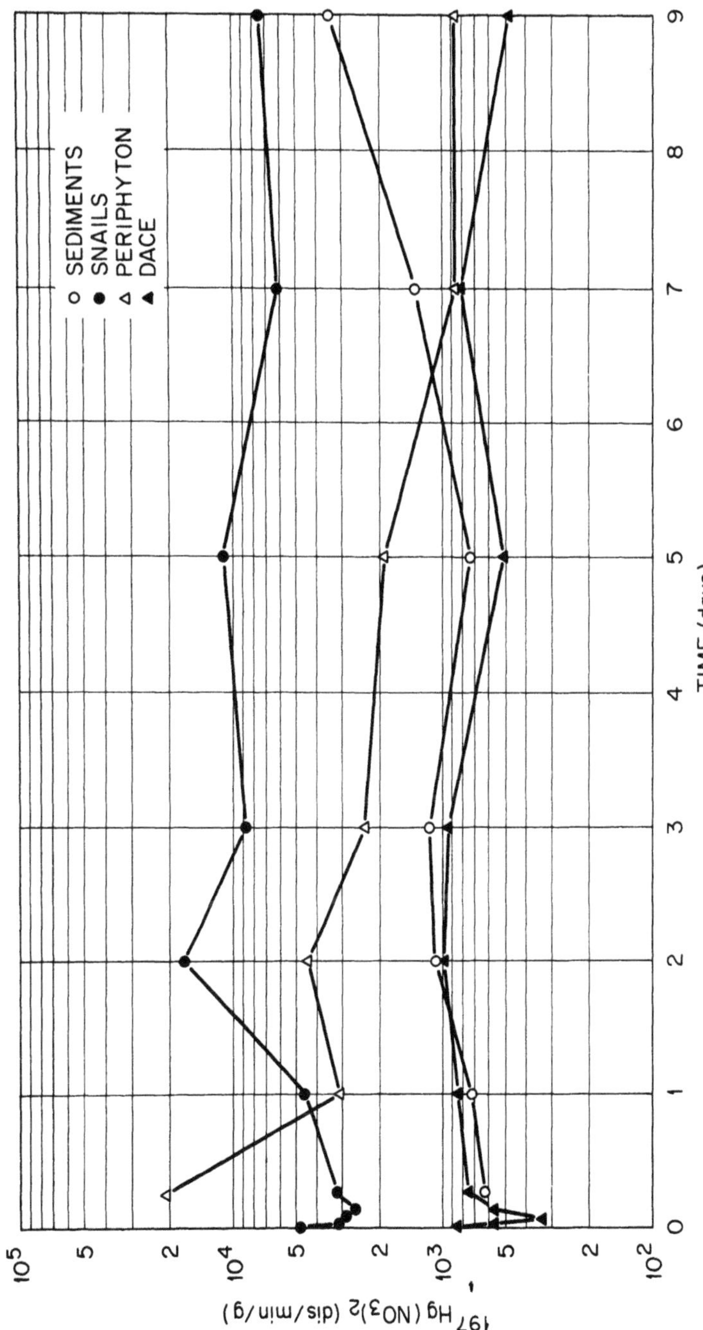

Figure 7. Activity (means, dis/min/g) in stream ecosystem compartments in Reach 3 of $^{197}Hg(NO_3)_2$ stream tag experiment. Composite samples collected for nine days. Sediments showed large variations in activity due to inhomogeneity of sediment distribution.

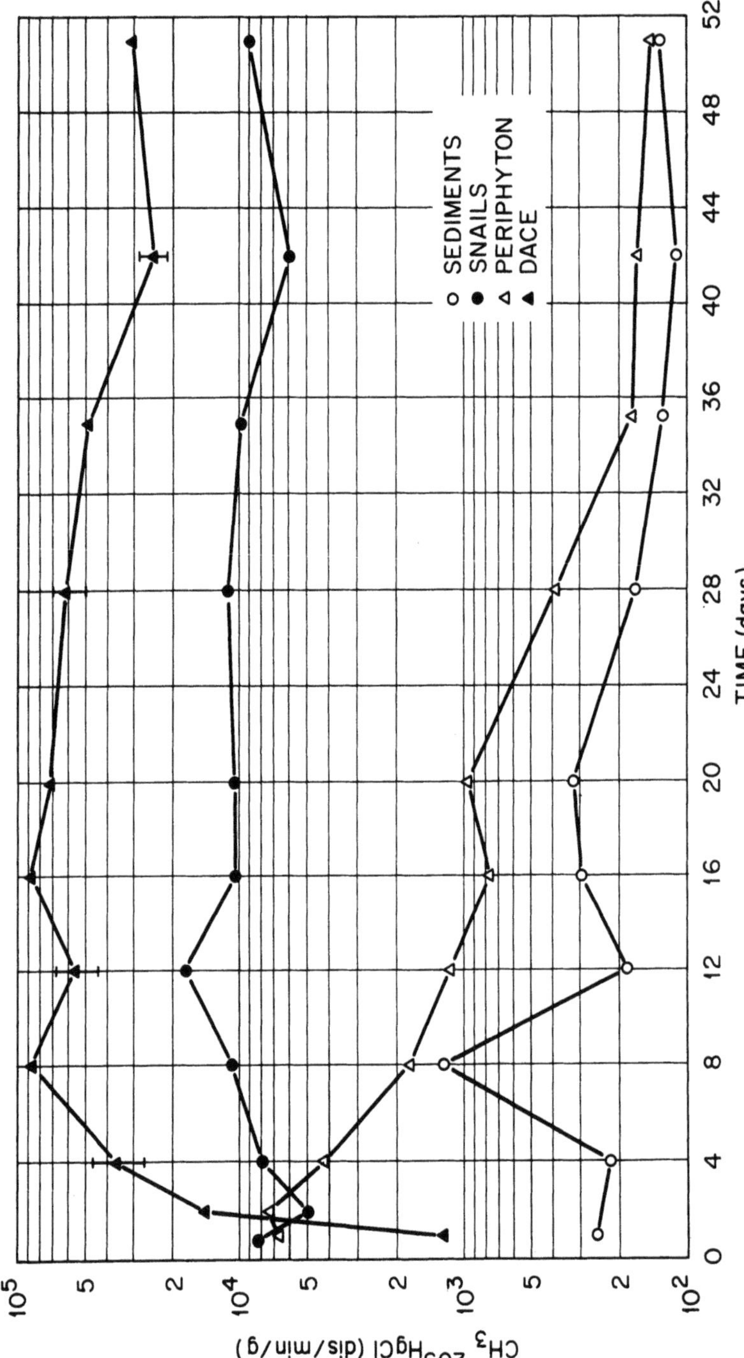

Figure 8. Activity (means, dis/min/g) in stream ecosystem compartments in Reach 3 of $CH_3^{203}HgCl$ stream tag experiment. Composite samples of sediments, snails, and periphyton. Error terms shown for dace when n = >3.

dis/min/g (wet weight) in stonerollers at approximately 300 m.

Cadmium Stream Tag Experiment

There seems to be no organic cadmium analog of methylmercury, but the itai-itai incident in the Jintsu River basin and the excessive concentrations of cadmium in the water of the Coeur d'Alene River in Idaho, point out not merely the potential but the fact of toxic levels of cadmium occurring in natural waters (9). The microcosm experiment described earlier indicates that cadmium enters the aquatic environment from terrestrial sources at a slow rate, but it yielded little information on the dynamics of cadmium transfer once in the aquatic system. In order to obtain information on the movement of cadmium in natural waters, a stream tag experiment was conducted in a similar manner to the mercury stream tag just described.

The isotope used was $^{109}CdCl_2$. The experimental parameters are shown in Table 10. Sampling procedures were the same as in the

Table 10. Experimental parameters of cadmium stream tag experiment using $^{109}CdCl_2$ at ORNL, Tennessee, August 12, 1972.

Parameter	$^{109}CdCl_2$ 12 Aug. 1972
Isotope added (mCi)	5
Total Cd added (mg)	2.97
Spike duration (min)	21
Maximum Cd concentration at 10-m (mg/ℓ × 10^{-4})	0.462
pH	9.2
Temperature (C)	17
Discharge (ℓ/m)	362
Dissolved O_2	saturated

mercury experiment, except no fish were taken at 200 m. Figure 9 shows the water activity during isotope flow-through at each sampling station. Cadmium-109 rather than ^{115m}Cd was used because ^{109}Cd yields better counting statistics. ^{109}Cd emits one gamma-ray accompanied by the emission of an x-ray. The x-ray is counted instead of the gamma-ray because it occurs in the energy region of maximum photopeak efficiency of the detector. This means that counting efficiency using the x-ray is usually better than that for the gamma-ray. Cadmium-115m emits several gamma-rays, but the branching ratio of all of them is very low (∿18%). In order to get enough ^{115m}Cd for good counting statistics, the concentration of cadmium may approach toxic levels for sensitive organisms.

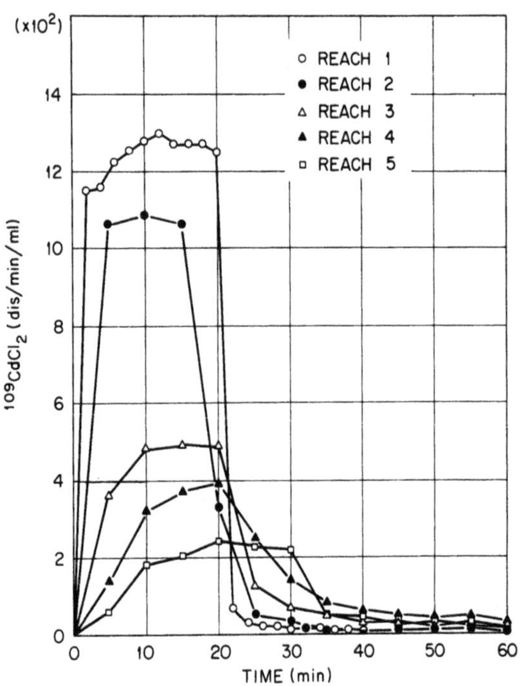

Figure 9. Activity in stream water at successive downstream sampling stations during tagging experiment with 5 mCi of $^{109}CdCl_2$ at ORNL, Tennessee, 1972. Sampling interval was 2 minutes at Reach I and 5 minutes at Reaches 2-5.

Cadmium proved to be less mobile than mercury when spiked into the stream, since 95% of the ^{109}Cd added was retained in the first 100 m below the input point. Unlike the two mercury compounds, the per cent of the ^{109}Cd activity associated with particulate matter suspended in the water did not change systematically with sampling station, but maintained at about 20% activity on > 0.45 μ particles, with about 80% activity remaining in solution. Figure 10 shows the transport of ^{109}Cd through the stream ecosystem compartments in Reach 3.

The concentration of ^{109}Cd in the stream sediments was quite variable, due probably to the turbulence of the water and the non-uniform distribution of the sediments. The concentration of ^{109}Cd in the periphyton samples at day 12 is probably in error. The

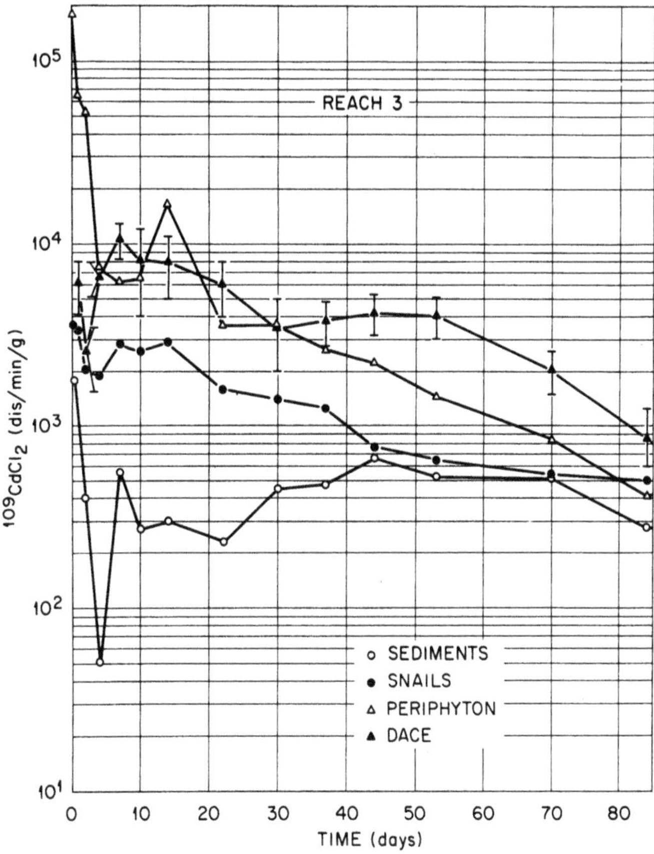

Figure 10. Activity (means, dis/min/g) in stream ecosystem compartments during $^{109}CdCl_2$ stream tag experiment at ORNL, Tennessee, 1972. Composite samples for sediments, snails, and periphyton.

elimination curve for the periphyton follows an exponential function. Both the snails and fish showed peak concentrations of ^{109}Cd at about one week. After this peak, the snails showed an apparent exponential loss of ^{109}Cd, but the fish showed what seems to be a bimodal elimination curve, the elimination rate increasing at about 3 months post-tag.

If the periphyton data point at day 12 is in error, the fish had higher average ^{109}Cd concentrations than periphyton after about one week, and significantly higher ^{109}Cd concentration after about 2 1/2 months. These results indicate that cadmium bioaccumulates in animals, but not nearly to the extent methylmercury does. Cadmium appears to be generally less mobile in the environment than mercury.

Food Chain Experiments

The accumulation of mercury in fish is one of the most important aspects of the mercury pollution problem, both in terms of ecological effects and human consumption (32,33,34). The Japanese experience at Minamata and Niigata emphasize the possible serious consequences of eating mercury contaminated fish, and the damage to wildlife due to widespread mercury distribution in Sweden has been well documented (33,34).

Fish acquire body burdens of mercury directly from the water by adsorption through the gills and through the food chain, but it is not known if the retention time of mercury differs according to the route of uptake (24,35). There have been many analyses of mercury in aquatic organisms, but the trophic dynamics of mercury in aquatic ecosystems remains obscure. Understanding this problem requires consideration of the behavior of both inorganic mercurials and methylmercury. It has been reported that 60-90% of the total mercury in fish tissue is the methyl form, even though the major industrial input to the environment is inorganic or metallic mercury (24,36). The conversion of inorganic and metallic mercurials to the methyl form by methanogenic bacteria has been demonstrated (37,38).

We are experimentally comparing the accumulation and retention of CH_3HgCl and $Hg(NO_3)_2$ at 25° acquired directly from the ambient water and from the food in the trophic chain: green algae - detritus → blood worm (_Chironomus_ sp. larvae) → mosquito fish (_Gambusia affinis_) → large mouth bass (_Micropterus salmoides_). Table II summarizes the data obtained so far.

Biological half-lives (T_b) are measured by exposing the organism (via water or food) and removing it to clean water and live-counting daily until an elimination curve can be determined.

The detritus--mostly the green algae _Chlorella_--was tagged with ^{203}Hg to a concentration of 3 ppm (air dried weight) in the case of $CH_3{}^{203}HgCl$, and 16 ppm in the case of $^{203}Hg(NO_3)_2$. The detritus was millipore filtered (0.45 μ) and placed into the water with _Chironomus_ larvae. The larvae fed for one hour, and were removed to elimination cages in clean water and ^{203}Hg activity counted for seven days.

For the _Chironomus_ → _Gambusia_ link, the _Chironomus_ larvae were tagged by placing them in small containers of ^{203}Hg until they acquired 0.04 ppm (wet weight) in the case of CH_3HgCl and 0.03 ppm in the case of $Hg(NO_3)_2$. These larvae were then fed to _Gambusia_ (∿10/hour) for one hour, and the Gambusia were live-counted for 30 days in the case of $CH_3{}^{203}HgCl$ and seven days in the case of $Hg(NO_3)_2$.

Table II. Preliminary determination of biological half-lives (T_b, days), mercury concentration (ppm), and assimilation efficiencies (A_e, %) of aquatic food chain components at 25°.

	$CH_3{}^{203}HgCl$			$^{203}Hg(NO_3)_2$		
	Chironomus	Gambusia	Micropterus	Chironomus	Gambusia	Micropterus
Direct Uptake						
T_b	48			0.5	14	27
Hg concentration (water)	0.4			0.4	0.5	0.05
Food Chain Uptake						
T_b	7	55		5.5	12	12
Hg concentration (food)	3	0.04		16	0.03	0.004
A_e	76	81.5		60	12	40

For the *Gambusia* → *Micropterus* link [$Hg(NO_3)_2$], *Gambusia* were placed in water with ^{203}Hg until they acquired ∿0.04 ppm (wet weight) and were then fed to the *Micropterus* (one *Gambusia* to each *Micropterus*). The CH_3HgCl *Gambusia* → *Micropterus* link is not yet completed, nor are the direct uptake parameters of CH_3HgCl for *Gambusia* or *Micropterus*.

The mechanism of uptake and retention of inorganic mercurials by fish differs from that of organic mercurials, as shown by Jernelov (24), Backstrom (39), Haga, *et al*. (40), and McKone, *et al*. (41). Organic mercurials are metabolized in the liver and the Hg^{2+} ions then excreted via the kidneys. The cleavage rate of the C-Hg bond is slow; consequently, the organic moieties are retained longer than inorganic ions, resulting in the deposition of organomercurials in the nervous system and in muscles (24). Since fish muscle protein has a relatively high per cent of sulfur-containing amino acids and nitrogen, both of which react readily with the mercury and other heavy metals to form stable chelates, fish muscle tends to contain relatively high mercury concentrations (39,42,43).

Hannerz (35) has compared direct uptake and food-chain uptake of several forms of mercury by both marine and fresh water fish. He concluded that food-chain accumulation of inorganic mercurials in fresh water fish seem negligible, and that his experiments led to an overestimation of the role of accumulation of mercury from food. The experiments contained an ambiguity, since some of the fish were not fed mercury-tagged natural food, while others were. Hannerz further states that his data cannot be used to make any conclusions as to the relative importance of mercury uptake from the water and from food in nature.

Jernelov (24) reported that in nature the transfer of methylmercury from benthic organisms to benthic-feeding fish is small, but that for piscivorous predators (pike), 60% of the mercury body burden is acquired through the food chain. These calculations are based on analyses of stable mercury from fish collected in Swedish waters. Our preliminary experimental data indicate that food chain accumulation can contribute significantly to the body burden of mercury in fish.

Cadmium Food Chain Experiments

Experiments to determine the accumulation and transfer of cadmium through an aquatic food chain have been initiated and some preliminary data are available. The experiments involve the use of radiotracer techniques and the same aquatic food chain as the mercury experiments.

Gambusia tagged with $^{109}CdCl_2$ were fed to largemouth bass and the retention time determined. The retention of cadmium from $CdCl_2$ was very similar to the retention of inorganic mercury by largemouth bass. Two days after the tagged Gambusia were fed to the largemouth bass, the bass retained about 7% of the cadmium. After thirty days the percentage had decreased to less than 2%. Direct uptake and food chain studies are continuing to determine the flow of cadmium through aquatic trophic chains.

SUMMARY

An holistic view of the transfer of heavy metals from terrestrial to aquatic environments can be achieved through a synthesis from many small scale experiments each of which considers a limited aspect of the overall problem. The information presented here is not yet an holistic statement, but is a first approximation from an experimental strategy intended to provide a unified picture of the transfer of heavy metals in the environment.

The measurement of background concentrations of heavy metals is a necessary step in determining the rates of transfer of these substances in the environment. In terrestrial situations, predatory organisms appear to have higher tissue concentrations of mercury than the prey species on which they feed.

An analysis of plants from the Great Smoky Mountains National Park (GSMNP), a putative low background area (i.e., little or no mercury pollution) showed no concentrations exceeding 0.2 ppm (air dried weight) with mosses generally having higher concentrations than vascular plants. Cadmium concentrations were as high as 0.4 ppm (oven dried weight) in moss, with vascular plants always having lower concentrations. These background levels should, of course, not be uniformly applied since the fectors influencing the uptake of heavy metals by plants vary from region to region.

Once the background levels of mercury in environmental samples for a region are known, the amount of mercury input via pollution can be deduced. Perhaps no source of regional mercury pollution is more important than the burning of mercury containing coal. Coal fly ash collected as it settled in the environment at Oak Ridge National Laboratory (ORNL) contained 0.98 ppm mercury. Samples of this fly ash was tagged with radioactive mercury and employed in cycling experiments in laboratory microcosms and natural field plots. Mercury from fly ash was shown to be leached from the particles and rendered available for biological uptake. Although the ^{203}Hg tagged fly ash was applied to the terrestrial portion of the microcosms, ^{203}Hg slowly accumlated in sediments, snails, and fish.

The transfer rate of mercury and cadmium from terrestrial to aquatic ecosystems is dependent upon the residence time in the terrestrial compartments. The terrestrial cycling of inorganic mercury and cadmium was investigated by tagging field plots with soluble ^{203}Hg and ^{109}Cd. At the end of one growing season, four months after application, the vegetation contained four times more cadmium than mercury (based on per cent of total input). Most (64%) of the remaining mercury was found in the soil, while most (59%) of the remaining cadmium was found in the liter. This data indicates that inorganic mercury is lost more readily from vegetation than is cadmium, in agreement with the analytical results of the Cades Cove vegetation.

A tree inoculation study at ORNL revealed that inorganic mercury was leached out of the crown to a greater extent than was cadmium. Most (82%) of the mercury that leached out of the tree crown was bound in the soil under the tree, while most (68%) of the cadmium that leached out was retained in the litter, in agreement with the field plot results. Cadmium would, therefore, be more available to organisms whose food base is the litter or herbaceous vegetation.

An uptake study with bean plants showed a large difference in the translocation rate and concentrations of inorganic and the more toxic methylmercury. The fruit concentrates methylmercury by a factor of 33 and inorganic mercury by a factor of 3.

A microcosm experiment with cadmium showed that 68 to 77% of ^{109}Cd added to the terrestrial portion was bound there in contrast to 42 to 54% of the ^{203}Hg bound in the terrestrial portion in the mercury microcosm experiment, again indicating the greater mobility of mercury in the environment.

The Great Smoky Mountains provide an opportunity to determine the background levels of trace metals in aquatic organisms, since several large watersheds are maintained in an undeveloped state in the higher elevations of the GSMNP. Geological and meterological considerations indicated that biota from these remote areas of the park would contain low concentrations of mercury. Fish of five species were collected from three of the larger streams and were found to contain a mean of 0.04 ppm mercury. There was no significant difference in mercury content in or among species from all three streams, nor was there any relationship of mercury concentration with fish weight. No differences in mercury concentration were noted with fish analyzed whole, with gastrointestinal tract removed, and axial musculature.

The transport of mercury in oligotrophic streams such as those in GSMNP was studied at ORNL by tagging experiments with $CH_3{}^{203}HgCl$ and $^{197}Hg(NO_3)_2$. Mercury-203 from methylmercury was shown to be cycled more slowly than ^{197}Hg from $^{197}Hg(NO_3)_2$. The elimination rate of ^{203}Hg from methylmercury was twice as long as that of ^{197}Hg from $^{197}Hg(NO_3)_2$ in periphyton, and five to six times as long in snails and fish. However, both mercurials were retained in the sediments for comparable lengths of time. About 80% of the total activity from methylmercury was retained in the first 100 m of the stream, while about 75% of the activity from mercuric nitrate was retained in the first 100 m. Twenty-five per cent of the ^{197}Hg from $^{197}Hg(NO_3)_2$ detected in the stream water was bound on suspended particles, while 4% of the ^{203}Hg from methylmercury was bound on the suspended particles.

Fish concentrated ^{203}Hg from methylmercury higher than any of the other biota, but they concentrated ^{197}Hg from $^{197}Hg(NO_3)_2$ less than did the other biota.

A similar stream tag experiment with cadmium revealed that ^{109}Cd from $^{109}CdCl_2$ is less mobile in the stream than mercury, since 95% of the total ^{109}Cd added was retained in the first 100 m. Twenty per cent of the ^{109}Cd in the water was found on the suspended particles. As was the case with mercury, the cadmium concentration in the sediments was highly variable, due to stream turbulence and sediment distribution patterns.

The cadmium elimination rate of periphyton followed an exponential function, but snails and fish showed a net uptake in cadmium for about one week post-tag after which they showed net elimination. After 40 days, the fish had a higher concentration of cadmium than the other biota and remained higher for the duration of the experiment (85 days). Cadmium is thus shown to bioaccumulate in fish, with the liver-kidneys-intestines containing the highest per cent of the body burden. Cadmium appears to be generally less mobile in the environment than mercury, although concentration factors of cadmium by aquatic animals are much lower than for methylmercury.

In aquatic environments, organisms acquire body burdans of heavy metals directly from the water via the gills and the food chain. The relative importance of these two routes of uptake has not been adequately assessed. The food chain green algae - detritus → blood worm → mosquito fish → largemouth bass is being studied using ^{203}Hg tracer. Preliminary results indicate that ^{203}Hg from methylmercury acquired via gills or food has a very long biological half-life in largemouth bass (> 1000 days) compared to mosquito fish (55 days). The efficiency of transfer of inorganic mercury from mosquito fish to largemouth bass was higher (40%) than for blood worm to mosquito fish (12%). The blood worm assimilated 60% of the inorganic mercury contained in the algae-detritus. Although these experiments are incomplete, it appears that food chain uptake can account for a significant per cent of the mercury body burden in fish.

ACKNOWLEDGEMENTS

Several colleagues and students were responsible for much of the work reported here. Cyrus Feldman and J. A. Carter of the ORNL Analytical Chemistry Division did the stable chemical analyses, and D. R. Matthews of the ORNL ACD tagged the fly ash with radiomercury.

N. A. Griffith of the Environmental Sciences Division aided in all of the studies described.

R. A. Stella, Oak Ridge Associated Universities (ORAU)-Emory University Undergraduate Research Participant, and R. G. Olmsted, NSF-ORAU trainee, performed the field tagging experiments with ^{109}Cd and ^{203}Hg.

LITERATURE CITED

1. Huckabee, J. W. (1972), Bioscience, 22(6), 336.

2. Borlaug, N. E. (1972), Bioscience, 22(1), 41-44.

3. Wallace, R. A., Fulkerson, W., Shultz, W. D., and Lyon, W. S. (1971), Mercury in the Environment: The Human Element, USAEC Rpt. ORNL-NSF-EP-1, Oak Ridge National Laboratory, Oak Ridge, Tennessee.

4. Harriss, R. C. (1971), Biological Conservation, 3(4), 279-283.

5. Goldschmidt, V. M. (1958), Geochemistry, London E. E.: Oxford University Press.

6. Pierce, A. P., Botbol, J. M., and Learned, R. E. (1970), Mercury content of rocks, soils, and stream sediments, Mercury in the Environment, U. S. Geol. Survey Prof. Paper 713.

7. Huckabee, J. W., Cartan, F. O., Kennington, G. S., and Franz, J. C. (1973), Bull. of Envir. Contam. and Toxicol., 9(1), 37-43.

8. Shacklette, H. T. (1970), Mercury in plants, Mercury in the Environment, U. S. Geol. Survey Prof. Paper 713.

9. Fulkerson, W. and Goeller, H. E. (1973), Cadmium: The Dissipated Element, ORNL-NSF-EP-21, Oak Ridge National Laboratory, Oak Ridge, Tennessee.

10. Huckabee, J. W., Mosses: Sensitive indicators of airborne mercury pollution, accepted for publication, Atmospheric Environment.

11. Smith, G. F. (1953), Analytica Chimica Acta, 8, 397.

12. Feldman, C., Carter, J. A., and Bate, L. C. (1972), Environment, 14, 48.

13. Joensuu, O. I. (1971), Science, 172, 1027-1028.

14. Bertine, K. K. and Goldberg, E. D. (1971), Science, 173, 233-235.

15. Billings, C. E. and Matson, W. R. (1972), Science, 176, 1232-1233.

16. Schlesinger, M. D. and Schultz, H. (1971), Analysis for Mercury in Coal, Bureau of Mines, Managing Coal Wastes and Pollution Program, Technical Progress Report-43, Bureau of Mines, Pittsburth, PA.

17. O'Gorman, J. V., Suhr, N. H., and Walker, P. L., Jr. (1972), Applied Spectroscopy, 26, 44-48.

18. Matti, C. A., Cycling of ^{203}Hg and ^{109}Cd in an oldfield ecosystem for one growing season, M. S. Thesis, University of Tennessee, in preparation.

19. Witherspoon, J. P., unpublished data.

20. Joint FAO/IAEA Coordination Meetings on Isotope Tracer Aided Studies of the Fate and Significance of Foreign Substances in (a) Food and (b) the Agricultural Environment, Ispra, Italy, Oct. 30-Nov. 10, 1972.

21. Witherspoon, J. P. and Gaskins, R., unpublished data.

22. Huckabee, J. W. (1972), Mercury concentrations in fish from Great Smoky Mountains National Park, ORNL-TM-3908, Nov.

23. Espenshade, G. H. (1963), Geology of some copper deposits in North Carolina, Virginia, and Alabama, Geol. Ser. Bull. 1142-I, Washington, D. C.: U. S. Government Printing Office.

24. Jernelov, A. and Lann, H. (1972), Oikos, 22, 403-406.

25. Zitko, V., Finlayson, B. J., Wildish, D. J., Anderson, J. M., and Kohler, A. C. (1971), J. Fish. Res. Bd. of Canada, 28(9), 1285-1291.

26. Gebhards, S., Cline, J. E., Shields, F., and Pearson, L. (1971), Mercury residue in Idaho fishes--1970, Environmental Monitoring in Idaho for Metallic Poisons Using the Materials Testing

Reactor, Special Research Issue No. 2, J. Idaho Acad. of Sci.

27. D'Itri, F. M., Annett, C., and Fast, A. W. (1971), Mar. Technol. Soc. J., 5(6), 10-14.

28. Nelson, D. J., Kevern, N. R., Wilhm, J. L., and Griffith, N. A. (1969), Water Research, 3, 367-373.

29. Elwood, J. W. and Nelson, D. J. (1972), Oikos, 23, 295-303.

30. Curlin, J. W. and Nelson, D. J. (1968), Walker Branch Watershed Project: Objectives, facilities, and ecological characteristics, ORNL-TM-2271.

31. Crutchfield, P. J. (1966), Nautilus, 79(3), 80-86.

32. Johnels, A. G., Westermark, T., Berg, W., Persson, P. E., and Sjostrand, B. (1967), Oikos, 18, 323-333.

33. Lofroth, G. (1970), Methylmercury, Bull. No. 4 (2nd ed.), Ecological Research Committee, Swedish Natural Science Research Council, Sreavagen 166 VIII, 5-113 46 Stockholm.

34. Nelson, N., ed. (1971), Environmental Research, 4, 1-69.

35. Hannerz, L. (1968), Inst. Freshwater Research Drottingholm, 48, 120-176.

36. Westoo, G. (1966), Acta Chem. Scand., 20, 2131-2137.

37. Jensen, S. and Jernelov, A. (1969), Nature, 223, 753-754.

38. Wood, J. M., Kennedy, F. S., and Rosen, C. G. (1968), Nature, 220, 173-174.

39. Backstrom, J. (1969), Acta Pharm. et Toxicol., 27, 5-103.

40. Haga, Y., Haga, H., Hagino, T., and Kariya, T. (1970), Bull. Jap. Soc. of Scient. Fish., 36, 225-231.

41. McKone, C. E., Young, R. G., Bache, C. A., and Lisk, D. J. (1971), Environmental Science and Technology, 5, 1138-1139.

42. Beveridge, J. M. R. (1947), J. Fish. Res. Bd. Canada, 7, 51-54.

43. Hughes, W. L. (1957), Ann. N. Y. Acad. Sci., 65, 454-460.

BIOLOGICAL EFFECTS OF HEAVY METAL POLLUTANTS IN WATER

ROLF HARTUNG

Department of Environmental and Industrial Health
The University of Michigan
Ann Arbor, Michigan 48104

I beg your indulgence if I begin my presentation with a brief review of toxicological principles. Toxicology covers the entire range of harmful interactions of chemical substances with the biota. This therefore expands the range from studying very acute effects down to the study of exceedingly subtle changes in behavior and function after long-term exposure for the determination of safety. The studies of functional deficits, carcinogenicity and teratogenicity are well established in toxicological investigations, at least for the safety evaluation of drugs, pesticides, and food additives for man. Studies on subtle behavioral changes and mutagenicity are gaining increasing acceptance. The types of measurements which have been noted most frequently for the study of heavy metals in water are lethality, growth, and reproduction. Behavioral changes and alterations in community structure are not normally investigated, even though they might have sizeable repercussions on the sustained survival of any species.

Toxicity studies are normally described by relating the exposure to a certain chemical to the effect that is observed in the form of dose-response curves. Such dose-response curves describe the response of a defined strain or species under a rigorously controlled set of experimental conditions to accurately define exposures of a toxicant. If the chemical form of the toxicant is changed, the dose response curve is likely to change. If the species or the experimental conditions are changed, the dose-response curve is also likely to change.

One of the most important variables in determining the biological effects of a heavy metal on organisms is the amount of the heavy metal that is absorbed, or the amount and form of the heavy metal that is available to do damage at the surface of epithelial cells. The assumption has usually been that heavy metals as determined by standard methods (A.P.H.A.) are an accurate presentation of the amount of absorbable heavy metal found in water. Likewise, the assumption has been that during experimental exposures of aquatic organisms to heavy metals, that the added toxicant was readily available to the aquatic organisms, and, moreover, that it could readily be related to environmental concentrations of heavy metals as determined by standard methods. The extent to which such assumptions are met is not very well understood, but there are many instances in which all of these assumptions have obviously not been satisfied. It has been found that heavy metals in natural waters are predominantly associated with particles suspended in water. The extent to which heavy metals associated with such particles are absorbable is not well known. It is commonly thought that unless particles can be incorporated into cells by phagocytosis, that a metal must first be solubilized and then be transported and absorbed in solution. The extent to which this occurs under natural conditions is not very well known, but whenever attempts have been made to determine the amounts of heavy metals in solution in natural waters, as compared with those adsorbed onto or part of particles, it has been found that only a minor percentage of the heavy metals in natural waters is in solution. In contrast, when natural water samples were analyzed according to standard methods, the samples are acidified and adsorbed heavy metals are solubilized to a high degree. While the concentrations found for such methodology may in general terms agree with some measure of exposure to aquatic organisms, it may also occur that heavy metal concentrations found in water by such methods correlate closely to turbidity instead.

Only recently has there been an effort to closely define the water supply in which toxicity studies are to be conducted. The literature abounds with descriptions of acute and sub-acute toxicity testing done under static conditions, where an appropriate amount of toxicant is initially added to the water and the organisms are introduced subsequently without any further attempts to renew the water supply. The exposure that the aquatic organism gets under such conditions is virtually impossible to quantify for the following reasons: even though the heavy metal may have been added in an ionic form, it may rapidly be changed to less soluble hydroxides, carbonates, sulfides, or phosphates which are then plated onto the walls of the container and are adsorbed onto the particles suspended in the water. This would result in a rapid decline of the effective concentration to which the fish are exposed. In addition, the amount of toxicant that is absorbed by the fish itself will reduce subsequent exposures. The dynamics of decline of ionized heavy metals in a static bioassay system are essentially impossible to predict. Thus, static toxicity tests will provide answers which will provide underestimates in the response of aquatic organisms to heavy metals because the actual exposure is usually considerably lower than the nominal exposure at the inception of the test. Since a bioassay run under static conditions provides essentially an undefined or, at best, a very poorly defined analytical system, it has become more accepted to undertake bioassays under dynamic conditions. While such a system does not do away with the formation of insoluble heavy metal salts and adsorption of heavy metals onto particles, it does do away with many of the problems inherent in static toxicity tests, and does expose the aquatic organisms to defined concentrations of heavy metal which may, however, be in an altered form than that originally intended. The actual amount of toxicant absorbed still depends, in addition to toxicant concentration, on the chemical form of the toxicant, the ligands which may be present for complexation of the heavy metal, the particle size, and nature of the particles that are available for adsorption in the water, the pH and alkalinity. In addition to the physical and chemical characteristics of the water itself, certain characteristics of the biological organism would also influence the rate at which a heavy metal is absorbed. Primary among these are the species of the organism, its age, its metabolic rate, and perhaps its previous health experience.

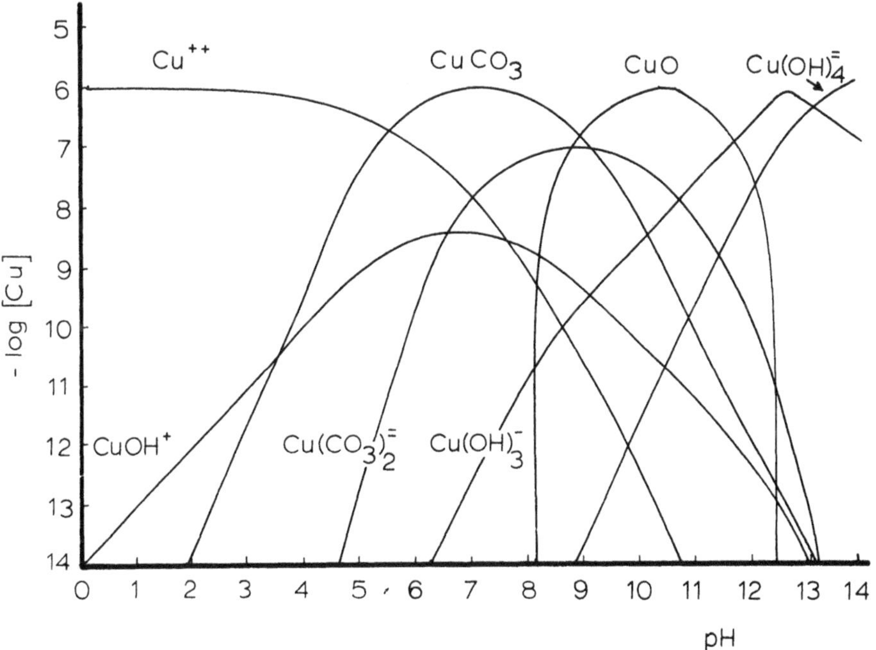

Figure I. Copper Carbonate and Hydroxide Equilibria. Total copper 10^{-6}M; total carbonate - 10^{-2}M. (Adapted from Black, _et al_., 1971)

Interconversions of chemical species of heavy metal can be illustrated by copper. The data presented in the above figure account for only a portion of the carbonate and hydroxide salts of copper. They do not take into account other anions which are known to be important in natural waters. It is therefore evident that unless a system is accurately described, that we will not know the particular heavy metal compound or combination of compounds that we are exposing the organism to. Since many of the carbonates of heavy metals are relatively insoluble, it is not surprising that the toxicity of many heavy metals decreases as the carbonate concentration increases. Doudoroff and Katz (1953), McKee and Wolf (1963), and Skidmore (1964) had reported on the apparent beneficial effects of water hardness on heavy metal toxicity. Stiff (1971) felt that the differences in toxicity between hard and soft water are actually due to differences in alkalinity which usually accompany differences in hardness.

At this point it is difficult to review the biological effects of various heavy metals in water for the simple reason that most of the experimental work in the past decade has involved the use of static bioassays, and has restricted itself to the analysis of acute or lethal effects. Much of the chronic work using dynamic bioassays over longer periods of time is still in progress and only a few reports have reached the literature. Studies on effects at the physiological or biochemical levels in aquatic organisms are exceedingly rare, so that mechanisms of the production of toxic effects are usually only poorly understood, or else must be inferred from mammalian studies. In spite of these shortcomings it may be appropriate to review some of the dose response relationships for heavy metals on aquatic organisms. I will try to emphasize chronic exposures since it is more likely that exposures of that type are encountered under natural conditions.

Cadmium

Ball (1967) found that the toxicity of cadmium to fish (*Salmo gairdnerii*) was relatively slow even at high concentrations. He found that the 24-hour LC_{50} was 30 ppm. When the duration of the test was extended, it became apparent that cadmium has a remarkably cumulative toxicity. A 7-day exposure resulted in an LC_{50} of 0.01 ppm. These data were essentially confirmed by Velsen and Alderdice (1967). *Daphnia magna* appear to be considerably more sensitive to cadmium than fish. Thus, Anderson (1948) found that the threshold concentration to produce immobilization in *Daphnia magna* after an exposure of 64 hours was 0.0026 ppm in a static bioassay.

In mammals cadmium at moderate exposure levels appears to produce kidney damage (Axelsson and Piscator, 1966) and hypertension (Schroeder and Buchman, 1967). It has also been implicated in the causation of ouch-ouch disease in Japan (Tsuchiya, 1969).

Chromium

Pickering and Henderson (1965) found in static bioassays that hexavalent chromium was more toxic in soft water than in hard water. The 96 hour LC_{50} values for various species were as follows:

96 Hour LC_{50} Values for Cr^{+6}
Static Bioassay

	In Soft Water	In Hard Water
Pimephales promelas (Fathead Minnow)	5.07 ppm	67.4 ppm
Lepomis macrochirus (Bluegill)	7.46 ppm	71.9 ppm
Carassius auratus (Goldfish)	4.10 ppm	
Lebistes reticulatus (Guppy)	3.33 ppm	

Duodoroff and Katz (1953) also found that bluegills (*Lepomis macrochirus*) were very tolerant to hexavalent chromium in hard water. They found that these fish could tolerate 45 ppm of hexavalent chromium for 20 days.

Chromium appears to be more toxic for algae. Hervey (1949) found that the growth of *Chlorococcales* was inhibited at 3.2 to 6.4 ppm and the growth of euglenoids was inhibited at 0.3 to 1.6 ppm. The growth of diatomes was inhibited at concentrations of 0.032 to 0.32 ppm.

Copper

The toxicity of copper is greater in soft water than in hard water (Mount and Stephan, 1969). However, when the pH of a body of water is increased, the toxicity of freshly added copper increases also. As the pH increases, the proportion of free copper is decreased and more of the copper occurs in the form of fine particles (Mancy, 1972). These data appear to indicate that finely divided particles of copper are more toxic than copper in solution. It may also be possible that finely divided particles of copper adhere readily on gills and can then be readily dissolved in the mucous coating on the gills, or that the absorptive behavior of the gill membrane is different at higher pH values or that the fish is more stressed at a high pH and, therefore, is more sensitive to these agents. A similar increase in toxicity in a heavy metal has also been found for zinc by Mount (1966). However, the situation is still unclear because Sprague (1964) found that suspended zinc was less toxic than dissolved zinc and Lloyd (1960) found that particulate zinc has the same toxicity as dissolved zinc.

Copper is known to be especially toxic for algae, aquatic plants, and molluscs. Copper sulfate has consequently been utilized extensively for the control of algae and for the control of snails when they are present as an intermediary host for swimmer's itch.

Lead

Lead also has a significantly different toxicity in hard water when compared to soft water. The static bioassays conducted by Pickering and Henderson (1965) indicate this relationship clearly.

96 Hour LC_{50} of $PbCl_3$
Static Bioassay

	In Soft Water	In Hard Water
Pimephales promelas (Fathead Minnow)	5.58 ppm	482 ppm
Lepomis macrochirus (Bluegill)	23.8 ppm	442 ppm
Carassius auratus (Goldfish)	31.5 ppm	
Lebistes reticulatus (Guppy)	20.6 ppm	

Crandell and Goodnight (1962) reported that 2 ppm of lead retarded the growth of guppies and delayed their sexual maturity. Weir and Hine (1970) demonstrated that 0.07 ppm of lead in soft water changed conditioned behavior in goldfish.

While it is not entirely clear from environmental data as to the extent to which lead can have an effect on fish, invertebrates, and plant life under natural conditions, it is quite clear that the presence of lead pellets from shot in sediments does have a detrimental effect on waterfowl. Under certain conditions waterfowl appear to mistake lead shot for either seed or grit and ingest it. The lead shot is then ground up under acid conditions in the presence of grit in the muscular stomach of the waterfowl. While some lead shot may be expelled early during the grinding process, it is clear that the digestive physiology of waterfowl favors the occurrence of high exposures to lead. It has been demonstrated that the ensuing lead poisoning may terminate either in recovery, in increased chance of predation, or in death.

The incidence of lead shot in gizzards of waterfowl averaged 6.6% in 18,454 ducks (Bellrose, 1951). Among the ducks which had been found to have ingested lead shot by x-ray fluoroscopy, approximately 68% were found to contain only one lead shot and about 18% contained more than two lead shots (Jordan and Bellrose, 1951). These epidemiological data have since been confirmed by many other workers. There also appear to be seasonal variations in the mortality of waterfowl and, in addition, there appear to be species differences. The LD_{50} dose in grain-fed waterfowl appears to be approximately 3 to 6 No. 6 shot pellets. The toxicity is greatly influenced by diet. Soft diets decrease the toxicity of the lead (Jordan, 1952). The extent of the direct mortalities due to lead poisoning in waterfowl is difficult to estimate, but it appears that the epidemiological data of the incidence of lead shot in the gizzards of waterfowl and the dose response curve for mortality for lead shot in waterfowl overlap. In addition, Bellrose (1959) has indicated that lead poisoning in waterfowl reduces their ability or their desire to move about, and apparently enhances the frequency with which waterfowl become prey to predators.

Mercury

Inorganic mercury in an ionic form is fatal to fish at concentrations of 1 ppm after 96 hours (Weir and Hine, 1970). That concentration is fairly high and the major concerns regarding the effects of mercury in water center around the methylated forms and the methylating process itself. Various aspects of the methylating process have been described previously by Jernelov (1972) and Landner (1971). From those references it is evident that mercury may be methylated under anaerobic as well as aerobic conditions. The various significant forms of mercury are represented diagramatically in Figure 2. The dynamics of the methylation process are influenced by the mercury concentration, the micro-organisms present, the water temperature, and the pH (Wood, *et al.*, 1968). Indications are, at this time, that only microorganisms possess methylating activity. Warm-blooded organisms have not been found to convert ionic mercury to methyl mercury to any significant extent (Clarkson, 1972). Fish are able to

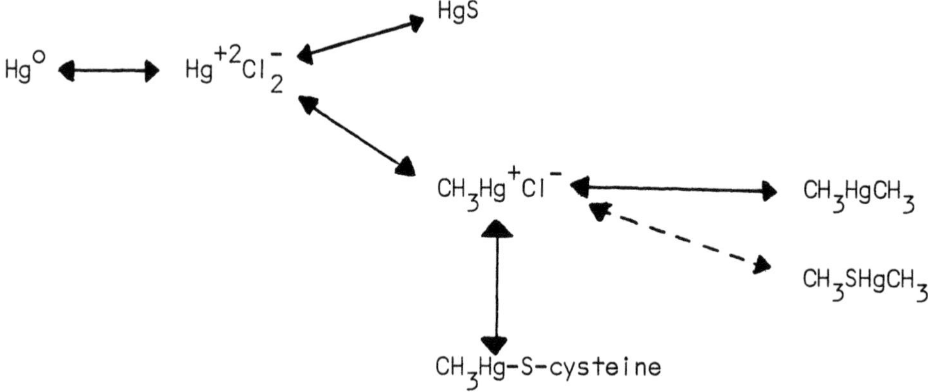

Figure 2. Important Transformations of Mercury.

concentrate methyl mercury either directly from the water through their gills or through the food chain (Johnels, et al., 1967). In freshwater fish essentially all of the mercury is in the form of methyl mercury. It is not clear to what extent food chains are important in determining the relative amount of methyl mercury present in freshwater fish, since differences between fish from various trophic levels tend to be less than one order of magnitude, and some species of herbivorous fish contain higher levels of methyl mercury than some species of carnivorous fish. The rate of loss of methyl mercury from fish after cessation of exposure is exceedingly slow. Miettinen and his associates (1971) have shown that methyl mercury in fish appears to have a half-life in the neighborhood of two years. At this time it is not entirely clear as to why freshwater fish contain such a high proportion of methyl mercury. In addition, it is not clear as to what the form, or what the absolute concentration of mercury in water is. Hannerz (1968) found that when pike are reared in water containing 0.1 ppb of methyl mercury, the pike became uncoordinated, lost weight, and experienced sporadic mortality.

The most significant hazards due to methyl mercury appear to be to fish-eating birds and mammals. The incidents of Minamata disease in Minamata and Nigata, Japan, are well documented (National Institute of Public Health, Sweden, 1971). In its early stages the disease is characterized by relatively nonspecific symptoms due to impairment of the central nervous system. When the disease becomes well established, sensory defects of hearing and visual field become evident. There is a loss of motor control and coordination followed by paralysis, mental deterioration, loss of consciousness, and death. All of the well established central nervous system defects appear to be irreversible (Dinman and Hecker, 1972). Takeuchi (1972) noted that cats appeared to be more sensitive than people to the CNS symptoms caused by methyl mercury. It is not entirely clear what the concentration of methyl mercury was that caused Minamata disease in Japan. Median concentrations in fish appeared to range from 1.6 to 7.7 mg/kg wet weight. The values varied depending on the species that were sampled and on the investigator reporting the analyses. The ranges of values found were considerably wider, ranging from 0 to 41 mg/kg (National Institute of Public Health, Sweden, 1971).

Johnels and Westermark (1969) reported that the concentration of mercury in the feathers of the osprey and the great crested grebe had been increasing steadily since the latter half of the 19th century, probably due to the increasing levels of general industrialization in Sweden. Dustman, *et al*. (1972) found elevated levels of mercury in a survey of wild animals living on or near Lake St. Clair. Fish-eating birds contained especially high concentrations of mercury. Concentrations in one heron reached 175 ppm in the liver and 23 ppm in the flesh. Maximum residues in a tern were 39 ppm in the liver and 7.5 ppm in the flesh. Similar values were reported by Henriksson, *et al*. (1966); Borg, *et al*. (1969); and Holt (1969) in birds that had died after being dosed experimentally with methyl mercury. Mercury in eggs from terns, grebes, and the mallard were also found to be elevated. Since no population surveys were made in the St. Clair area, it is not clear whether the population of fish-eating birds has been increasing, holding constant, or decreasing. In any event, it appears to be clear that fish-eating birds readily reach elevated levels of mercury.

Zinc

Cairns and Scheier (1968) found that zinc was considerably more toxic in soft water than in hard water. They found that bluegills (*Lepomis macrochirus*) demonstrated a 96-hour LC_{50} of 1.9 to 3.6 ppm of zinc in soft water and a 96-hour LC_{50} of 10.1 to 12.5 ppm of zinc in hard water. He found relatively little difference due to temperature. In contrast, Mount (1966) found that zinc was most toxic at a pH of 8.0 and a water hardness of 50 ppm and least toxic at pH 6.0 and hardness of 200 ppm. He found that the primary controlling factor appeared to be pH rather than hardness. Toxicity differences due to the presence of zinc either as a fine precipitate or in the ionic form have been discussed previously in this paper.

CONCLUSIONS

The findings cited above reiterate the difficulty in correctly analyzing an exposure of various aquatic organisms to heavy metals. In most instances, it has not been possible to define the exact chemical and physiological conditions under which the exposure was made, and it has become increasingly clear that these conditions have a major effect on the resulting biological response. The current methodology utilized for water analyses does not permit us to predict the absorbed dose or the effect that may result. Their indiscriminate use in regulatory processes is, therefore, questionable at least. It might be preferable to study the actual dose absorbed by determining body burdens of heavy metals in aquatic organisms and comparing that to background levels and to levels at which early effects have been noted.

The effects which have been studied for aquatic organisms have, in most instances, been limited to the study of survival under more or less poorly defined conditions. More subtle effects, such as growth, reproduction, or behavior have been utilized much less frequently. Mutagenicity, tercitogenicity, and carcinogenicity have been utilized almost exclusively for mammalian species. Their significance to most aquatic organisms has been virtually ignored.

We are thus in the early infancy of study of the biological effects of heavy metals on aquatic organisms. Safety evaluations and regulations made to protect aquatic organisms are assailable from many directions, and far more research is needed to upgrade the existing information.

LITERATURE CITED

Anderson, B. G. (1948), Trans. Amer. Fish Soc., 78, 96-113.

A.P.H.A. (1971), Standard Methods for the Examination of Water and Wastewater, 13th ed., A.P.H.A., Washington, D. C.

Axelsson, B. and Piscator, M. (1966), Arch. Env. Hlth., 12, 360-373.

Ball, I. A. (1967), Water Research, 1, 805-806.

Bellrose, F. C. (1951), Trans. N. Amer. Wildl. Conf., 16, 125-135.

Bellrose, F. C. (1959), Bull. Ill. Nat. Hist. Survey, 27(3), 235-288.

Black, J. A., Roberts, R. F., Johnson, D. M., Minicucci, D. C., Mancy, K. H., and Allen, H. E. (1971), Abstracts of Papers, Div. Water, Air & Waste Chem. 162nd Nat. Meeting of the Amer. Chem. Soc., Washington, D. C., Sept. 13-17.

Bork, K., Wanntorp, H., Erne, K., and Hanko, E. (1969), Viltrery, 6(4), 301-379.

Cairns, J., Jr. and Scheier, A. (1968), Progr. Fish Cult., 30(1), 3-8.

Clarkson, T. W. (1972), Biotransformation of organomercurials in mammals, Environmental Mercury Contamination, R. Hartung and B. D. Dinman, eds., Ann Arbor Science Publ., pp. 229-238.

Crandall, C. A. and Goodnight, C. J. (1962), Limnol. Oceanogr., 7, 233-239.

Dinman, B. D. and Hecker, L. H. (1972), The dose-response relationship resulting from exposure to alkyl mercury compounds, Environmental Mercury Contamination, R. Hartung and B. D. Dinman, eds., Ann Arbor Science Publ., pp. 290-301.

Doudoroff, P. and Katz, M. (1953), Sewage Ind. Wastes, 25(7), 802-839.

Dustman, E. H., Stickel, L. F., and Elder, J. B. (1972), Mercury in wild animals, Lake St. Clair, 1970, Environmental Mercury Contamination, R. Hartung and B. D. Dinman, eds., Ann Arbor Science Publ., pp. 46-52.

Hannerz, L. (1968), Fisheries Bd. of Sweden, Inst. for Freshwater Res. Drottningholm, Report No. 48, pp. 120-176.

Henrickson, K., Karppanen, E., and Helminen, M. (1966), Ornis Fennica, 43(2), 38-45.

Hervey, R. K. (1949), Bot. Gas., 11(1), 1-11.

Holt, G. (1969), Nord. Vet. Med., 21, 105-114.

Jernelöv, A. (1972), Factors in the transformation of mercury to methylmercury, Envirnomental Mercury Contamination, R. Hartung and D. B. Dinman, eds., Ann Arbor Science Publ., pp. 167-172.

Johnels, A. G. and Westermark, T. (1969), Mercury contamination of the environment in Sweden, Chemical Fallout, M. W. Miller and G. G. Berg, eds., C. C. Thomas Publ., pp. 221-241.

Johnels, A., Westermark, T., Berg, W., Persson, P. I., and Sjostrand, B. (1967), Oikos, 18, 323-333.

Jordan, J. S. (1952), Univ. of Mich. Ph.D. Thesis, 144 pp.

Jordan, J. S. and Bellrose, F. C. (1951), Biol. Notes No. 26, Ill. Nat. History Survey, Urbana, Ill.

Landner, R. (1971), Nature, 230, 452-453.

Lloyd, R. (1960), Ann. Appl. Biol., 48, 84-94.

Mancy, K. H. (1972), Copper toxicity to fishes in natural waters, Progress Report FWQA 14-12-591.

McKee and Wolf, H. W. (1963), Water Control Criteria, Calif. State Water Quality Control Board, Sacramento, Publ. 3A, 548 pp.

Miettinen, V., Blankenstein, E., Rissanen, K., Tillander, M., Miettinen, J. K., and Valtonen, M. (1970), FAO Tech. Conf. on Marine Pollut., Rome, Italy, Dec. 9-18.

Mount, D. I. (1966), Air Wat. Pollut. Ind. J., 10, 49-56.

Mount, D. I. and Stephan, C. E. (1969), J. Fish. Res. Bd. Can., 29, 2449-2457.

National Institute of Public Health, Sweden (1971), Methylmercury in fish, Nordish Hygienish Tidshrift, Suppl. 4, 364 pp.

Pickering, Q. H. and Henderson, C. (1965), Proc. 19th Ind. Waste Conf. Purdue Univ.; (1966), Air Water Pollut., 10(6/7), 453-463.

Schroeder, H. A. and Buckman, J. (1967), Arch. Env. Hlth., 14, 693-697.

Skidmore, J. (1964), Quart. Rev. Biol., 39(3), 227-248.

Sprague, J. B. (1964), J. Fish. Res. Bd. Can., 21, 17-26.

Stiff, M. J. (1971), Water Research, 5, 171-176.

Takeuchi, T. (1972), Biological reactions and pathological changes in human beings and animals caused by organic mercury contamination, Environmental Mercury Contamination, R. Hartung and B. D. Dinman, eds., Ann Arbor Science Publ., pp. 247-289.

Tsuchiya, K. (1969), Keio J. Med., 18, 181-211.

Velsen, F. J. P. and Alderdice, D. F. (1967), J. Fish Res. Bd. Can., 24, 1173-1175.

Weir, P. A. and Hine, C. H. (1970), Arch. Env. Hlth., 20, 45-51.

Wood, J. M., Kennedy, S. F. and Rosen, C. G. (1968), Nature, 220, 173-174.

AIR POLLUTION BY LEAD AND OTHER TRACE METALS

PAUL R. HARRISON

Department of Environmental Control
City of Chicago
Chicago, Illinois

INTRODUCTION

Ecological investigations on heavy metals and trace metals, in general, have increased markedly in recent years. The impetus for these investigations is such that we have come to realize that the biological function of most organisms is dependent upon the proper balance and availability of trace elements in their life span. For example, if we were to consider the response of an organism to exposures of various concentrations of any element, we would, in general, have four modes of response. A typical response curve would show an increasingly beneficial effect up to an optimal value. The curve would then turn downward into a tolerance region where the organism would tolerate more than its optimum requirements with little difficulty. We then enter into a "stress" region where side effects or injury to the organism would start occurring. The fourth region would be that of termination by a lethal dose. For some elements the tolerance limits are so extensive that it would be physically difficult to assault the organisms with lethal amounts. Some metals which are essential to human existence are iron, copper, manganese, vanadium, nickel, cobalt, and zinc. Other elements such as cadmium, lead, mercury, thallium, and beryllium, are known to be toxic at low to moderate concentrations. One of the difficulties in their diagnosis has been that many of them require long periods of time to accumulate and cause damage to certain selective cells within the organism. The sensitivity of analytical procedures has also been a real problem.

Thus, it is evident that while a certain amount of trace elements are necessary for the proper functioning and survival of the organism, the same trace metals can assault the organism in toxic levels. It is known that most enzyme systems can be dehabilitated by a sufficiently high concentration of a selected trace metal. This obviously leads to the termination of that function within the organism and probably the termination of the organism itself. It is for these reasons we have seen the recent increase in the investigations and analyses for trace metals, the more toxic of which belong to the group called "heavy metals."

LEAD

Classically, one of the most discussed heavy metals is lead. Lead is unique in many of its chemical and physical properties. It is relatively unreactive with most acids, is physically pliable and malleable in its pure state, is comparatively easily refined, and readily available to surface mining. It is also the final resting place for many decay products of fission.

Because of its usefulness and low melting point, lead has been used for centuries for vessels of containment, pipes, and ornamentation. It was also found to contain certain "medicinal" properties but did not gain the eminence of mercury in alchemy (although many an alchemist sought to transform lead to gold!).

In recent history, lead has been used for shieldings for varied electrical cables, pipes, and for sealing joints. At the advent of commercial paints, lead was found to be quite useful as an oxide to form a very efficient white base for outdoor use. It was attractive, inasmuch as it did not discolor except in industrial regions where hydrogen sulfide was generated. It seemed to replenish itself by forming an oxide on the outside surface and inhibited the growth of various fungi which could discolor or adversely affect the aesthetics of the structure. Much of these coatings contained over 50% lead as solids. Many of these structures still stand with their potentially lethal paint chips and are available to children with pica tendencies. These buildings are usually more than 40 years old and tend to be located in the older and more impoverished areas of cities. In most of these cases, the lead was in a solid form available only to those who inadvertently or unknowingly would ingest it by eating, or drinking as in the case of the Roman wine vessels.

In the early 1940's, the U. S. auto industry decided that higher compression engines were necessary for the American consumer. At the advent of these high compression engines, it became necessary to increase the octane "rating" of the gasoline (pure octane being equal to 100). It was known that certain additives could be added to the gasoline to increase its octane rating without actually increasing the more precious octane hydrocarbon. Lead was found to be the optimum additive for the stated purpose. It was subsequently found that lead tended to form deposits in the engines and in the exhaust system shortening the life of the engine. A compound commonly known as tetraethyl fluid was developed so that one could add the lead compounded with chlorine and bromine in such a manner that upon combustion the lead would perform its "anti-knock" function and subsequently be carried through the exhaust in the form of very small particles of lead-chloro-bromide. This development ushered the present day worldwide air pollution by lead aerosols. Hitherto, the airborne lead pollution was localized at the mines and in refineries.

Today lead is found throughout the globe, especially in the northern hemisphere. Even in remote areas of Greenland (1) there are detectable amounts of lead caused by long-range advection of these purposely manufactured small aerosols. In fact, one can trace the history of lead contributed by the automobile by the stratification of the snows over the last two to three decades.

OTHER ELEMENTS

Similar discussions can be made for other elements. The overriding fact becomes apparent that improper use and unwise distribution of these trace metals have become a serious environmental problem. The following discussion will be an overview of the sources of lead and some other trace metals, both man-made and natural, their ambient levels found in some populated regions, and their transport, fallout, and rainout into the surface biosphere in which man lives. A discussion concerning the possibility of a secondary emission of refloatation will also be made. Finally, a brief discussion as to the biological effects will be presented.

SOURCES OF LEAD AND OTHER TRACE METALS IN THE AIR

When using the word "trace," one is referring to the relative abundance of the element with respect to others. Table 1 represents the annual consumption of ten selected metals in the United States in 1969 (2). The range is from 600 tons of selenium up to 2.2 million tons of copper. In order to understand the distribution and pathways into the environment of these metals, one should be aware of their uses. A short representation of 22 metals and their more common uses is presented in Table 2 (3). Most of the uses for the more common metals are well known. However, the distribution and use of the toxic elements such as thallium, selenium, and arsenic, which represent the less common elements, are less known. It may be seen from Table 2 that many of the metals are used in bulk form and are bound up in the primary material to which it is added. Others, however, are readily available to the environment in a relatively accessible form. These usually become the problem elements.

Table 1. Consumption of metals in the U.S.A. in 1,000's of pounds (USDI, 1969).

Metal	Consumption
Arsenic	51,334
Cadmium	13,328
Chromium	2,632
Copper	4,194,000
Mercury	5,732
Lead	2,657,580
Zinc	2,667,398
Nickel	318,612
Beryllium	17,438
Selenium	1,216

Table 2. Common uses of metals.

Metal	Used in
Antimony	Alloys, metallurgical processes
Arsenic	In metallurgy to increase hardening and heat resistance, glassware, ceramics, tanneries, dyes, pesticides, and wood preservatives
Barium	Alloys, paints
Cadmium	Alloyed with copper, lead, silver, aluminum, nickel; electroplating; ceramics; pigments; photography; nuclear reactors
Cerium	Textile dyes, printing, arc lamps and neon tubes
Chromium	Metal plating, leather tanning, paints, dyes, explosives, ceramics, photography
Copper	Alloys, electrical equipment, pipes and roofing, textile processes, pigmentation tanning, photography, electroplating, insecticides and fungicides
Lead	Pipes, water storage chambers, dyes, paints
Magnesium	Light alloys, electrical and optical apparatus
Manganese	Steel alloys, dry-cell batteries, glass and ceramics, paints and varnishes, inks, dyes in matches, in fertilizers

Table 2. Continued.

Metal	Used In
Mercury	Dentistry, in scientific and electrical equipment, in chlor-alkali plants, medical products, detonators, pigments, photo-engraving
Molybdenum	Metallugry, electrical and electronic equipment, in glass and ceramic industry, and as part of the fertilizer for legumes
Nickel	Metal plating
Palladium	Photography, toning solutions, galvanocoatings
Selenium	Pigmentation in paints, dyes, and glass; in rectifiers, semiconductors, photoelectric cells, and in other electrical apparatus
Silver	Jewelry, silverware, alloys, electroplating, food processing, photography
Strontium	Polytechnic, signal lights, flares, matches
Thallium	Rat poison, dyes, pigments, in optical glass, depilatory
Thorium	Incandescent mantles and filaments for lamps
Tin	Mordants for reviving colors, fabric dyes, fingernail polish, lacquers, varnishes, fungicides, insecticides and antihelminthics
Vanadium	In metallurgy to increase hardness and malleability of steel, glass manufacture, photography, dye mordant
Zinc	Galvanizing pipes, alloys, electrical purposes, printing plates, dye-manufacturing, pharmaceuticals

McKee, J. E. and Wolf, H. W. (1963), Water Quality Criteria, 2nd ed., Publication 3-A, April, 1971, The Resources Agency of California.

Man-Made Sources of Lead

Use of lead is continuing to rise. Table 3 is a tabulation of the tonnage of lead used from 1966 through 1969 (4). Thus, we are consuming approximately 1.4 million tons of lead per year in the United States of which 0.27 million tons are used as gasoline additives, which leads to an annual consumption of lead for gasoline additive at approximately 2.6 pounds per person per year.

Table 3. Industrial lead consumption in the United States* (in short tons).

	1966	1967	1968	1969
Storage Batteries	472,492	466,665	513,703	558,121
Oil Refining & Gasoline	246,879	247,170	261,897	271,128
Pigments	119,888	103,190	109,734	101,886
Solder	78,898	68,883	74,074	65,126
Ammunition	78,435	78,766	82,193	77,805
Cable	66,491	63,037	53,456	54,078
Construction (caulking)	63,250	48,789	49,718	41,888
Printing (type metal)	30,421	28,554	27,981	26,469
Total Classified	1,156,754	1,105,004	1,172,756	1,196,501
Unclassified	167,123	155,512	156,034	178,699
Grand Total	1,323,877	1,260,516	1,328,790	1,375,200

*Lutz, et al. (1970), Lead Model Case Study, Battelle Memorial Institute Research Report.

Perusal of Table 3 will further reveal that over half the lead used is recycled as storage batteries and as the "type" metal used for printing. No attempt whatsoever is made to recycle the lead used for gasoline additives. This is especially remarkable since 98% of the airborne lead that can be attributed to combustion sources comes from the combustion of gasoline. Approximately 6.6 billion pounds of lead have been used in gasoline since 1920, most of which has resulted in airborne particulate! (5)

At this high rate of consumption, some parts of our environment are substantially contaminated. Examples are buildings using lead paint, and areas close to heavily trafficked roadways. Although lead paint accounts for 2 to 3 times as much lead as compared to the automobile in urban areas, the lead in the paint is relatively inaccessible except through peeling off of paint chips and their accidental ingestion by children.

It is illustrative to trace the history of lead aerosol production in the northern hemisphere by year. Table 4 shows its production during 1753, 1815, 1933, and 1966, respectively (6). It is interesting to note that the percentage of lead converted to aerosol was much less for smelters in 1966 than in previous years. In fact, smelting produces less total emissions today than it did

Table 4. Lead aerosol production in the northern hemisphere.

Year	Lead Smelted, Thousands of Tons	Lead Converted to Aerosols, Per Cent	Lead Aerosols from Smelters, Thousands of Tons
1753	100	2	2
1815	200	2	4
1933	1,600	0.5	8
1966	3,100	0.06	< 2

Year	Lead Burned as Alkyl, Thousands of Tons	Lead Converted to Aerosols, Per Cent	Lead Aerosols from Alkyls, Thousands of Tons
1753	--	--	--
1815	--	--	--
1933	10	40	4
1966	300	40	120

over 200 years ago! The record for lead additives, however, is not at all as reassuring, where 40% of lead burned as alkyl additives end up as aerosols. Table 5 presents further breakdown of lead emissions in the United States for 1968, of which gasoline combustion represented approximately 98% of the total (7).

In order to determine the amount of lead emitted to the atmosphere by the average automobile, a study was conducted in a tunnel on the Pennsylvania turnpike. Table 6 represents the emission rates of eight selected vehicles. This study returned an average emission rate of 0.1 grams of lead per mile of which one-half became airborne for more than a few minutes (8). The old adage of "it isn't what you do but the way you do it" applies in the case of automotive lead emissions. Table 7 presents the emission size distribution of particles emitted from those same cars used to gain the emission rates. Investigators have provided evidence that over 50% of those particles emitted by these automobiles are less than one micron (Table 8) (8). It is widely believed that particles less than 10 microns in size are easily respirable and reach the innermost parts of the lungs. This study is further substantiated by Figure 1 which indicates that the mass median equivalent diameter of these particles is approximately one-half micron (9) and, thus, over 60% of the aerosols captured

Table 5. Lead emission in the United States, 1968*.

Emission Source	Lead Emitted, Tons/Year
Gasoline combustion	181,000
Coal combustion	920
Fuel oil combustion	24
Lead alkyl manufacturing	810
Primary lead smelting	174
Secondary lead smelting	811
Brass manufacturing	521
Lead oxide manufacturing	20
Gasoline transfer	36
Total	184,316

*Data from National Inventory of Air Pollutant Emissions and Controls, a loose-leaf data bank on file at the Bureau of Air Pollution Sciences, Environmental Protection Agency, Research Triangle Park, North Carolina.

Table 6. Lead emission rate from production vehicles.

Vehicle	Mileage	Emission rate, grams of lead per mile	
		Total	Airborne
F-47	0 to 54,000	0.086	
C-76	0 to 55,000	0.106	
C-82	8,000		0.039
H-1	41,000		0.061
CW-1	21,000		0.047
CW-21	11,000		0.055
C-40	0 to 28,000	0.102	
C-83	21,000	0.14	
Average		0.108	0.050

Table 7. Lead particle size distribution from production vehicles.

Vehicle	Mileage	Emission rates, grams of lead per mile				
		Size in microns				
		Total	9	9 to 1	1	0.3
C-40	5,000	0.102	0.028	0.029	0.046	0.031
C-40	16,000	0.095	0.037	0.024	0.034	0.024
C-40	21,000	0.083	0.036	0.017	0.030	0.022
C-40	28,000	0.115	0.066	0.028	0.022	0.013
F-47	3,000		0.008	0.008	0.022	0.019
F-47	20,000	0.086	0.027	0.010	0.020	0.015
F-47	55,000		0.024	0.008	0.017	0.015
C-76	3,000		0.015	0.023	0.044	0.031
C-76	20,000	0.106	0.056	0.022	0.044	0.028
C-76	55,000		0.017	0.008	0.027	0.022
Average		0.098	0.031	0.018	0.031	0.022

Table 8. Size of atmospheric lead particles.

Investigator	Location Samples	Average size (mass median equivalent diameter), Microns
Robinson and Ludwig	59 urban areas	0.25
Lee, Robinson and Wagman	Downtown Cincinnati	0.2
	Suburban Cincinnati	0.4
Lundgren	Riverside, California	0.5
Habibi	Wilmington, Delaware	0.24
Pierrard	El Monte, California	0.2
Weighted Average		0.25

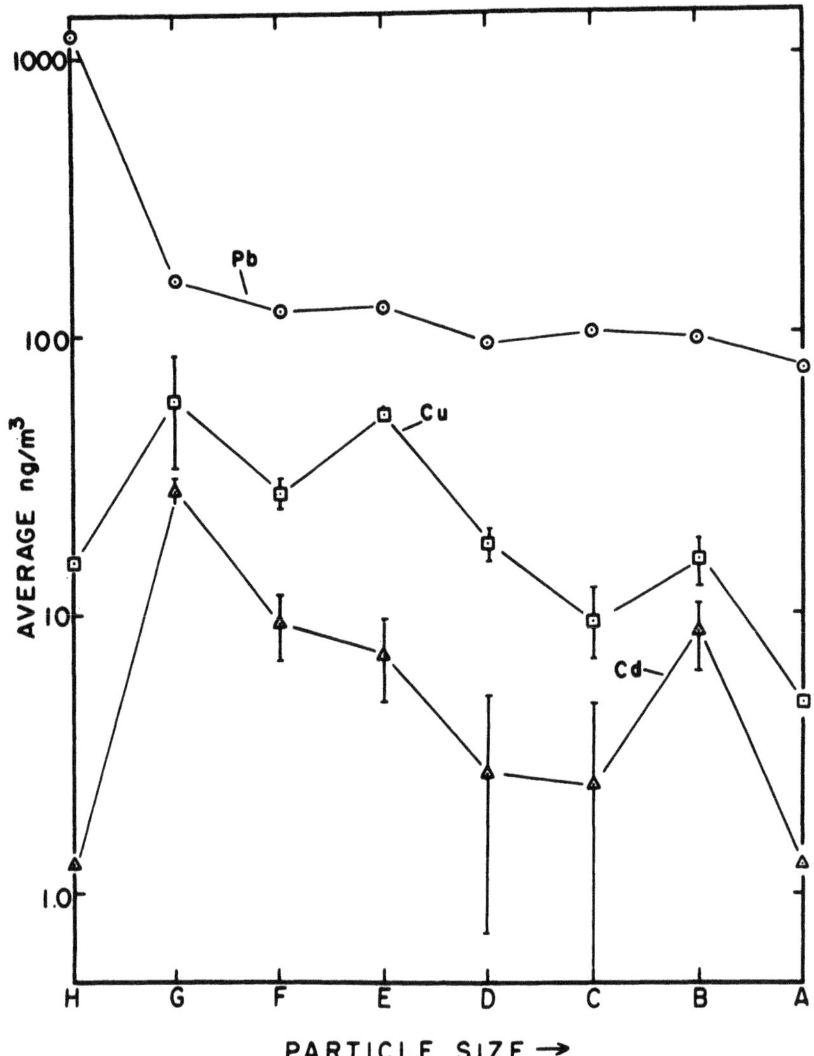

Figure 1. Average particle size distributions of Pb, Cd, and Cu.

during this experiment were less than one-half a micron in diameter.

NATURAL SOURCES OF TRACE METALS

Although man is responsible for much of his exposure to trace elements by his congested habits, the natural environment provides us with large quantities of these elements. Table 9 represents some of the elemental concentrations in crustal rock, soil, and sea water as residuals to the natural physical processes of erosion such as wind motion and sea spray (10).

Table 9. Elemental concentrations in crustal rock, soil, and sea water.

	Crustal Rock Abundance ppm*	Avg. Soil Abundance ppm**	Sea Water Abundance g/ton*	Sea Salt Abundance ppm*
Na	28,300	6,300	10,556	324,000
Mg	20,900	6,300	1,272	39,200
Al	81,300	71,300	0.1	3
Cl	130	100	18,980	584,000
K	25,900	13,600	380	12,000
Ca	36,300	13,700	400	12,000
Sc	22	7	0.00004	0.0012
Ti	4,400	4,600	0.005	0.15
V	135	100	0.005	0.15
Cr	100	200	0.002	0.06
Mn	950	850	0.0008	0.025
Fe	50,000	38,000	0.0034	0.105
Co	25	8	0.0001	0.003
Ni	75	40	0.005	0.15
Cu	55	20	0.002	0.06
Zn	70	50	0.015	0.45
Ga	15	30	0.0005	0.015
As	1.8	5	0.003	0.09
Se	0.05	0.01	0.004	0.12
Br	2.5	5	65	2,000
Ag	0.07	(0.1)	0.0002	0.006
In	0.1	---	---	---
Sb	0.2	---	0.0002	0.006
I	0.5	5	0.05	1.5
La	30	(40)	0.0003	0.009
Ce	60	(50)	0.0004	0.012
Sm	6.0	---	---	---
Eu	1.2	---	---	---
W	1.5	---	0.0001	0.003
Hg	0.08	0.01	0.00003	0.0009
Th	7.2	6	0.000005	0.00015

*Mason, B. (1966), _Principles of Geochemistry_, New York: Wiley.

**Vinogradov, A. P. (1959), _The Geochemistry of Rare and Dispersed Chemical Elements in Soil_, New York: Consultants Bureau, Inc.

Again, using lead as an example, we present Table 10 as an estimate of the concentration of lead in some naturally occurring materials (11). Although deep sea clays contain the highest concentrations, they are relatively unavailable to the ambient air. The most available aerosols of natural origin are from soils and coal which tend to be ejected by wind erosion and combustion. Tables 11, 12, and 13 present similar representations for cadmium, arsenic, and mercury (10).

Table 10. Concentrations of lead in some natural materials, ppm.

Type of Material	Range Usually Reported	Average
Continental crust	---	15
Basaltic igneous	2-18	6
Silicic igneous	6-30	18
Shales and clays	16-50	20
Black shales (high C)	7-150	30
Deep sea clays	---	80
Limestones	---	9
Sandstones	1-31	
Soils	1-150	15
Phosphorites	10-100	10
Coals	2-50	15

Table 11. Concentrations of cadmium in some natural materials, ppm.

Type of Material	Range Usually Reported	Average
Continental crust	---	0.2
Basaltic igneous	0.006-0.6	0.2
Silicic igneous	0.003-0.18	0.15
Shales and clays	0-11	1.4
Black shales (high C)	0.5-8.4	2.0
Deep sea clays	0.1-1	0.5
Limestones	---	0.05
Sandstones	---	0.05
Soils	0.1-0.5*	0.3*
Phosphorites	0-170	30
Coals	---	2

*Excluding soils from mineralized areas.

Table 12. Concentrations of arsenic in some natural materials, ppm.

Type of Material	Range Usually Reported	Average
Continental crust	---	2.0
Basaltic igneous	0.2-10	2.0
Silicic igneous	0.2-13.8	2.0
Shales and clays	---	10
Deep sea clays	---	13
Limestones	0.1-8.1	1.7
Sandstones	0.6-9.7	2
Soils	0-102*	5*
Phosphorites	0.4-2000	13
Sedimentary Fe ores	70-1100	400
Sedimentary Mn ores	Up to 1.5%	-

*Excluding soils from mineralized areas.

Table 13. Concentrations of mercury in some natural materials, ppm.

Type of Material	Range Usually Reported	Average
Continental crust	---	0.06
Basaltic igneous	0.002-0.5*	0.05*
Silicic igneous	0.005-0.4*	0.06*
Shales and clays	0.005-1.0*	0.16*
Black shales (high C)	0.03-2.8	0.5
Deep sea clays	0.02-2.0	0.4
Limestones	0.01-0.22*	0.04*
Sandstones	0.001-0.3*	0.05*
Soils	0.001-0.5**	0.05**
Phosphorites	0.001-0.95	0.050
Coals	0.05-13.3*	0.3*

*Not including many analyses, especially from the Donets Basin, Kerch-Taman area, and Crimea, USSR, that show contents 10-1000 times the average given.

**Excluding soils from mineralized areas.

Volcanic activity as a source is usually not thought of as a primary emitter of particulates because of its sparse geographical distribution and presence in remote areas. It can be shown that the amount of particulate blown into the air on a global scale by volcanic activity exceeds the total emissions of man's industrial activities with the primary difference being that man tends to emit his pollutants within his own living space. Volcanoes, on the other hand, emit effluents to great heights permitting upper air winds to transport and mix these materials over much greater volumes (Table 14) (11).

Table 14. Mercury in air and in volcanic emanations (in nanograms per cubic meter; 1 nanogram = 10^{-9} g).

Sample	Range	Average
Air		
Over Pacific Ocean, 20 miles off shore	0.6-0.7	---
Arizona and Calif., unmineralized areas	3-9	5
California	1-50	---
Chicago area	3-39	9.7
Moscow and Tula regions, USSR	80-300	---
400 ft. above porphyry copper deposits, USA	7-53	27
400 ft. above mercury deposits, USA	24-108	60
Over mercury deposit, USSR	200-1200	---
Air and Gases, volcanic regions		
Air, Kamchatka region, USSR	---	190
Air, Honolulu, Hawaii	40-910	---
Air, Sulfur Banks, Kilauea Volcano	21,400-23,300	---
Air, vent breccias of mud volcanoes, USSR	300-700	---
Gases, mud volcanoes, USSR	700-2000	---
Gases, Mendeleev and Sheveluch Volcanoes, USSR	300-4000	---
Gases from hot springs, Kamchatka and Kuriles	10,000-18,000	---

AMBIENT CONCENTRATIONS--LEAD AND TRACE METALS

Urban Concentrations of Some Trace Metals

During the summer of 1968, Harrison (12) measured lead, cadmium, and copper during six days around the southern Lake Michigan basin including Chicago and northwest Indiana. The first day was also used for the analysis for bismuth. Over 20 stations within the City of Chicago and 30 stations in northwestern Indiana were used. Figure 2 shows the locations of the sampling stations and their code designations. Tables 15 through 18 present the atmospheric concentrations of these elements as found on standard hi-volume sampler using glass fiber filters. The data are expressed in nanograms per cubic meter.

Lead showed high values from most of the stations. In fact, the lowest concentrations seen were approximately 100 micrograms per cubic meter, establishing a lower estimate of near urban or "proximate" location for a summertime situation. The variations were from 100 to 7000 nanograms per cubic meter, a factor of 70. Cadmium showed much lower concentrations from less than 5 nanograms per cubic meter up to 80 nanograms per cubic meter. Bismuth was found in extremely low concentrations, ranging from less than .05 nanograms per cubic meter up to 3 nanograms per cubic meter on the one day analyzed. Lead, cadmium, and bismuth seem to be from natural or diffused sources. Copper has a markedly different distribution as seen from Table 8. The variations are from less than 20 nanograms per cubic meter up to 10,000 nanograms per cubic meter. Since the isopleths show consistency in the northwestern Indiana region, we are convinced that this is an industrial emission emitting from a low-level source (short stack).

Data on the total suspended particulates measured on the selected days are recorded in Table 19. It follows from the data in Table 19 that lead on the average represents approximately two to three per cent of the total suspended particulate, indeed a significant source of particulate considering its wide distribution! Figures 3 and 4 are similar lead data obtained on two selected days in 1972 (13). The isopleths are in nanograms per cubic meter. These analytical data were obtained by atomic absorption spectroscopy, and compare well with the previous data in Table 15 obtained by anodic stripping voltametry. It is interesting to note that the concentration isopleths do appear to concentrate downwind of the expressways.

The same two days, analyzed for copper and cadmium, are similarly presented in Figures 5 through 8 (13). Again the data are in the same range of concentrations as presented by the previous analysis in Tables 16 and 17.

Figures 3 through 8 are presented also as a contrast to Figures 9 through 18 (14). It is obvious that the method of reporting data is critical to maximize the amount of information available in each figure. In the latter data set the concentrations are presented as percentage composition of total particulates with isopleths drawn as per cent of maximum concentrations. From these studies one can approximate the actual trace metal concentration by assuming that

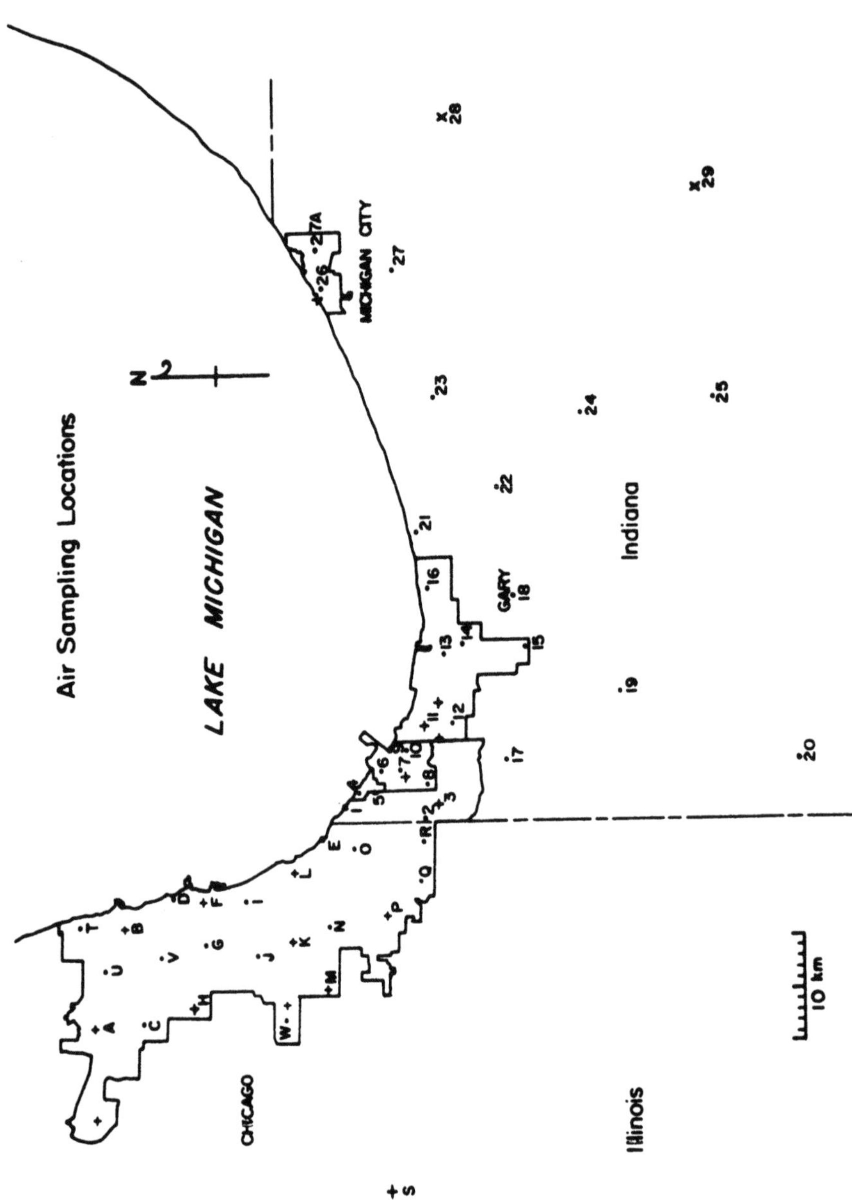

Figure 2. Key to high volume air sampling locations.

Table 15. Atmospheric concentration of lead, ng/m^3.

Station	May 21/22	June 6	June 20	July 9	Aug. 8	Aug. 29
1	2000	800	1000	100	1000	1000
2	----	2000	3000	2000	3000	2000
3	3000	1000	3000	2000	2000	2000
4	3000	1000	2000	800	2000	2000
5	1000	900	2000	400	1000	1000
6	2000	900	4000	600	4000	2000
7	----	----	3000	700	3000	2000
8	1000	700	2000	1000	1000	1000
9	2000	500	600	2000	2000	500
10	2000	300	800	1000	700	600
11	2000	1000	2000	----	----	----
12	4000	700	2000	----	----	----
13	6000	2000	3000	----	----	----
14	----	----	----	----	----	----
15	1000	500	2000	----	----	----
16	3000	800	----	----	----	----
17	1000	500	2000	900	2000	700
18	400	300	800	500	100	2000
19	300	400	400	----	100	3000
20	900	700	1000	1000	600	1000
21	----	----	1000	500	600	----
22	----	----	1500	400	1000	----
23	----	----	700	100	500	700
24	----	----	1500	500	300	500
25	----	----	500	300	200	----
26	1000	----	1000	700	800	500
27	----	----	----	300	400	800
A	7000	1000	5000	2000	3000	4000
B	4000	2000	2000	800	4000	4000
C	5000	300	2000	600	2000	3000
D	3000	2000	300	2000	4000	3000
E	4000	----	----	----	1500	5000
F	6000	2000	3000	----	3000	4000
G	4000	2000	4000	1500	----	3000
H	2000	1000	3000	700	3000	4000
I	----	----	----	----	----	2000
J	6000	900	2000	----	2000	----
K	5000	1000	2000	500	2000	2000
L	3000	1500	2000	200	4000	2000
M	6000	1000	2000	400	2000	4000
N	5000	1000	3000	600	2000	2000
O	6000	1000	2000	500	2000	1000
P	5000	1000	4000	600	2000	3000
Q	4000	700	2000	700	----	2000
R	4000	----	2000	300	2000	----
T	3000	1000	2000	800	----	----
U	4000	1500	3000	1000	3000	----
V	4000	1000	4000	1000	3000	2000
W	4000	800	2000	----	2000	3000
SHIP	----	----	----	700	----	----

Individual data points are relaible to a factor of 2.

Table 16. Atmospheric concentrations of cadmium, ng/m^3.

Station	May 21/22	June 6	June 20	July 9	Aug. 8	Aug. 29
1	50	5	30	10	**	10
2	--	**	20	20	10	20
3	10	**	**	50	**	30
4	30	**	**	5	10	8
5	40	10	10	50	**	7
6	10	8	20	10	10	10
7	--	--	8	60	10	10
8	15	**	5	10	8	**
9	30	7	10	20	**	**
10	20	**	3	6	**	**
11	10	40	6	--	--	--
12	70	10	**	--	--	--
13	30	30	**	--	--	--
14	--	--	--	--	--	--
15	15	**	**	--	--	--
16	7	9	--	--	--	--
17	10	**	**	**	**	**
18	**	20	**	10	**	20
19	9	**	**	--	**	40
20	10	6	80	20	5	20
21	--	--	30	40	**	--
22	--	--	30	10	**	--
23	--	--	30	30	**	**
24	--	--	20	7	20	10
25	--	--	30	**	**	--
26	5	--	**	7	10	**
27	--	--	--	**	10	20
A	**	**	20	20	10	30
B	30	30	**	10	30	30
C	30	6	10	**	20	**
D	M	20	**	20	**	20
E	60	--	--	--	15	40
F	20	10	10	--	10	5
G	50	10	**	10	--	15
H	**	**	20	8	30	40
I	--	--	--	--	--	**
J	80	**	10	--	8	--
K	20	**	8	10	10	10
L	30	15	**	**	20	30
M	40	6	9	**	20	**
N	10	8	20	9	10	**
O	50	10	10	**	10	7
P	5	20	9	20	**	30
Q	30	10	6	15	--	6
R	20	--	7	**	10	--
T	10	10	9	**	**	--
U	9	**	10	**	20	--
V	40	7	40	6	30	30
W	40	9	10	--	**	**
SHIP	--	--	--	20	--	--

Individual data points are reliable to a factor of 2. ** = ≤ 5

Table 17. Atmospheric concentrations of bismuth, ng/m^3.

Station	May 21/22	June 6	June 20	July 9	Aug. 8	Aug. 29
1	**					
2	--					
3	3.					
4	1.					
5	2.					
6	3.					
7	--					
8	0.7					
9	2.					
10	1.5					
11	0.4					
12	0.6					
13	0.4					
14	--					
15	0.3					
16	**					
17	--					
18	--					
19	--					
20	--					
21	--					
22	--					
23	--					
24	--					
25	--					
26	**					
27	--					
A	M					
B	**					
C	**					
D	M					
E	0.8					
F	1.					
G	0.1					
H	M					
I	--					
J	**					
K	0.6					
L	**					
M	M					
N	**					
O	**					
P	0.3					
Q	0.4					
R	0.4					
T	0.2					
U	**					
V	**					
W	0.8					
SHIP	--					

Individual data points are reliable to a factor of 2. ** = $\leq$0.05

Table 18. Atmospheric concentrations of copper, ng/m^3

Station	May 21/22	June 6	June 20	July 9	Aug. 8	Aug. 29
1	70	300	400	150	150	300
2	---	100	1000	150	200	150
3	300	200	2000	200	80	600
4	900	2000	200	80	600	400
5	1000	3000	600	60	500	200
6	2000	7000	(7000)	2000	2000	4000
7	---	---	1000	300	1000	600
8	5000	9000	5000	2000	3000	9000
9	7000	5000	9000	7000	10000	10000
10	4000	5000	1500	2000	(9000)	10000
11	300	700	400	---	---	---
12	***	300	500	---	---	---
13	80	700	200	---	---	---
14	---	---	---	---	---	---
15	200	200	150	---	---	---
16	100	200	---	---	---	---
17	150	200	150	200	30	70
18	300	200	300	200	***	100
19	400	700	400	---	70	100
20	150	200	100	200	40	100
21	---	---	100	70	70	---
22	---	---	300	70	70	---
23	---	---	90	***	100	100
24	---	---	500	60	100	70
25	---	---	200	60	90	---
26	80	---	300	200	400	200
27	---	---	***	300	300	200
A	70	***	600	100	100	300
B	300	200	70	200	300	150
C	100	100	100	20	***	100
D	150	70	30	100	200	200
E	400	---	---	---	500	1000
F	200	***	80	---	100	100
G	400	100	70	200	---	200
H	500	20	80	20	100	***
I	---	---	---	---	---	200
J	200	100	***	---	200	---
K	200	***	***	200	100	100
L	200	100	70	***	800	200
M	80	***	60	***	600	90
N	100	80	300	60	300	100
O	300	200	200	100	200	200
P	300	100	200	200	200	200
Q	900	30	100	200	---	200
R	500	***	200	---	200	---
T	200	200	200	30	***	---
U	200	200	200	20	300	---
V	600	100	400	80	200	200
W	200	40	100	---	200	200
SHIP	---	---	---	100	---	---

Individual data points are reliable to a factor of 2. *** = ≤20

Table 19. Atmospheric concentrations of total suspended particulate, $\mu g/m^3$.

Station	May 21/22	June 6	June 20	July 9	Aug. 8	Aug. 29
1	115	170	117	32	198	148
2	---	218	344	64	161	92
3	81	131	256	139	148	167
4	126	143	128	54	152	145
5	140	169	278	66	156	242
6	142	201	382	125	256	233
7	---	---	332	193	191	174
8	62	108	210	105	91	122
9	83	115	134	148	122	109
10	78	109	198	---	100	96
11	76	153	228	---	---	---
12	128	179	249	---	---	---
13	120	187	255	---	---	---
14	---	---	---	---	---	---
15	46	136	100	---	---	---
16	80	154	85	---	---	---
17	66	109	117	142	---	---
18	98	86	107	64	17	---
19	55	98	70	---	46	---
20	78	96	88	126	62	---
21	---	---	77	44	40	65
22	---	---	104	55	49	65
23	---	---	66	27	39	57
24	---	---	---	32	38	81
25	---	---	81	---	43	185
26	68	---	64	63	79	45
27	---	---	---	49	84	69
A	88	174	141	71	138	107
B	113	212	116	105	239	214
C	114	55	150	49	132	139
D	125	261	19	60	243	188
E	144	---	---	---	305	952
F	153	206	169	---	185	218
G	134	181	172	152	---	143
H	50	176	236	86	182	186
I	---	---	---	---	---	181
J	166	205	259	---	166	---
K	154	139	245	103	158	134
L	180	229	180	43	225	218
M	139	130	183	98	170	195
N	142	115	211	86	154	136
O	163	205	263	108	174	178
P	141	153	286	51	177	214
Q	195	161	408	109	---	235
R	143	---	241	110	130	---
T	92	152	81	67	132	---
U	75	172	121	74	146	---
V	114	186	190	100	196	148
W	131	155	216	---	161	185

All data furnished by the local agencies.

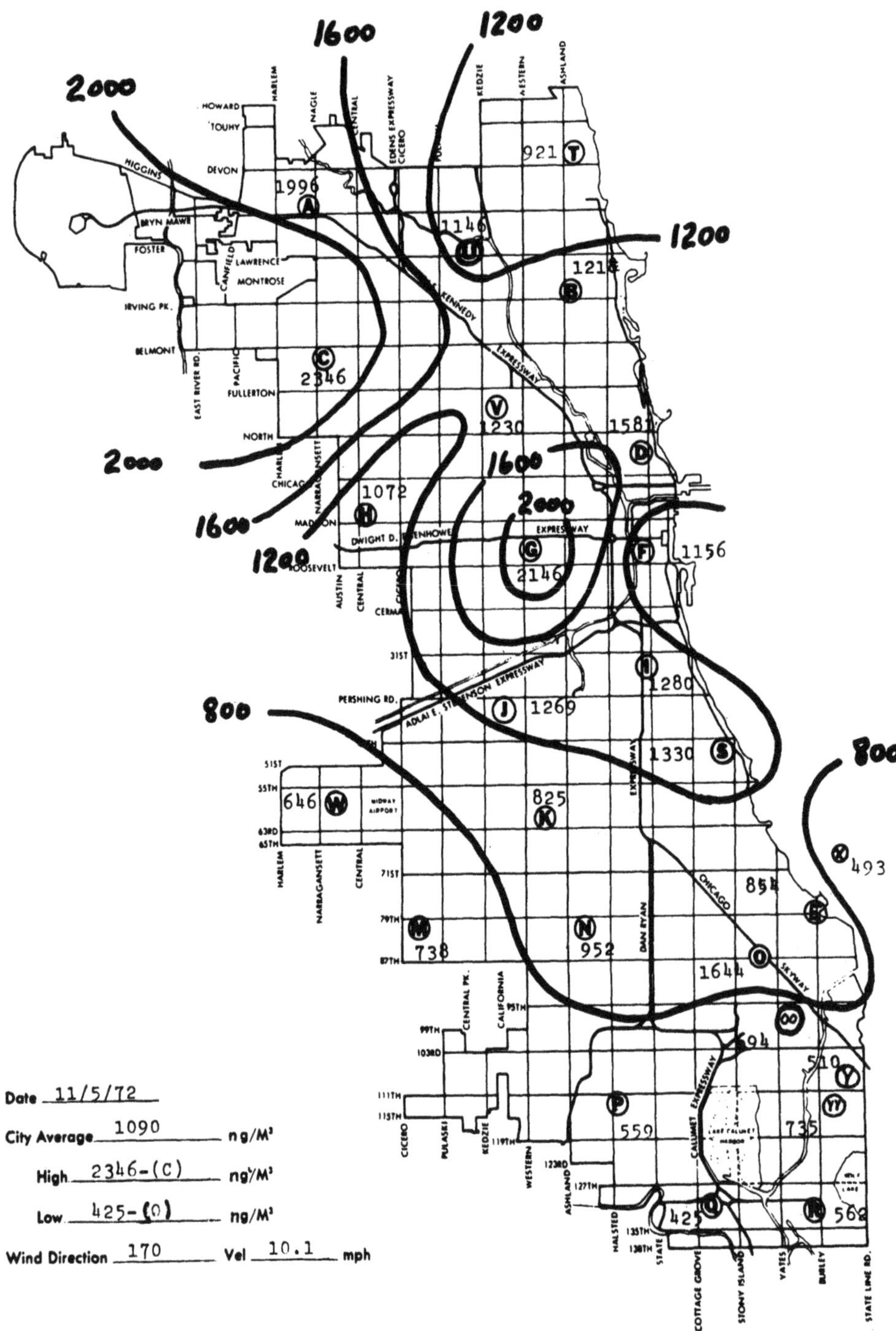

Figure 3. Ambient air metal concentrations--LEAD (in ng/m^3).

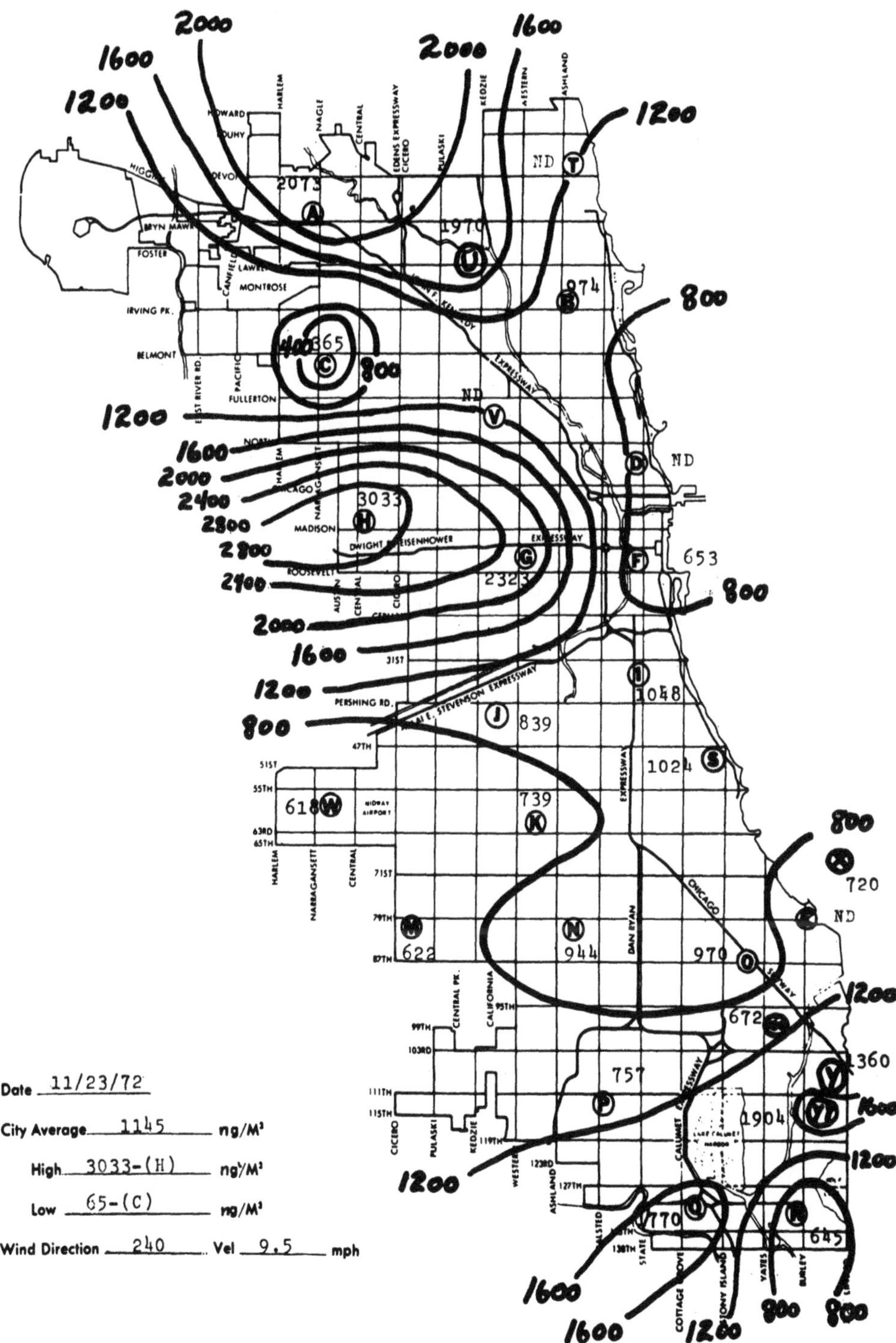

Figure 4. Ambient air metal concentrations--LEAD (in ng/m^3).

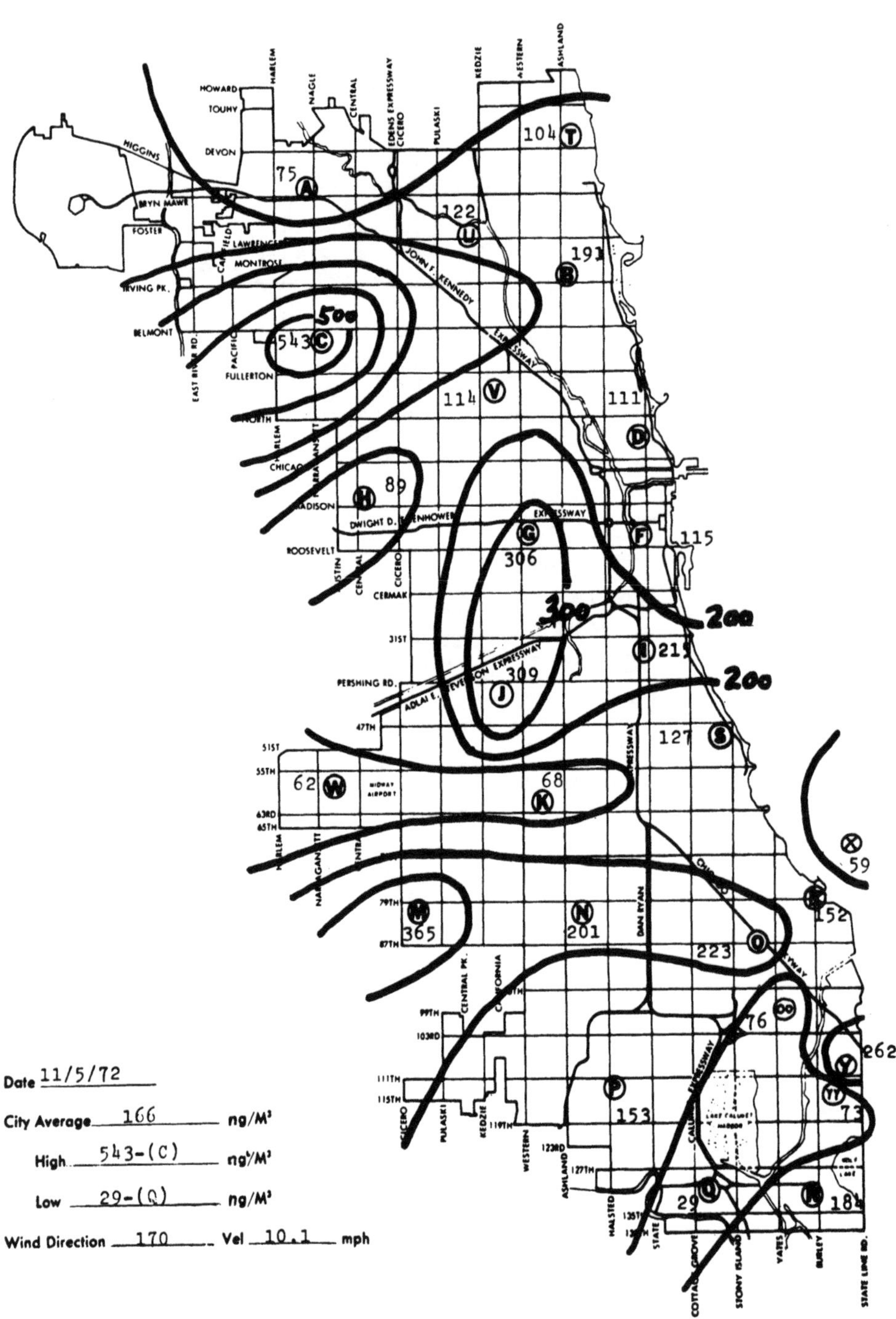

Figure 5. Ambient air metal concentrations--COPPER (in ng/m^3).

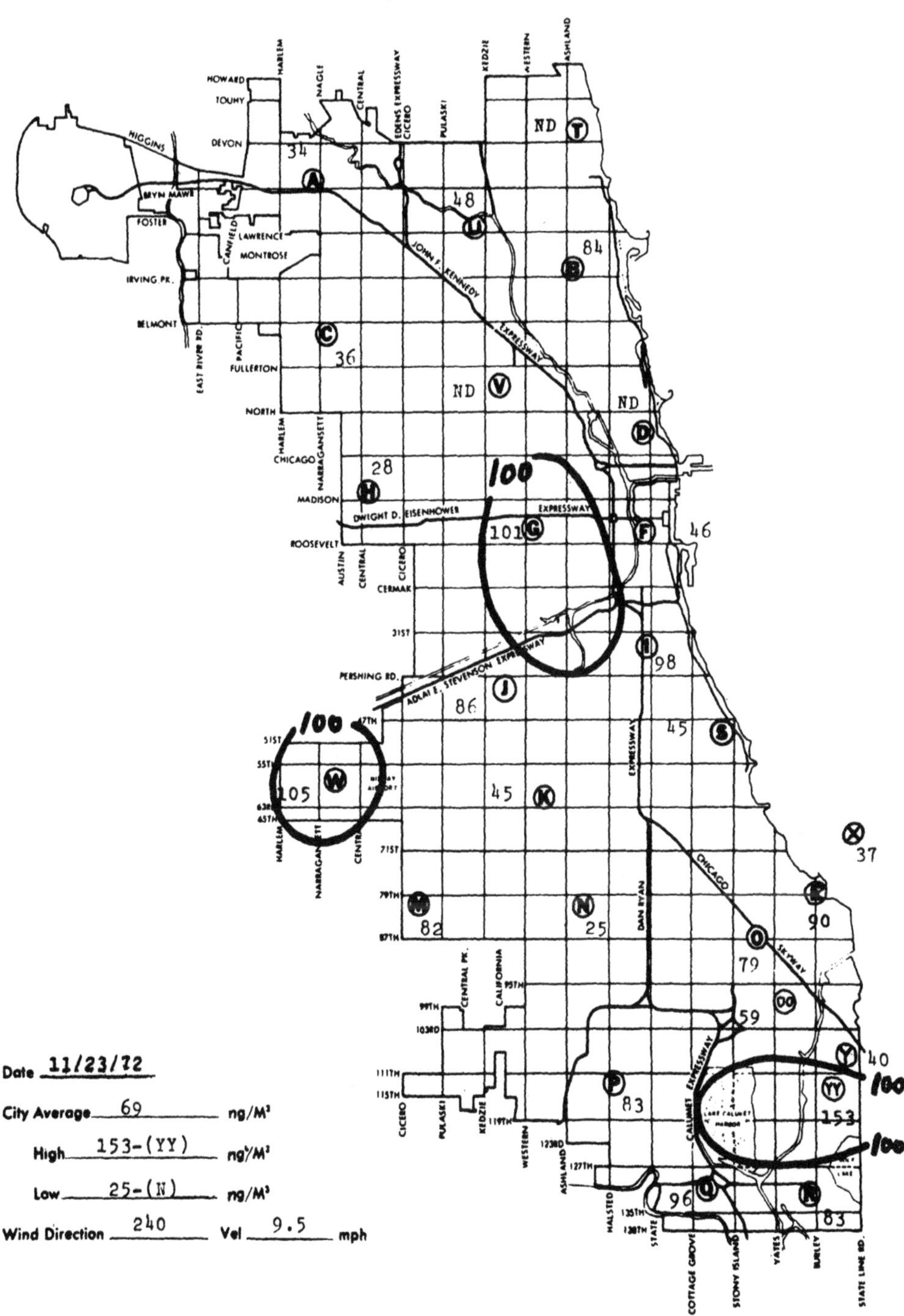

Figure 6. Ambient air metal concentrations--COPPER (in ng/m^3).

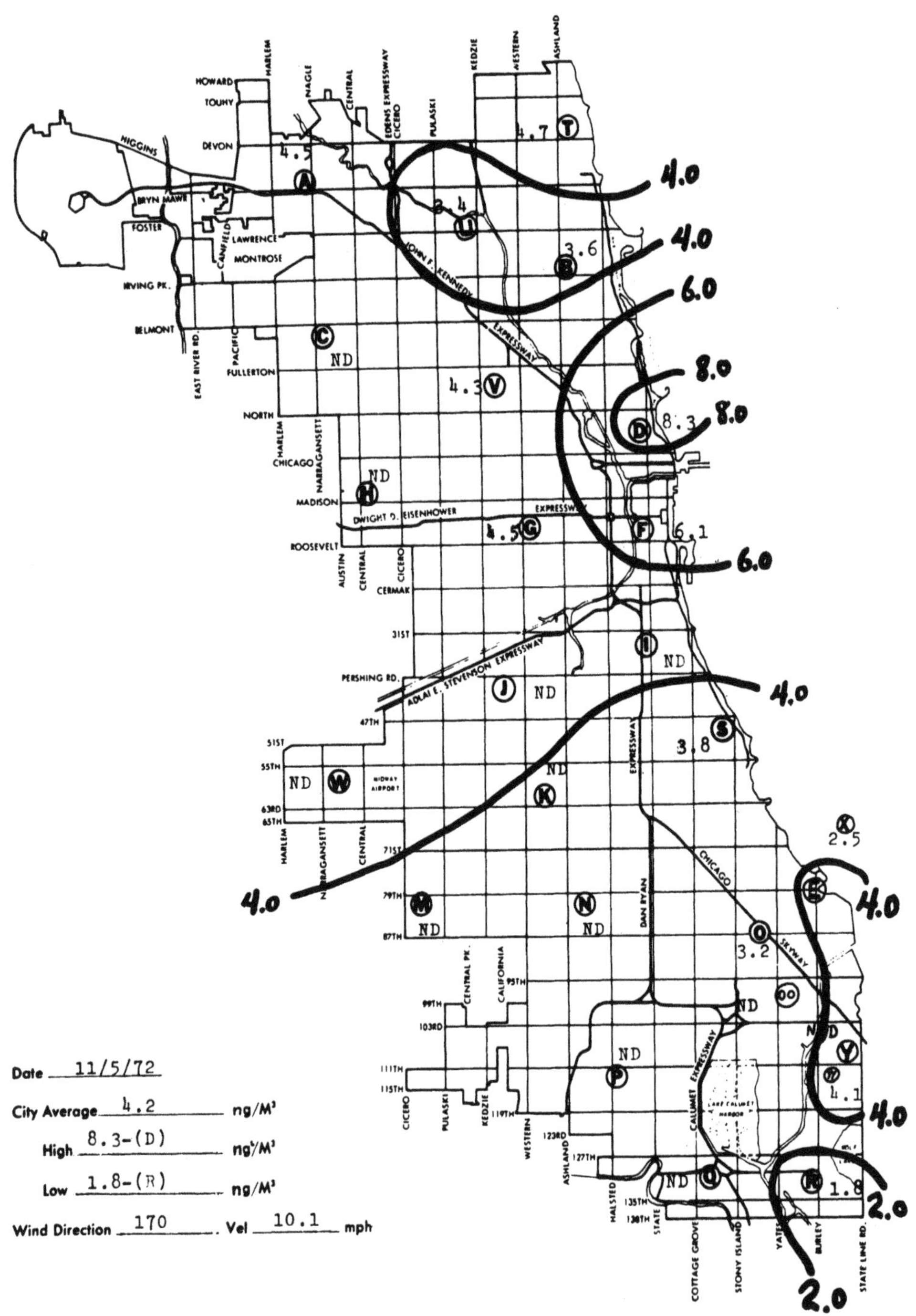

Figure 7. Ambient air metal concentrations--CADMIUM (in ng/m^3).

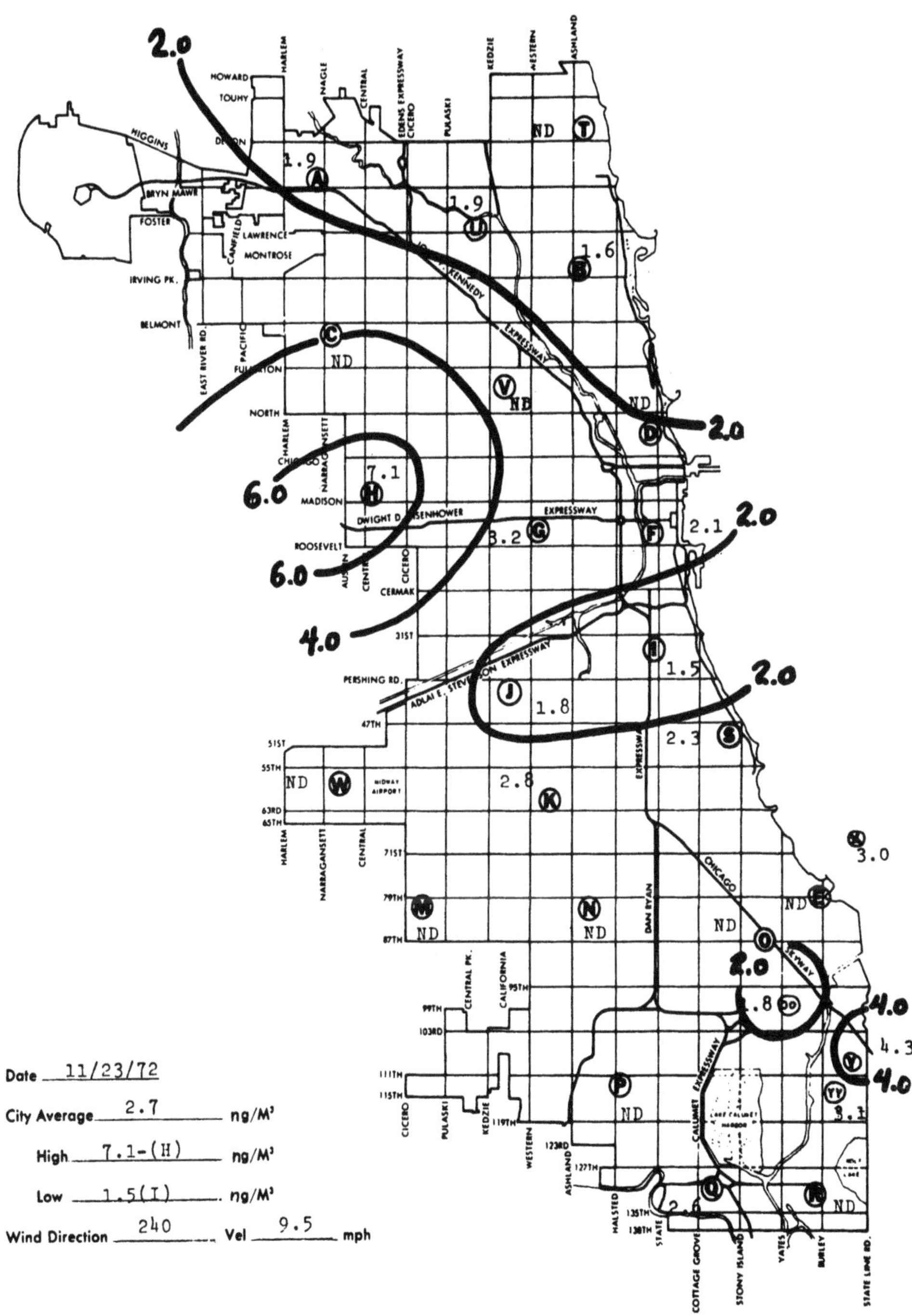

Figure 8. Ambient air metal concentrations--CADMIUM (in ng/m^3).

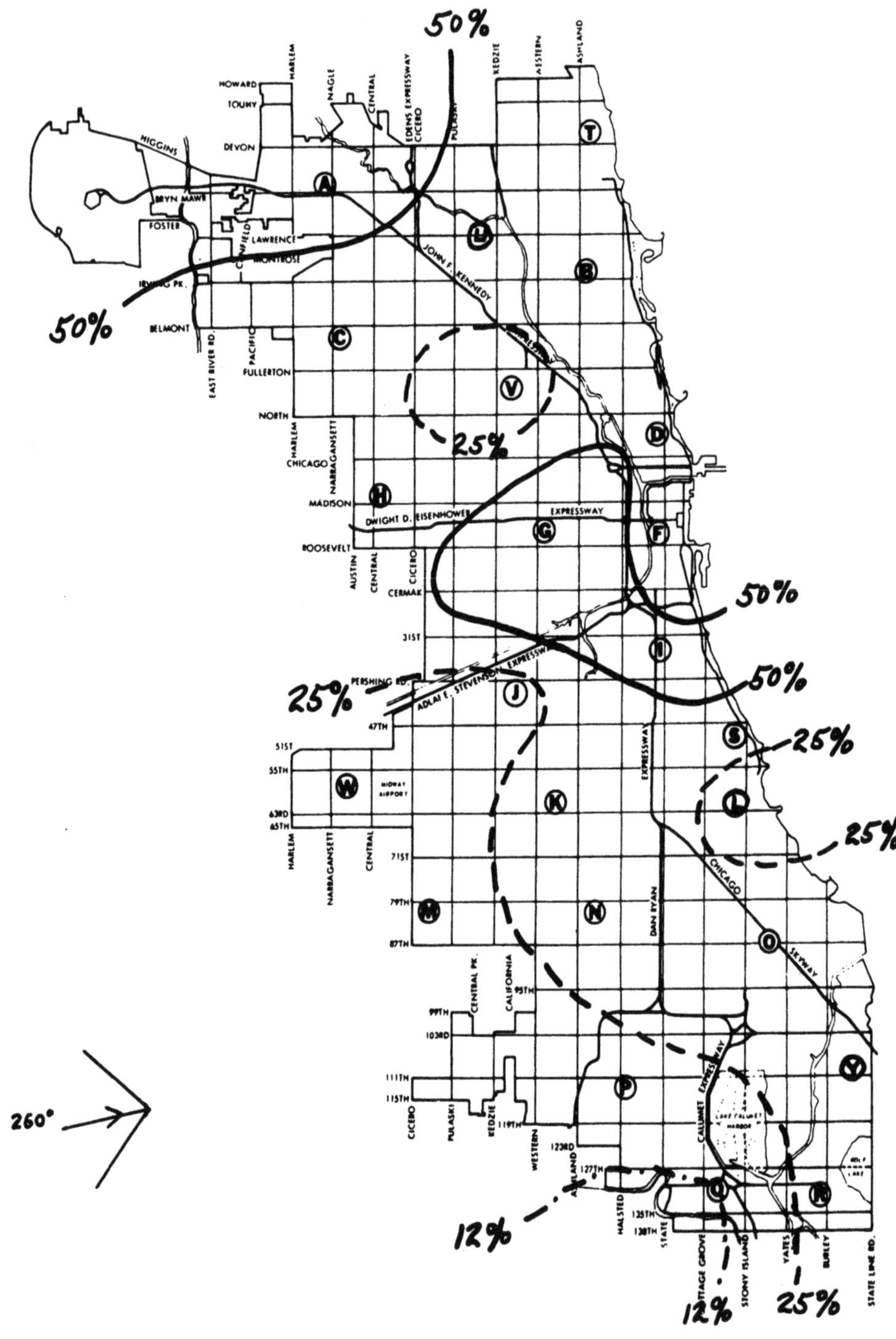

Figure 9. Distribution of lead. January 5 and 7, 1971.

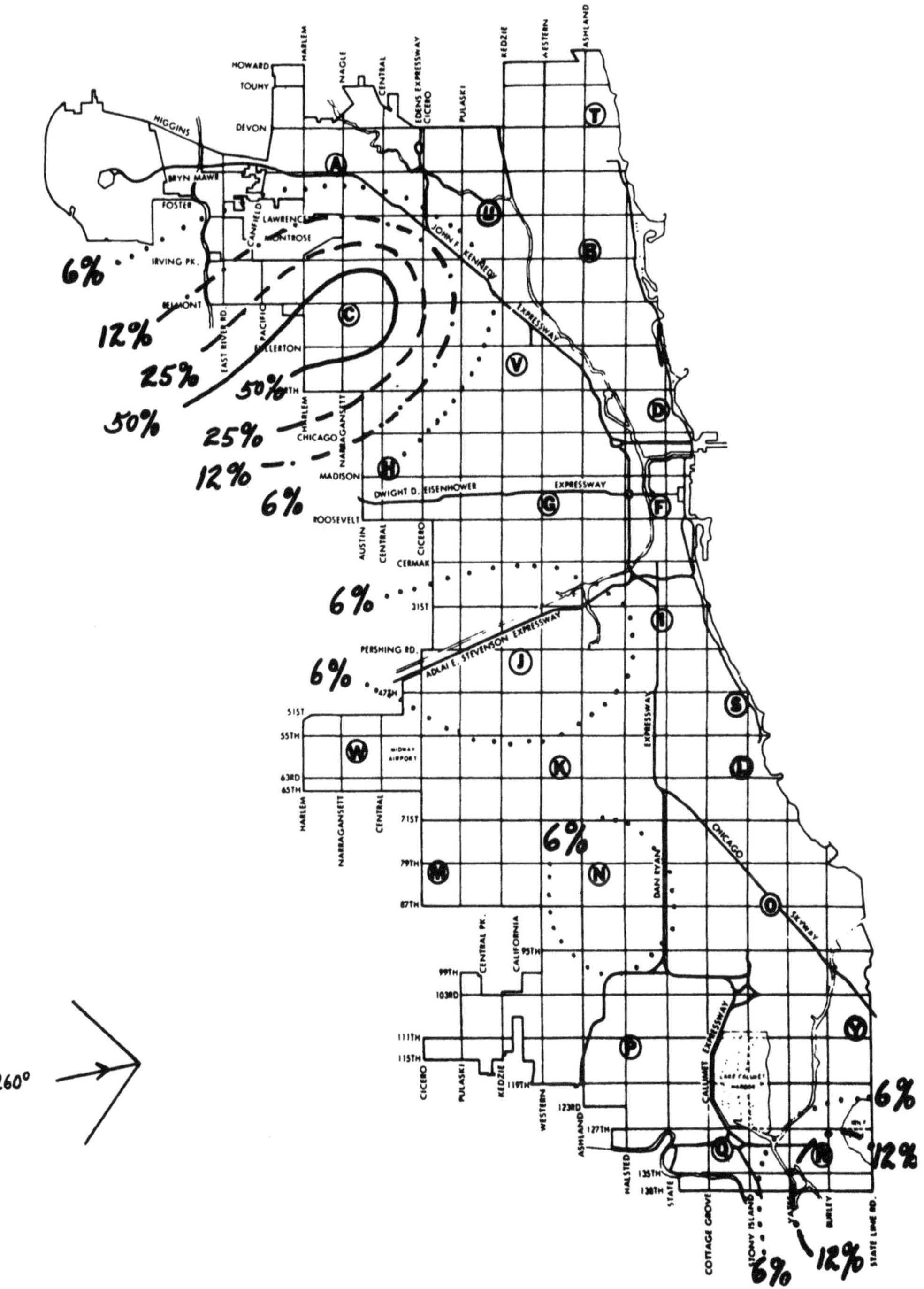

Figure 10. Distribution of copper. January 5 and 7, 1971.

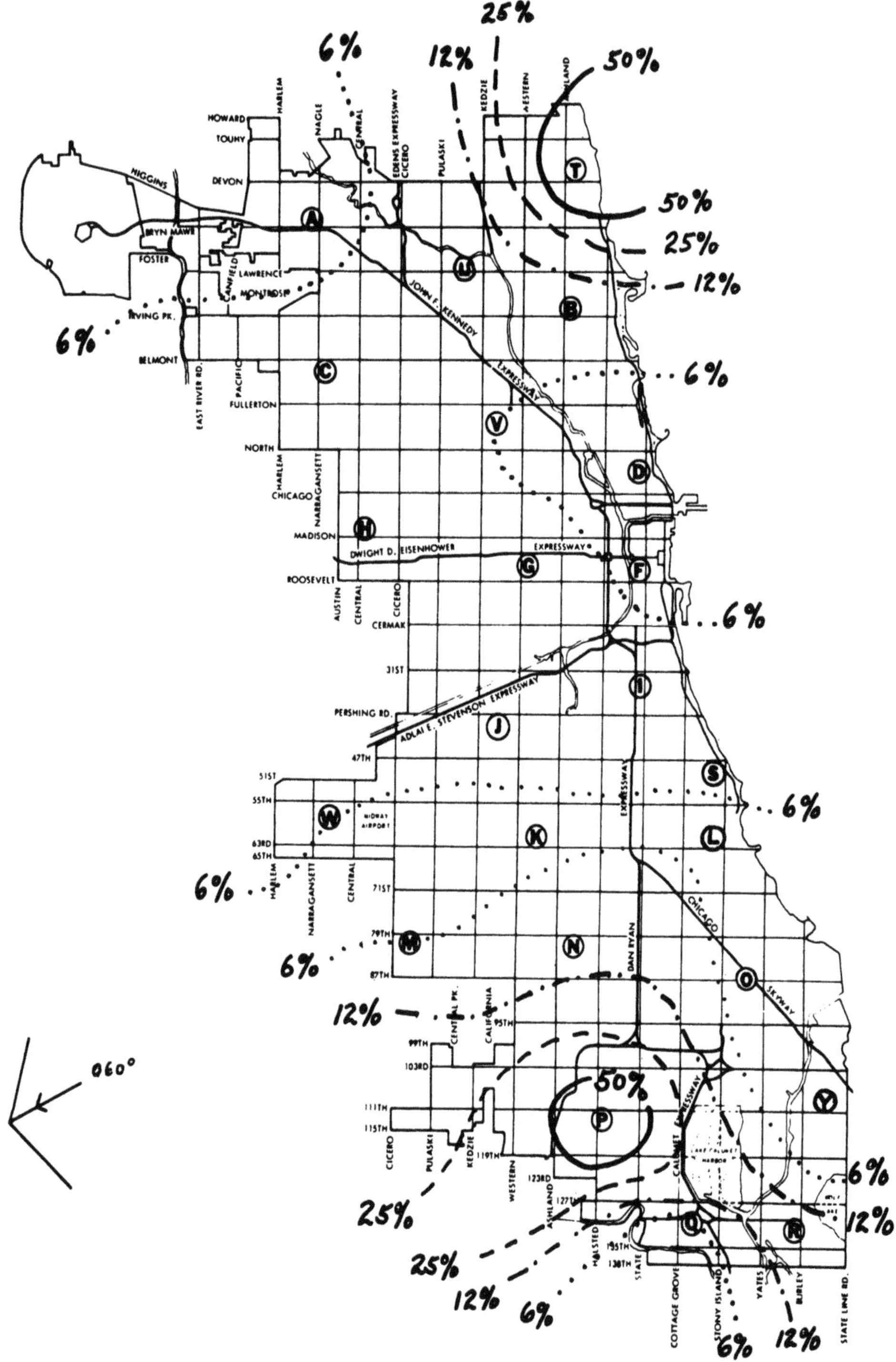

Figure 11. Distribution of copper. April 18 and April 20, 1971.

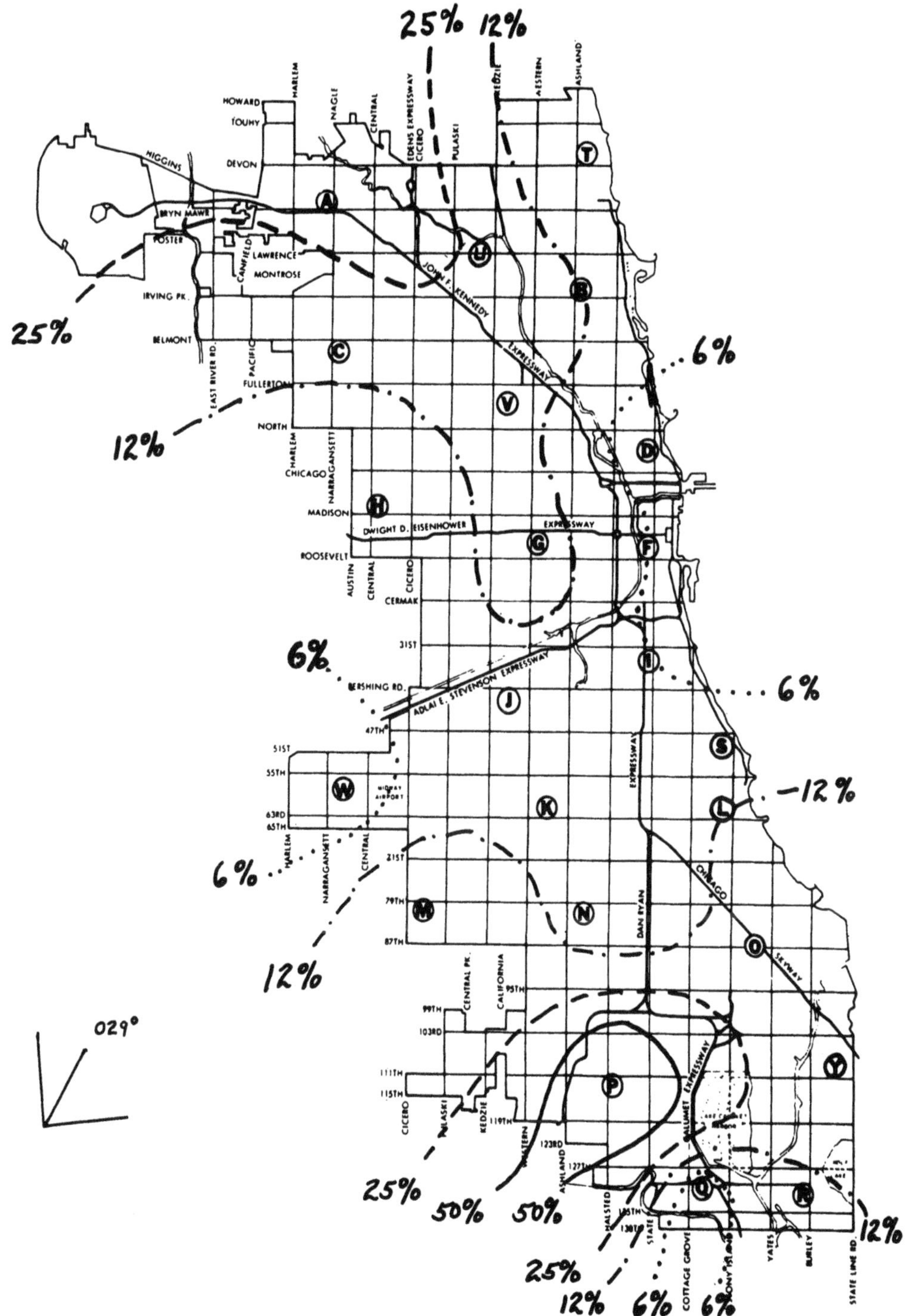

Figure 12. Distribution of copper. April 22 and May 6, 1971.

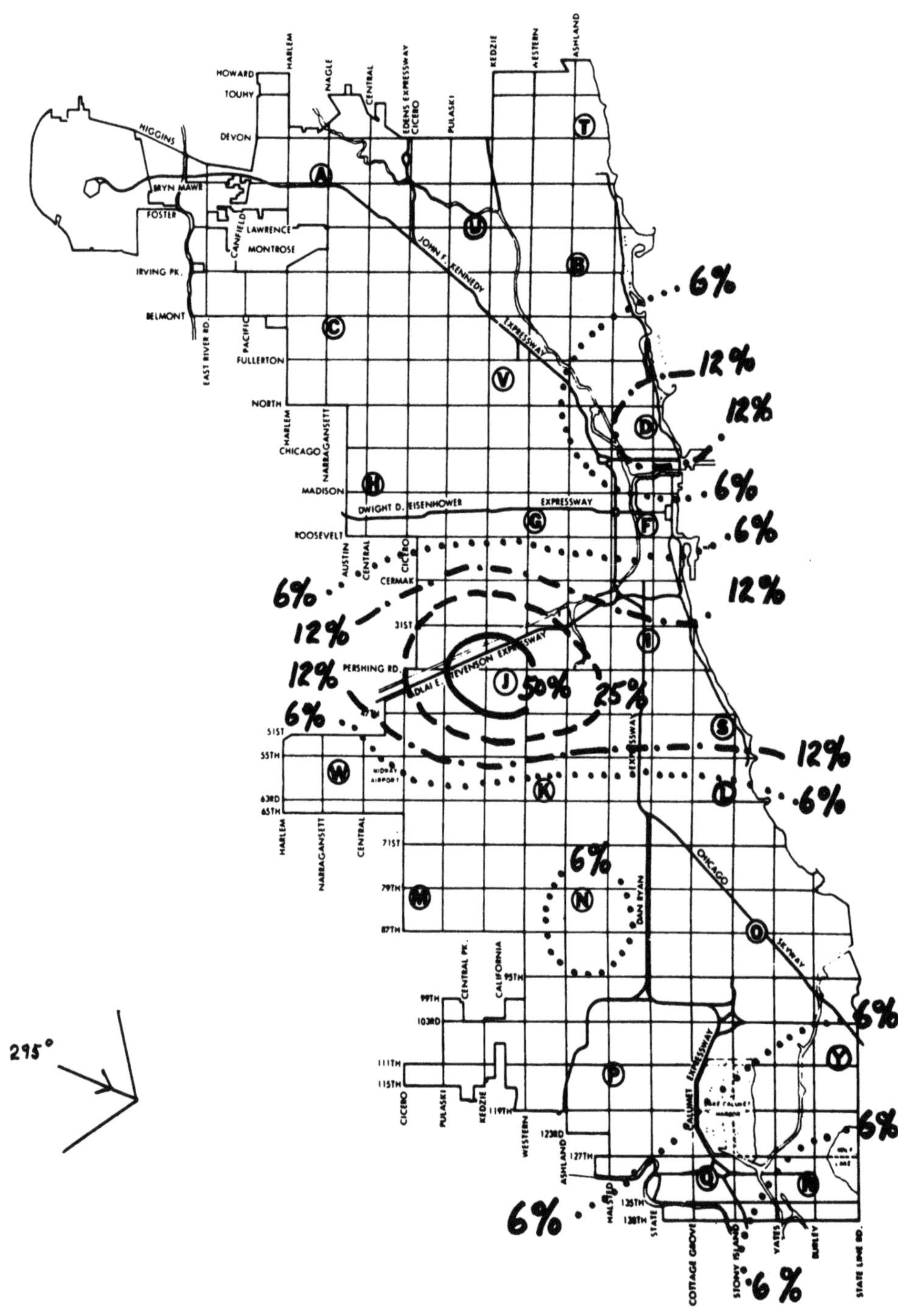

Figure 13. Distribution of zinc. March 7 and 23, 1971.

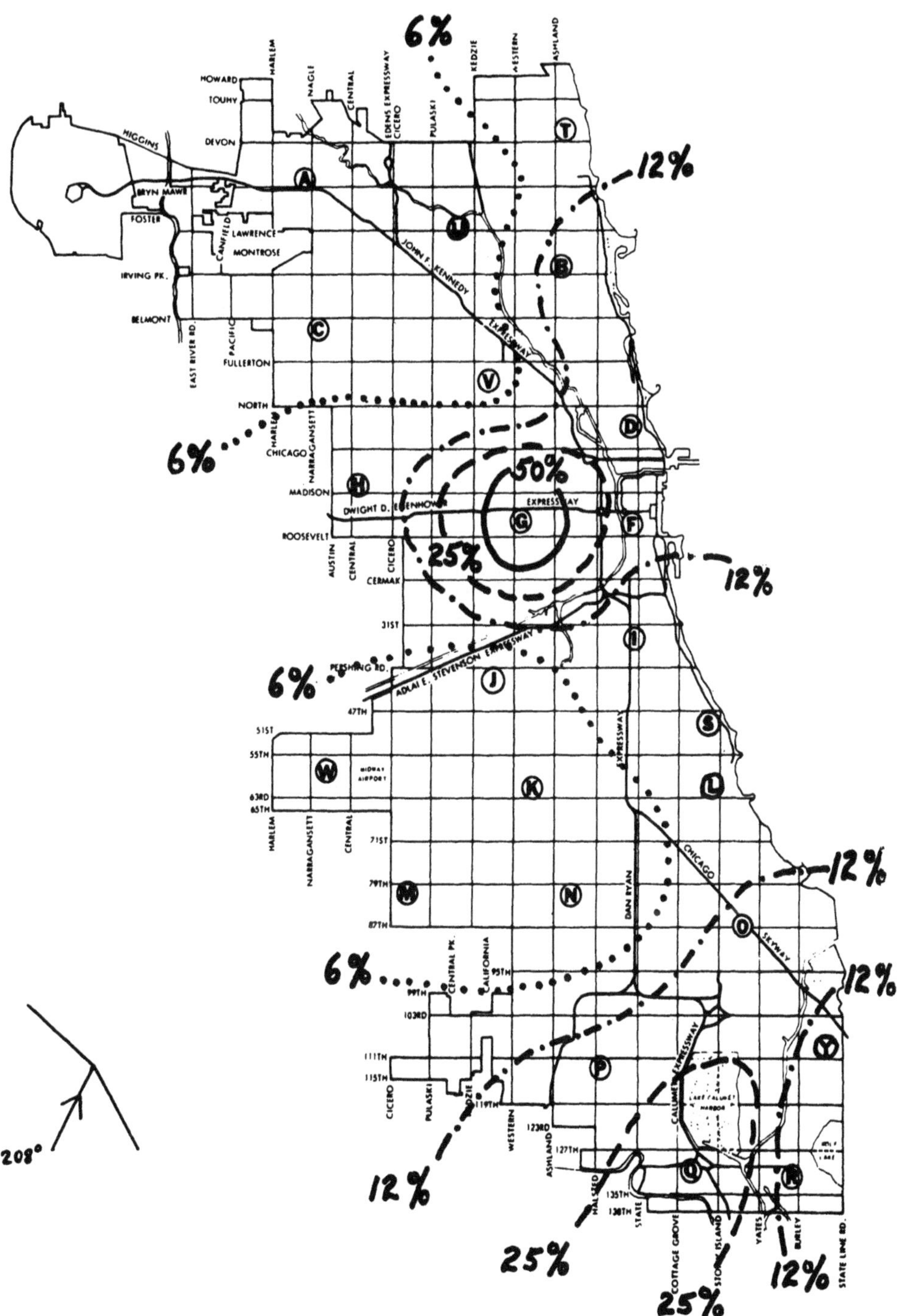

Figure 14. Distribution of zinc. April 11 and June 29, 1971.

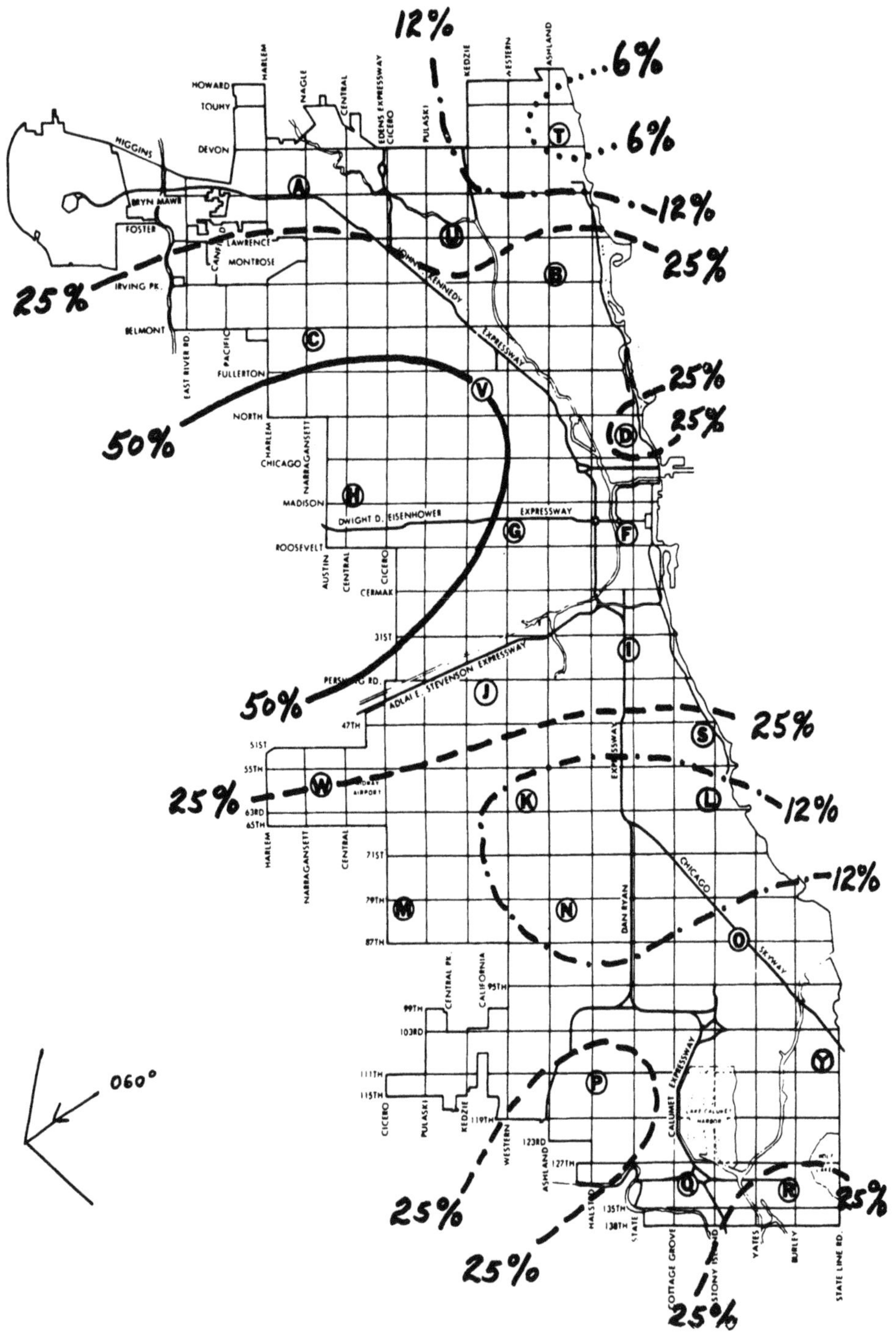

Figure 15. Distribution of zinc. April 18 and 20, 1971.

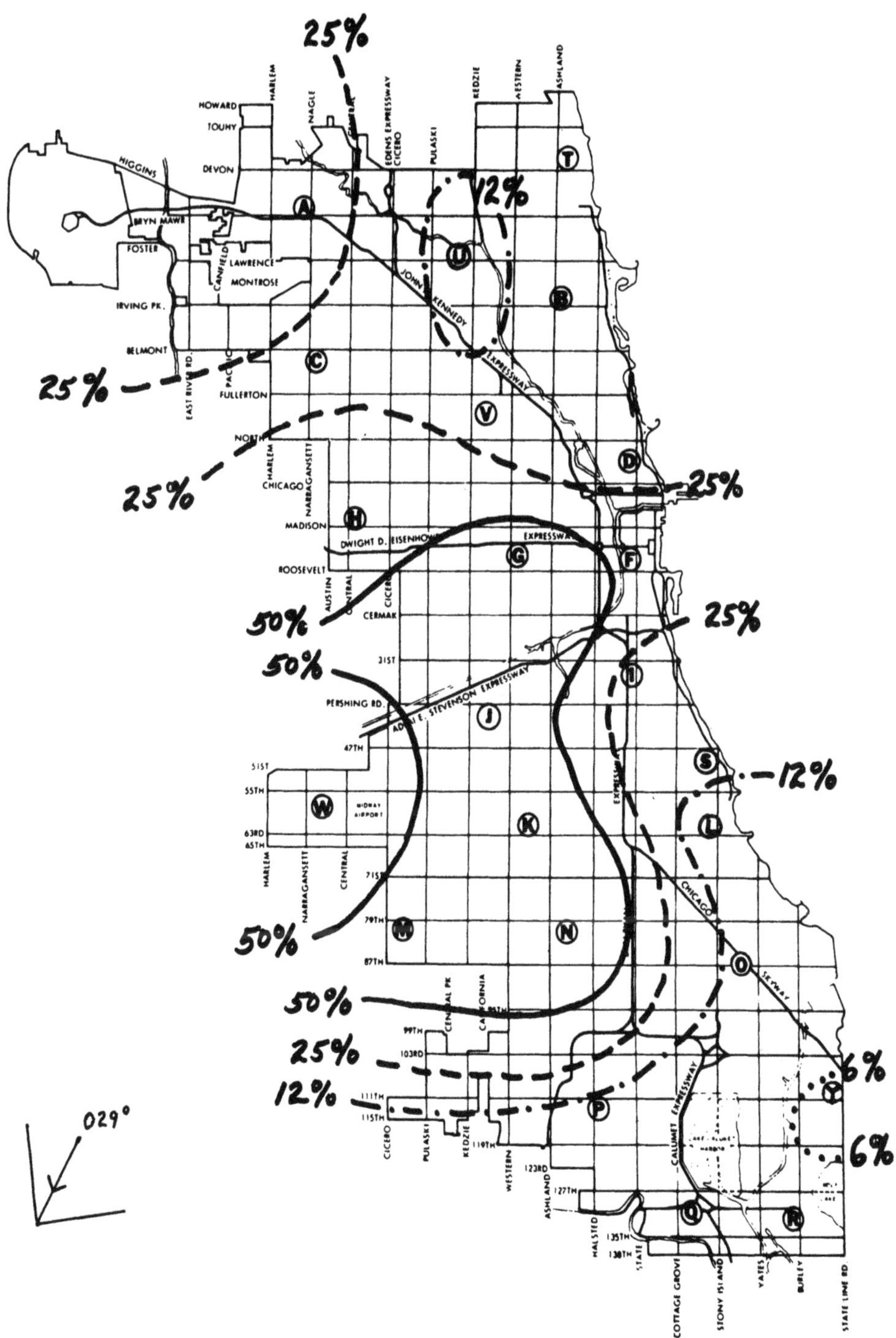

Figure 16. Distribution of zinc. April 22 and May 6, 1971.

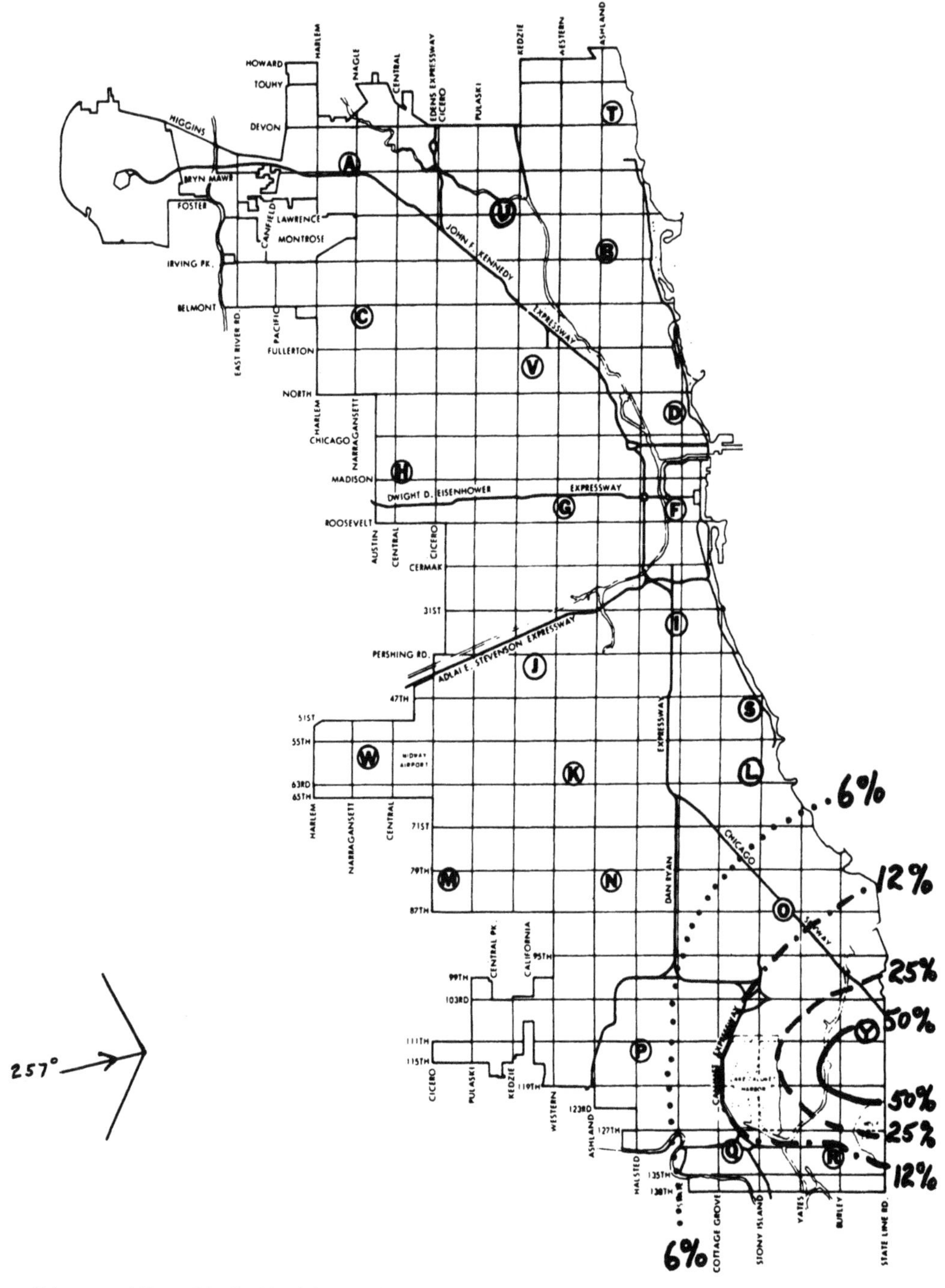

Figure 17. Distribution of manganese. February 9, 23, and 28, 1971.

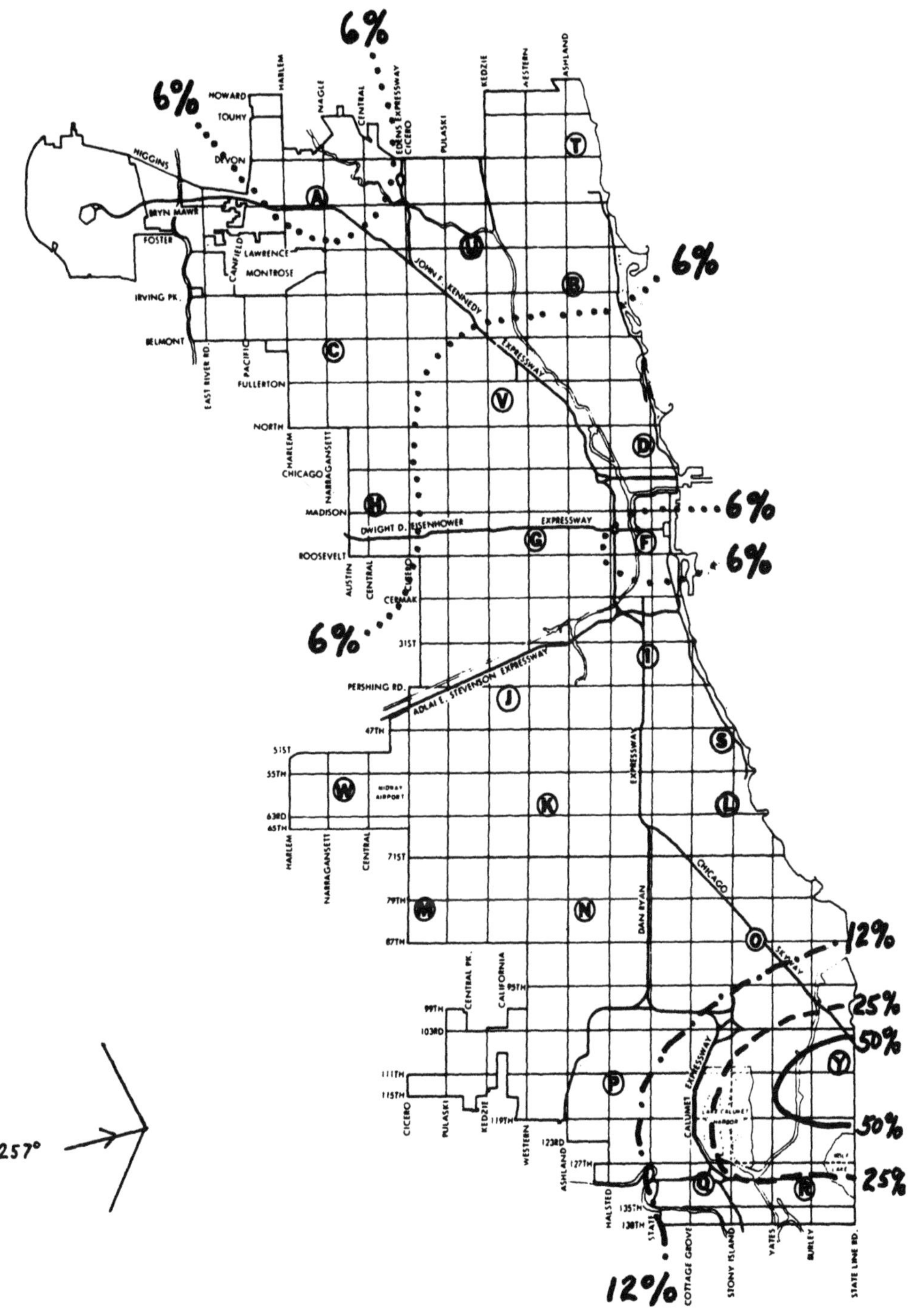

Figure 18. Distribution of iron. February 9, 23, and 28, 1971.

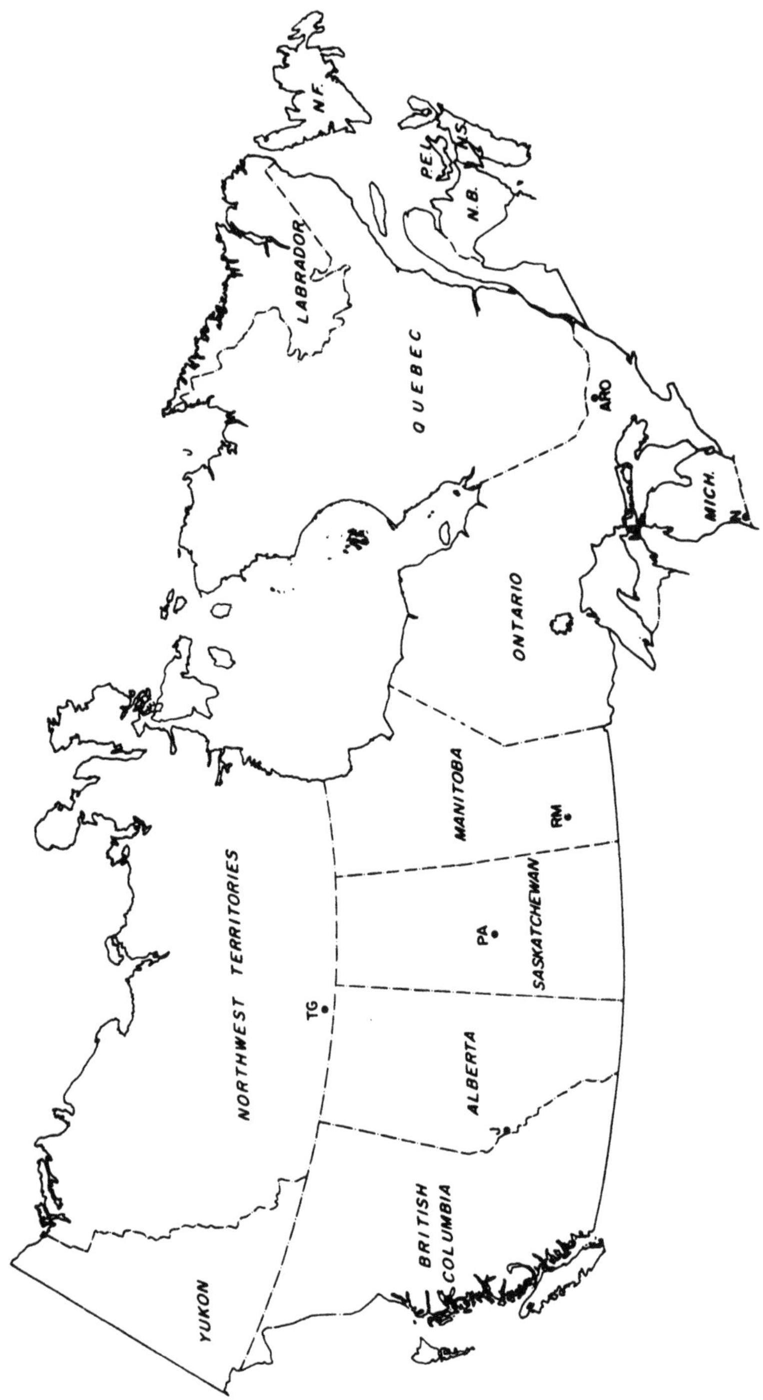

Figure 19. Sampling locations - Canadian experiments.

each station samples in the range of approximately 100 to 125 micrograms per cubic meter. The data cover lead, copper, zinc, and iron, and the method of analysis employed was atomic absorption spectroscopy.

Determining the approximate source location of the various elements mapped in Figures 3-18 requires tracing the wind direction upwind from the reference station in each figure to a general area of probable sources. "Hot spots" and associated wind direction in Figures 11 and 12 indicate the probable source area within the heavily industrialized southwest sector of Chicago. Another source location technique also traces the wind direction upward; however, instead of using one reference station as previously mentioned, it references two separate stations and their respective wind directions. In Figures 13 and 14 at stations "J" and "G," respectively, the two wind directions are traced upward to a point of intersection which describes the center of a probable source area, namely, a point north of the Sanitary and Ship Canal and west of the city limits. The source area is defined to be within a few square miles of this point.

Concentrations of Trace Elements in Remote Regions

During the summer of 1970, Rahn (10) conducted experiments in Michigan and the southern provinces of Canada and as far north as the southern region of the Northwest Territory. His purpose was to gain an understanding of the occurrence of trace metals in remote air mass source regions of relatively clean maritime-continental air sufficiently far from industrial sources. Figure 19 is a map showing the sampler locations for this "Canadian Experiment." Tables 20 and 21 show the average trace metal concentrations in nanograms per cubic meter of the summer portion of the Canadian experiment and the seasonal concentrations (summer and winter) for three selected stations, southern Northwest Territories (TG), eastern Ontario (ARO), and Sault Ste. Marie, Michigan (MI). These two tables, along with Table 22, show that much of the material is of soil origin. Some of the elements such as chlorine and mercury, show relatively larger aerosol concentrations. The chlorine is probably contributed from the sea salt spray advected over large distances (a known phenomenon). The mercury probably accumulates from its volatility and solar insulation on the soils. Origins of other element concentrations such as selenium are difficult to explain. Subsequent experiments conducted by Rahn lead us to believe that much of this enrichment factor comes from long range advection from industrial sources. In order to determine these source regions, Rahn further broke the elements into size distribution classifications as shown in Figures 20 and 21. Figures 22 and 23 are examples of the "M"-type elements found in the Northwest Territory.

Table 20. Total elemental concentrations (ng/m^3) - Canadian Summer Experiment

	TG	J	PA	RM	ARO	MI	N
Na	18	36	43	56	69	44	120
Mg	16	53	60	130	40	470	160
Al	66	145	150	330	240	230	580
Cl	9	13	11	28	4	35	46
K	54	106	112	175	170	150	340
Ca	40	150	130	360	160	1150	650
Sc	0.004	0.082	0.12	0.16	0.14	0.12	0.49
Ti	5	8	9	12	15	11	35
V	0.21	0.33	0.42	0.73	1.9	1.8	3.6
Cr	0.59	0.32	1.1	0.92	1.9	0.91	3.8
Mn	1.5	5.3	5.9	8.7	12	9.2	41
Fe	71	220	180	270	310	250	950
Co	0.042	0.059	0.085	0.11	0.16	0.17	0.34
Ni	<2	<2	<2	<2	5	<3	<7
Cu	0.9	3.8	0.9	4.4	7.9	10	15
Zn	3.8	5.2	13	15	40	22	130
Ga	0.026	0.042	0.035	0.056	0.14	0.15	0.35
As	0.31	0.27	0.32	0.45	4.7	3.2	4.6
Se	0.043	0.033	0.063	0.18	0.63	0.67	0.89
Br	0.54	2.0	2.9	3.1	5.7	7.2	94
Ag	<0.15	<0.15	<0.2	<0.2	<0.4	<0.5	<0.7
In	0.0013	<0.003	0.0020	0.0029	0.039	0.024	0.017
Sb	0.13	0.13	0.16	0.21	0.60	0.40	1.9
I	0.20	0.21	0.13	0.18	0.27	<0.2	<0.5
La	0.091	0.12	0.10	0.19	0.30	0.17	0.76
Ce	0.24	0.25	0.32	0.31	0.69	0.41	1.6
Sm	0.013	0.017	0.018	0.035	0.051	0.030	0.11
Eu	0.0017	0.0037	0.0036	0.0082	0.0090	0.0080	0.019
W	0.016	0.035	<0.03	<0.05	0.041	0.30	0.12
Hg	0.06	0.17	<0.3	0.41	0.19	0.38	0.61
Th	0.052	0.036	0.040	0.058	0.056	0.018	0.11

Table 21. Seasonal concentration comparisons - TG, ARO, MI

	TG1 Summer	TG2 Winter	TG1/TG2	ARO1 Summer	ARO2 Winter	ARO1/ARO2	MI3 Summer	MI4 Winter	MI3/MI4
Na	49	290	0.17	65	99	0.66	48	96	0.50
Mg	41	37	1.1	47	33	1.4	370	70	5.3
Al	170	38	4.5	240	250	1.0	250	180	1.4
Cl	7.8	310	0.025	6.9	21	0.33	15	11	1.4
K	83	41	2.0	140	44	3.2	190	75	2.5
Ca	120	48	2.5	120	38	3.2	880	200	4.4
Sc	0.065	0.016	4.1	0.11	0.033	3.2	0.12	0.089	1.3
Ti	14	2.7	5.2	12	5.1	2.4	12	12	1.0
V	0.32	0.37	0.87	1.3	2.5	0.52	1.4	0.88	1.6
Cr	2.5	1.1	2.3	0.88	1.3	0.68	1.1	1.2	0.9
Mn	3.1	0.73	4.2	9.5	7.1	1.3	8.1	7.1	1.1
Fe	96	810	0.12	230	300	0.77	240	250	1.0
Co	0.087	0.037	2.4	0.11	0.086	1.3	0.4	0.26	0.54
Ni	---	---	---	---	---	---	---	---	---
Cu	0.84	2.2	0.38	4.9	13	0.38	1.8	8.7	0.21
Zn	4.1	2.8	1.5	36	30	1.2	23	32	0.72
Ga	0.026	0.18	0.14	0.12	0.11	1.1	0.19	0.095	2.0
As	0.22	0.77	0.29	4.7	2.2	2.1	5.1	8.0	0.64
Se	0.038	0.164	0.59	0.43	0.34	1.3	0.38	1.2	0.32
Br	1.1	2.7	0.41	2.9	3.7	0.78	6.0	5.7	1.1
Ag	---	---	---	---	---	---	---	---	---
In	<0.018	0.0039	<0.46	0.074	0.071	1.0	0.011	0.061	0.18
Sb	0.078	0.085	0.92	0.71	0.28	2.5	0.54	0.38	1.4
I	0.24	0.35	0.69	0.11	0.26	0.42	0.17	0.37	0.46
La	0.20	0.10	2.0	0.33	0.065	5.1	0.29	0.17	1.7
Ce	0.45	0.11	4.1	0.58	0.16	3.6	0.35	0.38	0.92
Sm	0.024	0.020	1.2	0.050	0.012	4.2	0.035	0.021	1.7
Eu	0.0039	0.0064	0.61	0.0086	0.0040	2.1	0.0058	0.0047	1.2
W	<0.027	0.12	<0.23	0.043	0.045	1.0	0.062	0.038	1.6
Hg	0.34	0.048	7.1	0.19	0.19	1.0	0.46	0.90	0.51
Th	0.076	0.030	2.5	0.044	0.0071	6.2	0.033	0.028	1.2

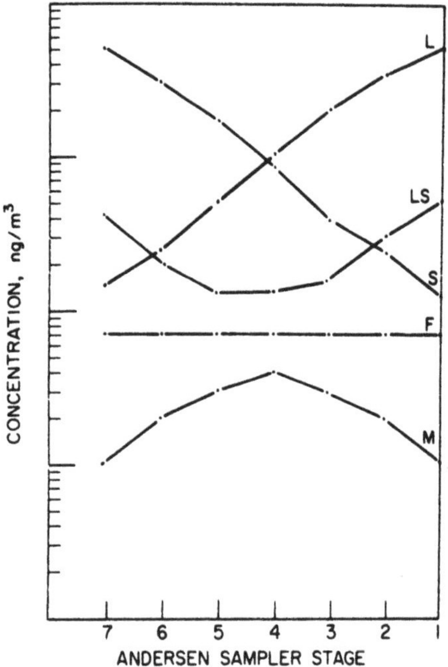

Figure 20. Idealized particle size distribution shapes.

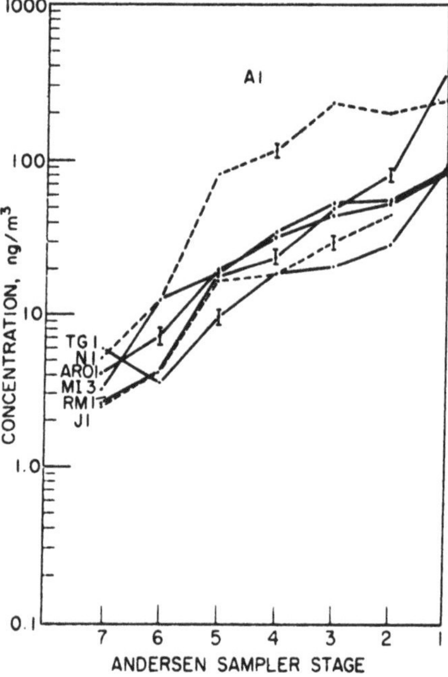

Figure 21. Particle size distributions of Al, Canadian Summer Experiment.

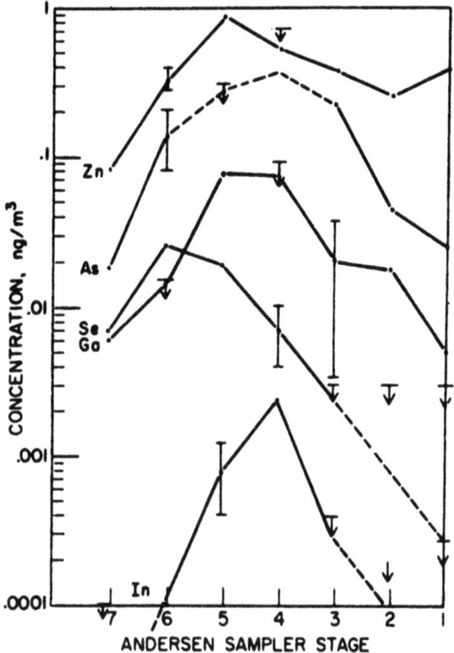

Figure 22. Andersen sample TG2 - "M"-type elements.

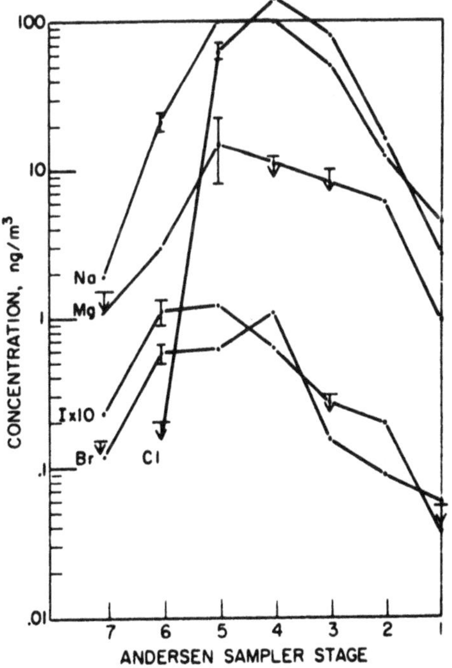

Figure 23. Andersen sample TG2 - "M"-type elements.

Table 22. Aerosol-soil normalized enrichments - Canadian Summer Experiment.

	TG	J	PA	RM	ARO	MI	N
Na	1.5	0.98	1.4	1.3	1.3	1.1	0.76
Mg	1.3	1.4	2.0	2.9	0.8	11	1.0
Al	0.49	0.34	0.46	0.65	0.42	0.45	0.32
Cl	47	22	23	41	5	51	18
K	2.1	1.3	1.7	1.9	1.5	1.7	1.0
Ca	1.5	1.9	2.0	3.8	1.4	13	1.9
Sc	3.3	2.1	3.6	3.2	2.5	2.6	2.8
Ti	0.58	0.29	0.41	0.38	0.40	0.42	0.30
V	1.1	0.57	0.87	1.1	2.4	2.7	1.4
Cr	1.6	0.28	1.1	0.68	1.2	0.67	0.76
Mn	0.95	1.1	1.5	1.5	1.8	1.6	1.9
Fe	1	1	1	1	1	1	1
Co	2.8	1.3	2.2	2.0	2.5	3.0	1.7
Ni	<26	<8.6	<10	<7.4	15	<11	<6.8
Cu	24	33	9.7	32	47	76	30
Zn	40	17	54	44	100	67	100
Ga	0.46	0.24	0.24	0.27	0.58	0.70	0.48
As	33	9.3	13	13	110	97	37
Se	2300	570	1300	2600	7700	10,000	2700
Br	58	69	120	88	140	220	760
Ag	<790	<260	<420	<290	<500	<760	<280
In	6.8	<5.2	4.2	4.1	47	36	6.8
Sb	340	110	170	150	370	300	380
I	21	7.2	5.4	5.3	6.5	<6	<4
La	1.2	0.52	0.54	0.65	0.92	0.63	0.76
Ce	2.5	0.86	1.3	0.88	1.7	1.2	1.3
Sm	1.2	0.48	0.62	0.82	1.0	0.75	0.72
Eu	0.74	0.53	0.62	1	0.92	1.0	6.4
W	5.8	4.0	<4.2	<5	3.3	30	3.2
Hg	3200	2900	<6200	5900	2200	5800	2400
Th	4.6	1.0	1.4	1.4	1.2	0.45	0.72

To further describe the behavior of these aerosols, Rahn conducted a two-and-one-half day study for several elements as presented in Figure 24. It is interesting to note that a classical dependence on wind speed does not apply at the Niles, Michigan rural location. This is due to the fact that most of this material is well mixed and comparatively evenly distributed and should not obey the classical, near-source plume diffusion regime. It is thought

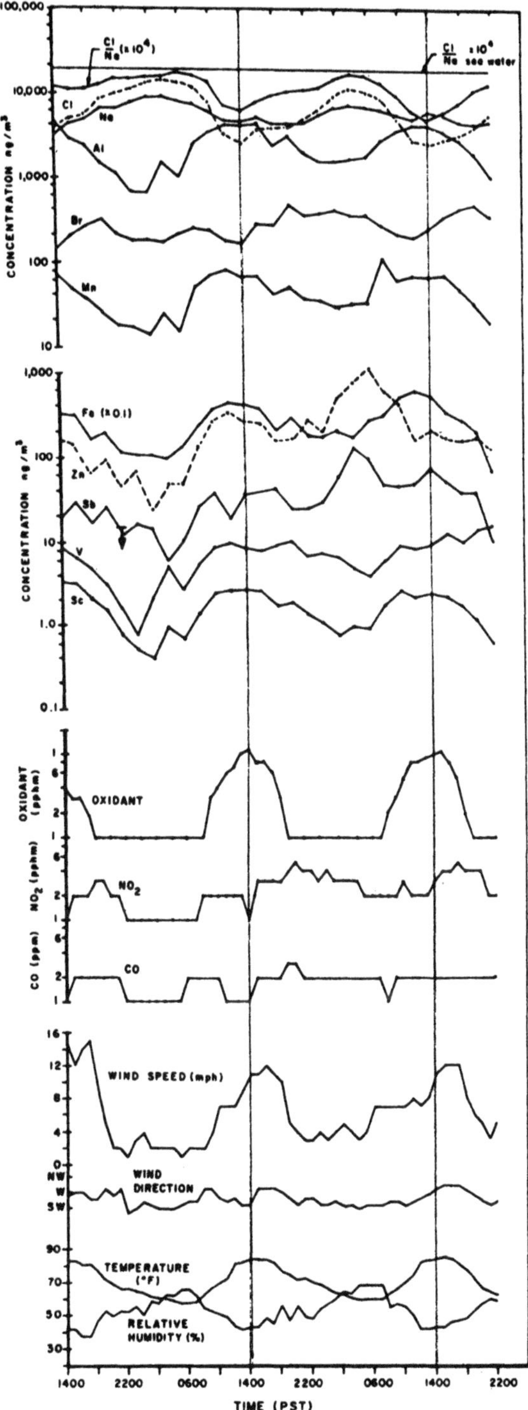

Figure 24. Results of measurements of diurnal variations of aerosol trace element concentrations.

that the iron and zinc, and other meterials presented, originate partially from the large steel-making operations in the southern Lake Michigan basin located due west of Niles. For further discussion the reader is referred to the thesis by Rahn (10).

Figures 25-28 represent a study conducted at Ann Arbor, Michigan, by Harrison, et al. over a 42-hour time period (15). This study relates average mass distribution with particle size for Pb, Cu, and Cd. These measurements are also related to short-time variations with regard to wind direction and source region. Figure 25 indicates the location of the sampling area in relation to Detroit. Temperature (T), relative humidity (RH), and dew point (Td) are graphed with respect to the sampling time in Figure 26. By inspection of the wind direction plotted in Figure 27 and the total concentrations plotted in Figure 28, one can see a definite coincidence with the currents of the wind direction coming from the "Detroit Sector" as shown by the dashed lines. With approximately a two-hour time lag the minimum occurs at 2:00 p.m., Saturday, April 27, when the source region had been determined to exist toward a direction containing virtually little or no industry. Towards the end of the experiment it is obvious that the diffusion characteristics, as well as change in wind direction, are functioning. These data represent a good example of moderate (10 to 100 miles) advection of automobile and industrial derived pollutants. It is likewise interesting to note that the lead minimum was approximately 800 nanograms per cubic meter, whereas the copper and cadmium levels continued below blanks for a period of time.

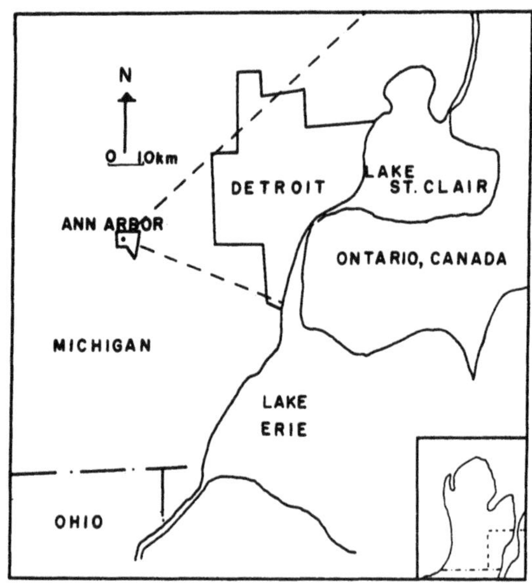

Figure 25. Location of sampling area in relation to Detroit.

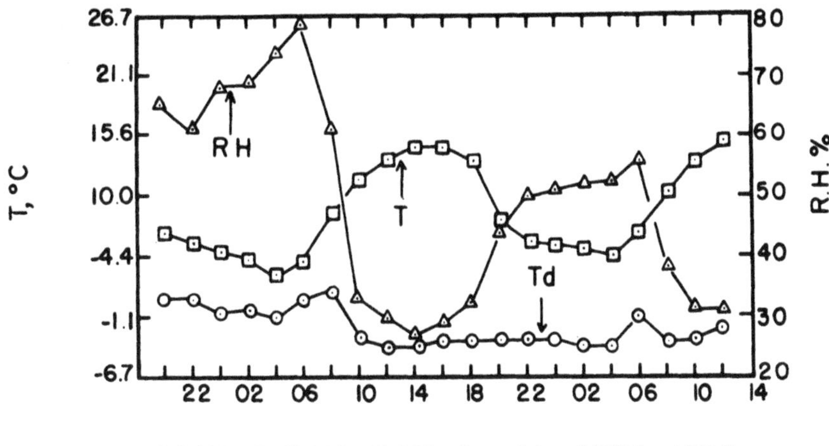

Figure 26. T, RH, Td versus sample time.

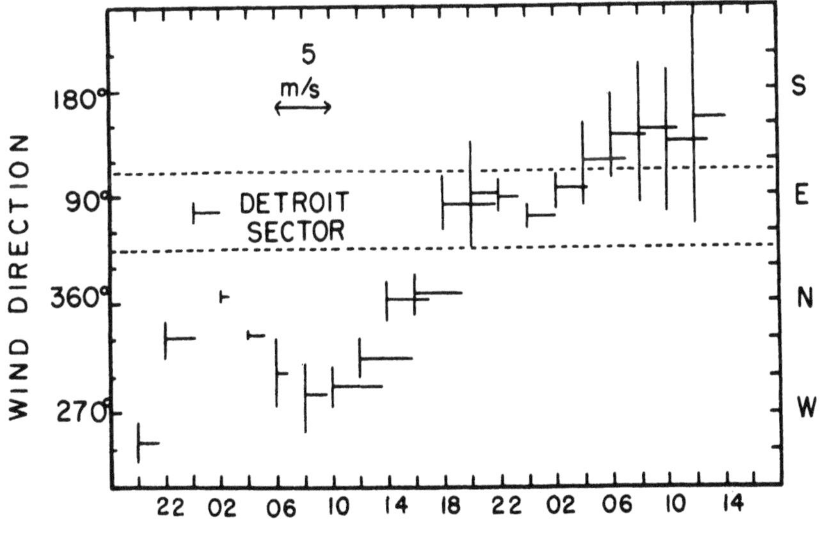

Figure 27. Wind speed and direction versus sample time.

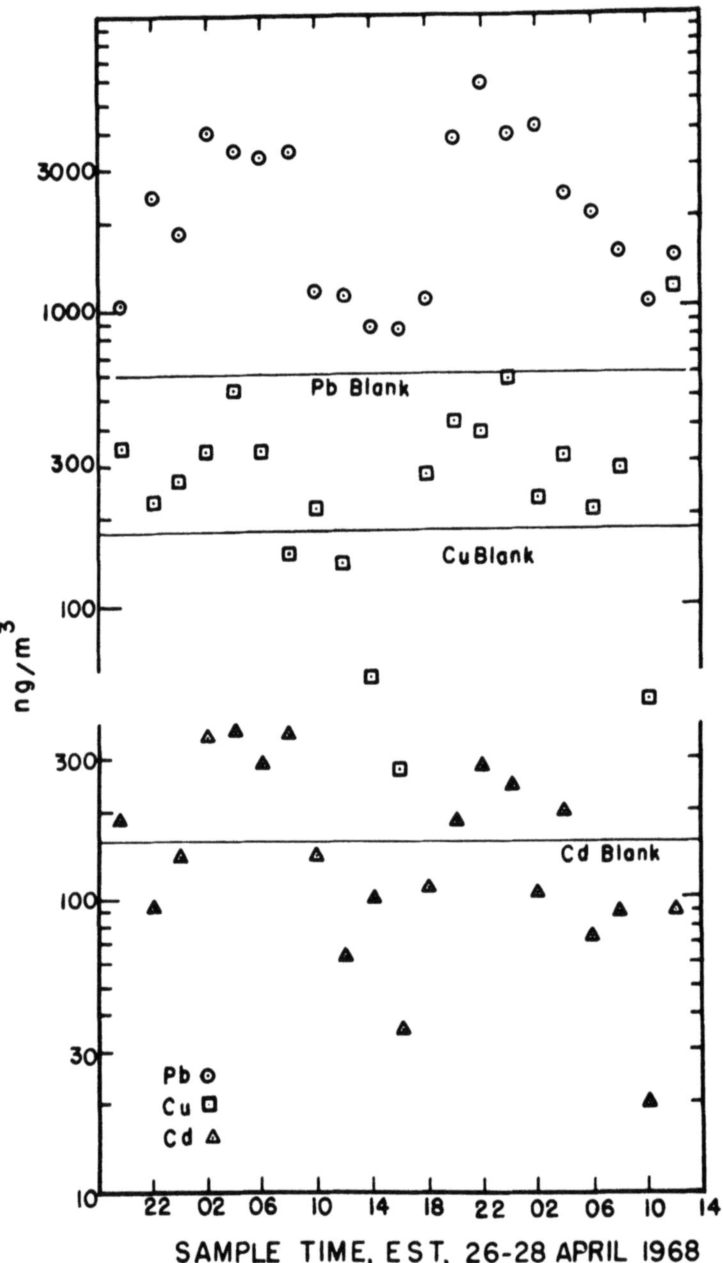

Figure 28. Total concentrations of Pb, Cd, and Cu versus time.

SURFACE DISTRIBUTION OF LEAD FROM AIRBORNE AEROSOLS

Advection and Fallout

We have established that airborne lead comes primarily from automobile exhaust systems. This lead is purposely encouraged to remain in small particles by chemical means. The tetraethyl fluid containing chlorine and bromine forms lead-chloro-bromide compounds upon combustion. Within the exhaust system, coagulation and attachment occur within the products of combustion materials. The resulting distribution is usually bimodal with those particles emitted directly from the exhaust being very small while those which are impacted or have attached themselves to the exhaust system and later emitted being larger in size. This bimodal distribution results in both a short range and extremely long range transport of these particles. About half of the particulate matter containing lead from automobile exhausts falls out from the air within a few hundred feet of the roadways in a matter of minutes. The remaining particles have a mean residence time of weeks to months increasing with decreasing size. Beyond several hundred feet from the roadway soil lead levels are not detectable from the background or soil content (1).

On a larger scale there is a logarithmic increase in atmospheric lead as measured from mid-ocean areas, high mountains, seashores, and then to suburban and urban environments in that order.

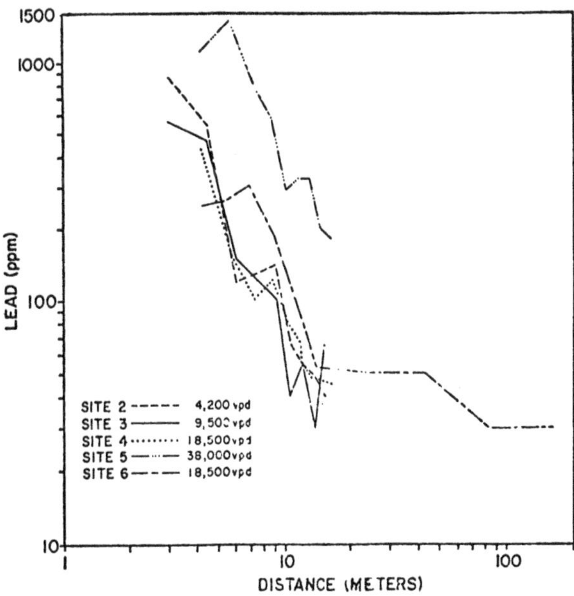

Figure 29. Surface soil lead concentration as a function of distance away from the highway.

Figure 29 is a representation of studies measuring the surface soild lead in ppm as a function of distance from the edge of the highway.* The results are seen to be nearly log-normal (5).

Referring back to Figure 25 through 28, we can estimate that significant amounts of lead are advected for decades of miles according to the source region and wind direction.

As the small lead particles "mature," they form conglomerates with other particles and either "fall out" or are precipitated out as rain or snow. Table 23 represents an analysis of five dustfall

Table 23. Dustfall for August, 1969 (mg/m^2) - month station

	Station					Soil	Grass
	6	7	8	9	10	(9)	Roots (9)
Pb							
Insoluble	5.2	29.	14.	16.	5.6		
Soluble	0.94	0.36	0.34	0.26	0.68		
Total	6.1	29.36	14.34	16.26	6.3	0.7 mg/gm	0.4 mg/gm
Per Cent	2.6	7.2	7.6	6.8	3.5		
Cu							
Insoluble	0.76	4.1	3.7	2.0	0.31		
Soluble	0.24	0.20	0.21	0.10	0.16		
Total	1.0	4.3	3.9	2.1	0.47	0.16 mg/gm	0.58 mg/gm
Per Cent	0.43	1.1	2.1	0.89	0.26		

stations located in East Chicago, Indiana (15). They are presented as insoluble and soluble fractions. It is important to note that most of the lead is insoluble. This important fact will be briefly discussed later. This was an exceptionally low dustfall month for the area. From this information we can estimate deposition rates of 40 to 300 ng per square meter per minute for copper and lead, respectively.

Figure 30 is the plot of the wind rose for that month. Noting the prevailing winds are from the southwest, we can see from Figures 31 and 32 that the surface isopleths of these metals line up with the prevailing wind. It should be noted that most of this area does not contain high densities of automobile traffic but contains large steel and petroleum industries. Most of the residential and automobile traffic is to the southwest. Previously presented information

*See also pp. 287-288, Editor.

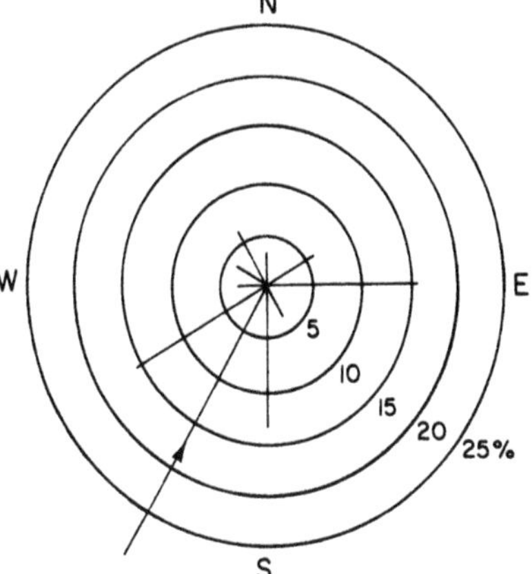

Figure 30. Wind rose for August, 1969.

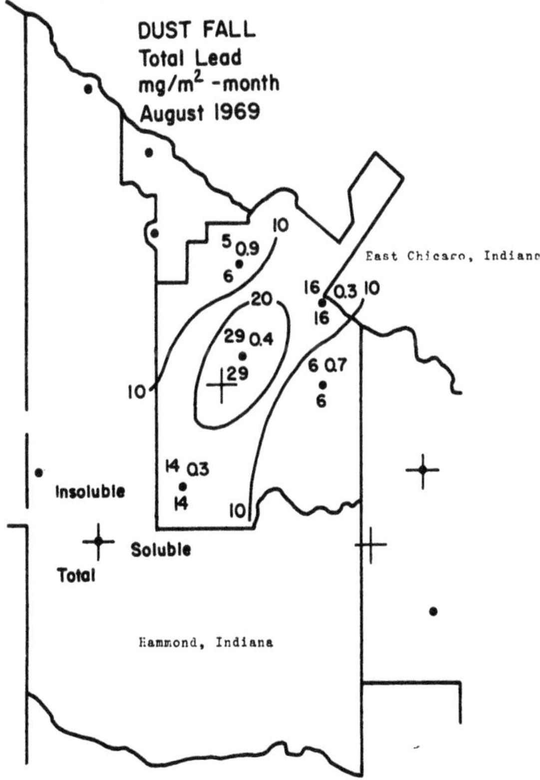

Figure 31. Lead fraction of total dustfall.

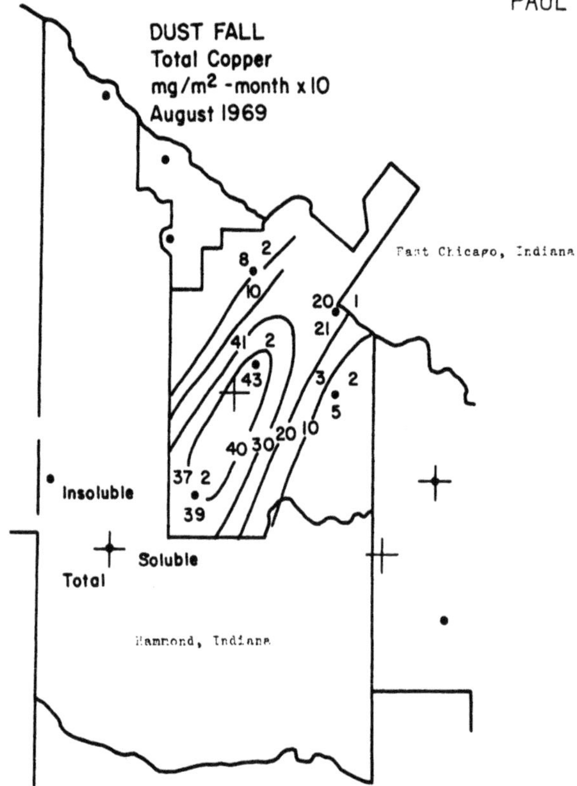

Figure 32. Copper fraction of total dustfall.

concerning potential copper sources leads us to believe that a very high source of copper particulate exists in the far southwest corner of the city which results in these high values of copper particulate fallout.

Lead then is a serious contaminant close to roadways as seen by elevated concentrations as compared to that of natural soils. At the same time, the small size distributions of the lead particles result in a virtual worldwide dispersion of lead aerosols from automobile exhausts. As mentioned previously, in the arctic regions, both north and south, we see measurable traces of lead in the snow covers. From these "frozen histories" one can measure the increased automotive activity within each hemisphere in a chronological manner. These ice records show that lead concentrations vary from 0.0005 micrograms per kilogram of ice at eight centuries B.C. to 0.2 micrograms per kilogram in 1965 (16). Since 1940, the levels have increased by a factor of 4. These are obviously accelerating rates.

In urban areas the saturation point seems to have been reached since the lead levels have not substantially changed in the last fifteen years. This is apparently due to petroleum industries using as much lead as possible for the high grade gasolines, and the urban areas having reached a near saturation point in the number of automobiles they can accommodate. As previously mentioned, lead painting accounts for two or three times the amound of lead as compared to the automobile but is bound up in the paints, whereas lead from automobilies is readily available either by direct emission or by refloatation of the street dusts and dirt.

PRECIPITATION PROCESSES

There have been recent discussions concerning the possibility of lead and lead-halide (in particular, iodine) as a condensation nuclei for snow. Recent investigations by a group at Colorado State University have shown that the effect of lead is minor as compared with other materials in their ice nucleation capabilities (5). Even at -22°C the effect of other automotive exhaust constituents is superior in forming snow crystals as compared to lead particles, both in the presence and absence of iodine. Assuming these experiments are valid, we can safely say that atmospheric lead is comparatively insoluble and does not seem to significantly enter into the precipitation process.

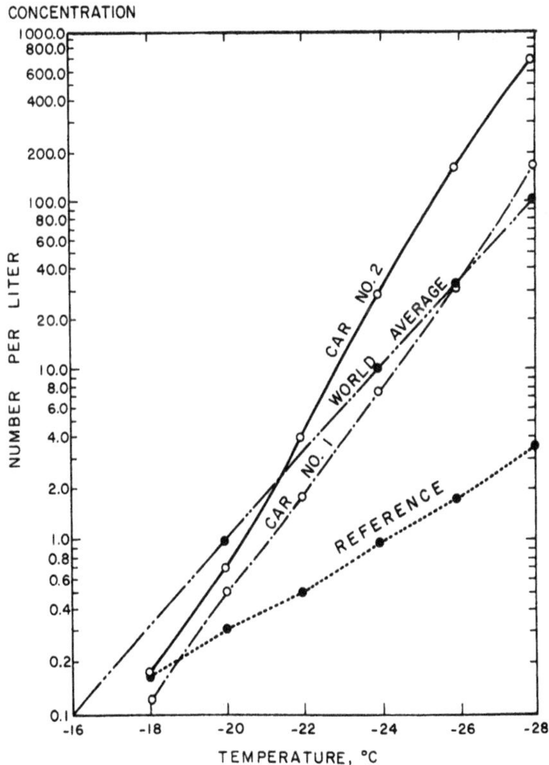

Figure 33. Ice nuclei activation spectra for auto exhaust from two cars. Fort Collins, Colorado, December, 1970.

Figure 33 shows that the average activation spectrum of aerosols from auto exhausts are approximately log-normally distributed (5). It should be pointed out that rain and snow do contain lead particles even though they may not be the prime nucleator. Table 24 is the

Table 24. Rainfall - June 26, 1968.

	ng/10 ml			ng/total rain		
	Cd	Pb	Cu	Cd	Pb	Cu
South Chicago	1932	453	437	77300	18200	17500
Gary - East	430	245	387	30100	17200	27100
LaPorte	246	209	109	20900	17800	9300

result of a rainfall catch of a frontal storm passing over Chicago from the northwest to southeast through northwest Indiana and Gary, Indiana (15). It can be seen that the lead levels expressed in nanograms per total washout, obtained by multiplying the number of liters of rain caught by the concentration per liter, stayed comparatively constant. Cadmium and copper decreased as distance away from the source area increased.

The lack of solubility of lead in the atmospheric aerosols suggests that the pollution of our major waterways occurs either by mechanical surface runoff or direct fallout into the large bodies of water. Table 25 is an estimation by Winchester and Nifong of the fallout of three metals into Lake Michigan as compared to concentrations of these elements in the river inflow. A 10% fallout from the regional emission inventory was assumed (17).

Table 25. Comparison of air pollution inventory with river inflow of trace elements to Lake Michigan, metric units.

	Cu	Ni	Zn
Air pollution emissions inventory, tons/yr.	3200	1000	3900
Lake Michigan rivers approx. mean conc., 1963-64, μg/liter	90	40	30
Inflow to Lake Michigan, tons/yr.*	2700	760	∿500

*Estimated for total Lake Michigan drainage basin from contents and flow rates of individual rivers.

THE IMPORTANCE OF REFLOATATED AEROSOLS IN MECHANICAL RUNOFF IN URBAN AREAS

Soil and street dusts in cities contain large amounts of lead. The gradient of lead levels in the total environment decreases rapidly away from urban areas, but within the urban areas the concentrations are relatively high and uniform except around the most heavily traveled streets and expressways. This blending occurs from the wind movements of the street dust and particles throughout these urban areas. Table 26 presents data concerning the constituents of street dirts as taken from various cities throughout the

Table 26. Major constituents of street surface contaminants.

Measured Constituents	Weighted Means for All Samples (lb/curb mile)
Total Solids	1400
Oxygen Demand	
BOD_5	13.5
COD	95
Volatile solids	100
Algal Nutrients	
Phosphates	1.1
Nitrates	0.094
Kjeldahl nitrogen	2.2
Bacteriological	
Total coliforms (org/curb mile)	99×10^9
Fecal coliforms (org/curb mile)	5.6×10^9
Heavy Metals	
Zinc	0.65
Copper	0.20
Lead	0.57
Nickel	0.05
Mercury	0.073
Chromium	0.11
Pesticides	
p,p-DDD	67×10^{-6}
p,p-DDT	61×10^{-6}
Dieldrin	24×10^{-6}
Polychlorinated biphenyls	1100×10^{-6}

United States. It is seen that the heavy metals do not constitute a large percentage of the total weight, although they do add up to around 1.6 pounds per curb mile, lead and zinc being the primary heavy metals available. A further breakdown is presented in Table 27 for the inventory for each city and location studied (18).

Table 27. Heavy metals loading intensities (lb/curb mile)

	Zinc	Copper	Lead	Nickel	Mercury	Chromium
San Jose-I	1.4	0.49	1.85	0.19	0.20	0.10
San Jose-II	0.28	0.020	0.90	0.085	0.085	0.14
Phoenix-II	0.36	0.058	0.12	0.038	0.022	0.029
Milwaukee	2.1	0.59	1.51	0.032	---	0.047
Baltimore	1.3	0.33	0.47	0.077	---	0.45
Atlanta	0.11	0.066	0.077	0.021	0.023	0.011
Tulsa	0.062	0.032	0.030	0.011	0.019	0.0033
Seattle	0.37	0.075	0.50	0.028	0.034	0.081
Arithmetic Means	0.75	0.21	0.68	0.060	0.080	0.12

We are concerned mainly with atmospheric aerosols in this discussion, but we must point out that the street dirts do represent a heavy burden on the water treatment plants and streams in the areas.

Table 28 is a presentation of the size distribution of these particles by element. The data are presented as percentage by weight in each distribution. Particles less than 100 microns can

Table 28. Fraction of heavy metals associated with each particle size range (% by weight)

	Particle Size (μ)				
	2,000	840 → 2,000	246 → 840	104 → 246	<104
Chromium	26.1	13.6	16.3	16.3	27.7
Copper	22.5	20.0	16.5	19.0	22.0
Zinc	4.9	25.9	16.0	26.6	26.6
Nickel	26.2	14.2	15.3	17.2	27.1
Mercury	16.4	28.8	16.4	19.2	19.2
Lead	1.7	2.6	8.7	42.5	44.5
Average	16.3	17.5	14.9	23.5	27.8

be easily refloatated and thus become available for respiration either in the upper or lower parts of the pulmonary tract. Approximately 28% of these heavy metal particles are less than 100 microns. Again, it is significant that approximately half of the lead particles (or particles with lead association) are less than 100 microns. We can infer that many of these are even less than 50 microns which can easily be blown about and refloatated to reasonable "peopled" heights.

Recent data from the city of Chicago particulate monitoring network has shown that, indeed, a large part of the particulate "seen" at street level are of refloatated origin. In fact, more than 50% in most instances (19).

Thus, street dusts are also a significant source of air pollutants.

EFFECTS OF LEAD ON THE BIOSPHERE

The most pertinent and reoccurring conclusion coming from recent studies into lead is that the effect of airborne lead on most organisms are relatively small, at least in the short term.

Except in cities, inhaled lead amounts to about one-half the amount ingested from the diet. When atmospheric levels increase above 2 to 3 micrograms per cubic meter, the blood lead levels start to increase. There seems to be sinergistic effects linked with other stresses such as iron deficiency anemia and elevated carbon monoxide levels. The lead levels in blood and tissue increase exponentially when the airborne lead concentrations increase beyond 2 to 3 micrograms per cubic meter (16). The "Threshold Limit Value" (TLV) suggested by the National Institute for Occupational Safety and Health is 0.15 micrograms per cubic meter, an 8-hour occupational standard (20).

Chronic signs of lead poisoning appear at blood lead concentrations around 80 micrograms per 100 grams of whole blood. Inhibition of the formation of red blood cells takes place below these levels. Usually these values are not nearly reached from exposure to airborne lead. In most cases, these blood lead levels are caused by ingestion of lead bearing foods, paint chips, or other sources of accidental intake.

A positive note is seen in that cases of lead ingestion by children are slowly decreasing, and so are industrial cases of lead poisoning.

The most significant thing about atmospheric lead aerosols is their small size distribution. Since particles smaller than 10μ can be inhaled in the respiratory track, it can be shown that although the amount of lead ingested in the diet is nearly twice that acquired by inhalation, the percentage of the lead absorbed through inhalation is greater by a factor of nearly three in comparison to that ingested. Thus, lead is contributed as much to the total body burden by inhalation as by ingestion. Our conclusion is that this short term, high concentration effect of exposures to lead are well known, but more needs to be known about the long term chronic exposures. We especially need to determine the chronic effect of atmospheric lead aerosols on young children. We do know that as one gets older their total body burden of lead increases; however, the blood lead does not. Although the lead ingestion by children is slowly decreasing, we must carefully assess this effect to make sure that those most stressed individuals in these older and usually failing neighborhoods are not assaulted by still one more unnecessary stress.

EFFECTS ON VEGETATION, BACTERIA, AND ANIMALS

The small effects on vegetation are detected primarily close to roadways having high traffic densities. As mentioned before, the levels become indistinguishable from the background soil concentration several hundred feet from the roadway. There is no evidence that atmospheric lead levels affect livestock or wild life. Atmospheric lead does not appear to be a hazard to pets or aquatic animals or plants. The primary physical or chemical reason we can site for this general lack of effects is that lead as found in soil and water is generally unavailable to plants due to its insolubility. Figures 34 and 35 are schematic diagrams of the atmospheric parameters and their influence on lead transport through the environment (1, 21). Figure 35 is an estimation of the pathways and amounts through various stages of the suggested models (gms/month/sq mi) as suggested by the University of Illinois Researchers. Figures 36 and 37 are similar schematics of the suggested physical and conceptual models in humans and animals. This model is taken from that suggested by the Illinois group, except a dash line has been added between the respiratory track and the gastro-intestinal track. We must insist on this addition since the upper respiratory track acts as an initial impactor of the larger particles (i.e., greater than 1 to 10 microns). Most people will swallow about 3,000 times daily which will result in ingestion of mucous and phlegm from their sinuses and respiratory track. Their ingestion rate of lead is thus increased in addition to the retention in the lower respiratory track of smaller particles.

Finally, Table 29 is a reproduction of a table showing the relationship of adult blood levels to airborne lead exposures (22). The exponential increase is obvious. Currently most researchers

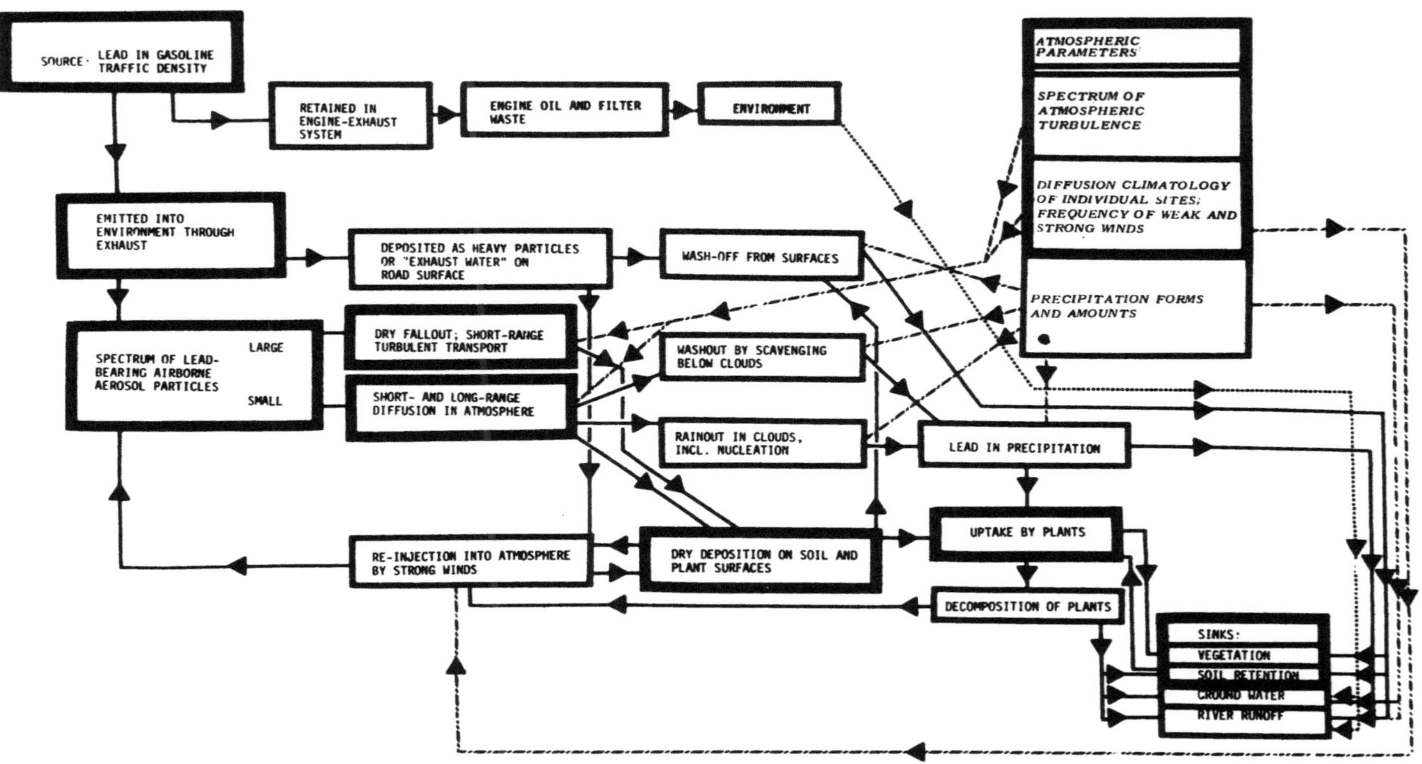

Figure 34. Schematic diagram of the influence of atmospheric parameters on the lead transport through the environment.

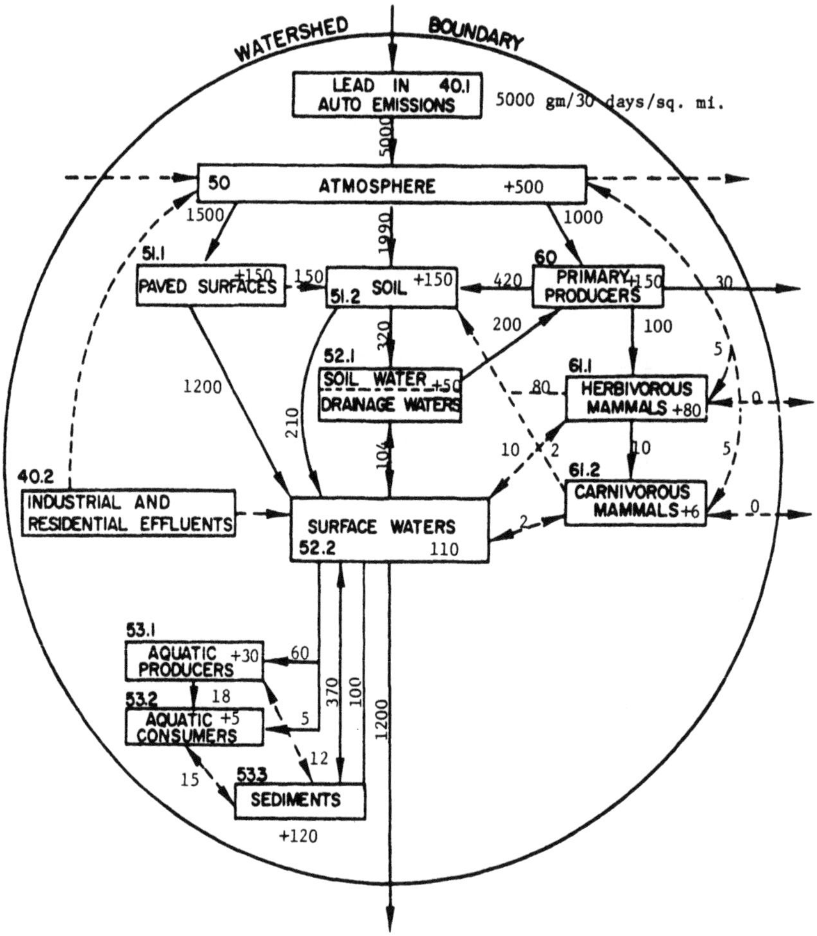

WATERSHED BOUNDARY
LEAD IN AUTO EMISSIONS 40.1
5000 gm/30 days/sq. mi.
5000
50 ATMOSPHERE +500
1500
1990
1000
51.1 PAVED SURFACES +150
150
51.2 SOIL +150
420
60 PRIMARY PRODUCERS +150
30
200
100
320
52.1 SOIL WATER DRAINAGE WATERS +50
1200
210
80
61.1 HERBIVOROUS MAMMALS +80
10
5
104
10
2
10
5
61.2 CARNIVOROUS MAMMALS +6
0
40.2 INDUSTRIAL AND RESIDENTIAL EFFLUENTS
SURFACE WATERS 52.2 110
2
53.1 AQUATIC PRODUCERS +30
60
53.2 AQUATIC CONSUMERS +5
18
5
370
100
1200
12
15
53.3 SEDIMENTS
+120

Figure 35. Basic system model.

Air Inhaled
Ingestion
Air Exhaled
RESP. TRACT
G. I. TRACT
FECES
Excretion
BLOOD
LIVER
BONE
MUSCLE
SPLEEN
KIDNEY
URINE
Excretion

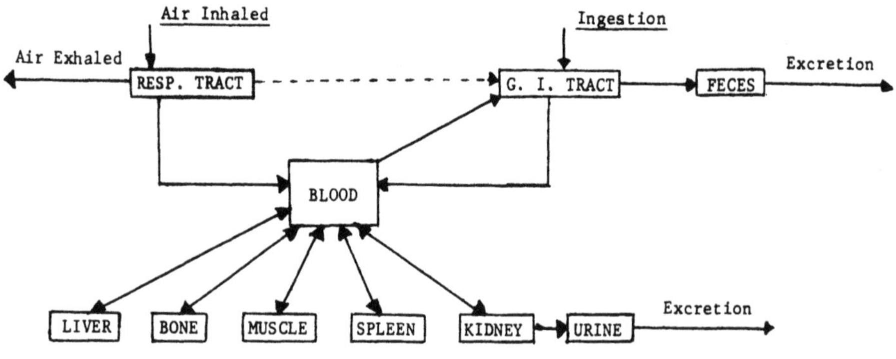

Figure 36. Physical flow.

Table 29. Relationship of adult blood lead levels to airborne lead exposure.

Air Lead Exposure	Daily Lead Absorption µg/day			Expected Blood Lead[3]	
µg/m^3	Air[1]	Diet[2]	Total	µg/100g	Relative Excess in Blood Lead (%)[4]
2.0	13.8	30	43.8	21.3	0%
2.5	17.3	30	47.3	22.8	7%
3.0	20.7	30	50.7	26.3	15%
3.5	24.2	30	54.2	25.8	23%
4.0	27.6	30	57.6	27.3	30%
4.5	31.1	30	61.1	28.8	38%
5.0	34.5	30	64.5	30.3	47%
10.0	69.0	30	99.0	40.0	137%
20.0	138.0	30	168.0	53.8	284%
50.0	345.0	30	375.0	71.6	780%
100.0	690	30	720.0	87.2	1550%

[1] Assumes inhalation of 23 m^3/day and 30% lung retention.

[2] Assumes 10% gastrointestinal absorption of the average adult daily total dietary intake (300 µg) from food and water.

[3] Computed from regression formula: Blood lead = -69.2052 + 54.7605 × log µg Pb absorbed daily as given in Chapter 3 of Airborne Lead in Perspective by the National Research Council, National Academy of Sciences.

[4] Relative excess in blood lead is associated with ambient air exposure above 2.0 µg Pb/m^3.

Reproduction of a table in the April 11, 1972 Revision of "Health Hazards of Lead" (Environmental Protection Agency, 1972b).

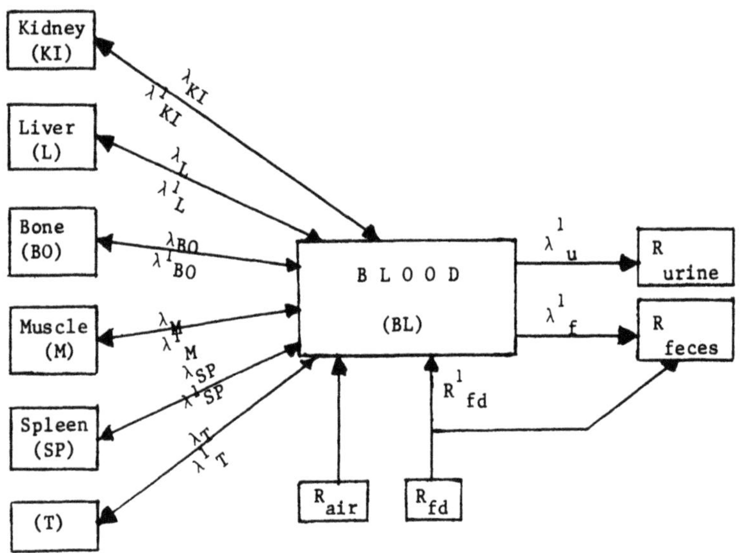

Figure 37. Conceptual model.

are suggesting an ambient air standard of 2 micrograms per cubic meter averaged arithmetically over a year.

CONCLUSIONS

The above discourse has carried us through inventories of natural, industrial, and human produced lead aerosols as well as those of some trace metals. Lead has been the "star of the show" for several reasons: It is widespread, it is concentrated "in peopled" areas in sufficient quantities, it serves no biologically useful purpose, and has caused death when ingested in large quantities. The primary source of atmospheric lead is additives to gasoline.

The only arguments for use of lead in gasoline are that unleaded gasolines are more expensive and replacement by more aromatic compounds may have greater adverse health effects creating a greater hazard. Other arguments concerning the potential inconvenience to the automobile engine and control devices are rather perfunctory compared to that of health considerations. Many investigators imply that there are no known adverse health effects due to atmospheric lead aerosols which is the crux of the problem of skipping priorities. The arguments against the use of lead are that there are human health hazards, and it serves no biologically useful function. It is also possible that there are disruptions in the ecological development of certain organisms due to these lead aerosols. The argument that lead aerosols cause weather modification has yet to be validated although they may still exist.

There is a potential threat in that bacteria can take up substantial quantities of lead with no apparent deleterious effects on cell growth rate and viability. The implication that lead could be transferred through the food chain in an increasing concentrated manner is still a possibility.

A final item that must also be included is the fact that certain compounds of some of these trace metals are many times more deleterious to the ecology (and man) than that in the pure state. Mercury presents an excellent example. Methylmercury is several times more toxic than raw purified mercury. A recent writer in _Scientific American_ (23) suggested that with the right incentives, he could swallow up to a pint of pure mercury and survive. Orders of magnitude less amounts of methylmercury would certainly be fatal. Currently in the Chicago area, the average 24-hour exposure to atmospheric mercury is around 28 nanograms per cubic meter, both as particulate and as vapor, representing no (known) health hazard (24).

According to the exposure, trace metals are potentially life saving as well as toxic. It becomes obvious that further investigations as to the effects of these trace elements on organisms (man, animal, and plant) are mandatory if man is to "optimize" his environment. Extensive research on lead indicates that the ecology is being assaulted unnecessarily by large amounts of lead as aerosols; however, no adverse health effects seem to have been detected in adults by exposure in this form. The effect on young children remains a question.

ACKNOWLEDGEMENTS

The author expresses sincere appreciation to Dr. H. W. Edwards, Engineering Research Center, Colorado State University; Dr. K. A. Rahn, Department of Meteorology and Oceanography, University of Michigan; and Dr. K. J. Yost, Bionucleonics Department, Purdue University, whose comments and suggestions contributed significantly to the development of this work.

REFERENCES

1. Edwards, H. W. (Principal Investigator) (1971), Impact on Man of Environmental Contamination Caused by Lead, NSF Grant GI-4, Interim Report, Colorado State University, Fort Collins, Colordao, June.

2. United States Department of the Interior (1969), Bureau of Mines, Minerals Yearbook, Vols. I-II, Metals, Minerals, and Fuels, Washington, D. C.: U. S. Government Printing Office.

3. McKee, J. E. and Wolf, H. W. (1963), Water Quality Criteria, 2nd ed., Publication 3-A, April, 1971, The Resources Agency of California.

4. Lutz, G. A., Levin, A. A., Bloom, S. G., Nielsen, K. L., Cross, J. L., and Morrison, D. L. (1970), Lead Model Case Study, Final Report, Vol. III, U. S. Department HEW, Contract #CPS-69-005, Columbus, Ohio: Battelle Memorial Institute, June.

5. Edwards, H. W. (Principal Investigator) (1972), Impact on Man of Environmental Contamination Caused by Lead, NSF Grant IRPOS GI-4, RANN GI-34813X, Interim Report, Fort Collins, Colorado: Colorado State University, June.

6. Murozumi, M., Chow, T. J., and Patterson, C. (1969), Geochim. et Cosmochim. Acta, 33, 1247-1294.

7. National Academy of Sciences, Committee on Biological Effects of Atmospheric Pollutants (1972), LEAD: Airborne Lead in Perspective, Washington, D. C.: National Academy of Sciences, 320 pp.

8. Cantwell, E. N., Jacobs, E. S., Kunz, Jr., W. G., and Liberi, V. E., SAE Paper No. 720672, Presented at National Meeting of the Soc. Automotive Engineers, May 26, 1972.

9. Harrison, P. R., Matson, W. R., and Winchester, J. W. (1971), Atmospheric Environment, 5, 613-619.

10. Rahn, K. A. (1971), Sources of Trace Elements in Aerosols--An Approach to Clean Air, U. S. AEC (Argonne) Contract No. AT(11-1) -1705, The University of Michigan, May.

11. Fleishcher, M. (1973), Natural Sources of Some Trace Elements in the Environment, Cycling and Control of Metals, Proceedings of Environmental Resources Conference, Columbus, Ohio, Oct. 31- Nov. 2, 1972; Cincinnati, Ohio: National Environmental Research Center, Feb.

12. Harrison, P. R. and Winchester, J. W. (1971), Atmospheric Environment, 5, 863-880.

13. Unpublished data produced from a cooperative program between Purdue University and the Department of Environmental Control, City of Chicago, NSF-RANN Program No. GI-35106, 1973.

14. Spittler, T. M., Tenny, K. S., Kowalski, T. A., and Harrison, P. R. (1972), The First Chicago Spectroscopy Forum, Paper #35, June.

15. Harrison, P. R. (1970), Ph.D. Thesis, The University of Michigan.

16. National Academy of Sciences, Committee on Biological Effects of Atmospheric Pollutants (1972), LEAD: Airborne Lead in Perspective, Washington, D. C.: National Academy of Sciences, 320 pp.

17. Winchester, J. W. and Nifong, G. D. (1971), Water, Air, and Soil Pollution, 1, 50-64.

18. Sartor, J. D. and Boyd, G. B. (1972), Water Pollution Aspects of Street Surface Contaminants, U. S. Environmental Protection Agency, Contract No. 14-12-921, Project 11034 FUJ, EPA-R2-72-081, Washington, D. C.: U. S. EPA, Nov.

19. Harrison, P. R. (1972), Paper #73-161, Presented at the 65th Annual Meeting of the Air Pollution Control Association, June 18-22, Miami Beach, Florida.

20. American Conference of Governmental Industrial Hygienists (1970), Threshold Limit Values of Airborne Contaminants and Intended Changes, Cincinnati, Ohio.

21. Environmental Studies Program, An Interdisciplinary Study of Environmental Pollution by Lead and Other Metals, NSF IRPOS GI-26, RANN GI-31605, Progress Report from May 1, 1971 to April 30, 1972, Graduate College, University of Illinois at Urbana-Champaign.

22. Environmental Protection Agency (1972), Health Hazards of Lead, Revised April 11.

23. Goldwater, L. J. (1971), Scientific American, 224(5), May, pp. 15-21.

24. Wroblewski, S. C., Spittler, T. M., and Harrison, P. R. (1973), Paper #73-152, Presented at the 66th Annual Meeting of the Air Pollution Control Association, June 24-28, Chicago.

HEAVY METALS--TOXICITY AND ENVIRONMENTAL POLLUTION

JACK SCHUBERT

Department of Radiation Health
Graduate School of Public Health
University of Pittsburgh
Pittsburgh, Pennsylvania 15261

CONTENTS

I. INTRODUCTION

The screaming Japanese cats of Minamata, the dying birds of Sweden, and the staggering pigs of New Mexico should have taught mankind a fundamental fact of environmental biology. Unfortunately, the lesson to be constantly relearned from those mercury-poisoned animals is that all organisms--from man with 10^{13} cells to the mindless single cell--are programmed by the same basic substance, deoxyribonucleic acid (DNA). With such a shared relationship, it would be expected that the response of all organisms to chemicals would qualitatively be the same. Yet, it is curious that to this very day, a lethal or debilitating action of a chemical to animals which may not have been manifested or observed in man, is often disregarded or jocularly dismissed--usually by those who have an economic interest in continuing the use of the chemical in question.

It cannot be overemphasized, however, that evidence of biological harm to any living organism by a chemical must, or should, automatically lead to a careful and persistent examination of the toxicity to man--especially when a substantial fraction of the population is exposed.

An ubiquitous group of chemicals being released to the environment in ever-increasing amounts are the heavy metals. The practical and fundamental toxicological problems involved in the evaluation of their actual and potential toxicity to man are basically identical to those required to evaluate the actual and potential toxicity of any chemical, whether organic or inorganic. However, certain of these heavy metals, namely lead, has reached such high levels in the environment and posed such a serious health problem that, in my opinion (S13) and others (N1), its use in gasoline (Table I) should be <u>immediately</u> discontinued.

I will devote most attention to three heavy metals--lead, mercury, and cadmium--because of their widespread distribution and the fact that relative to their toxicity, they are present in the body in the highest concentration (Table II). As Schroeder

(S4) pointed out, the toxicity of nonessential or abnormal trace metals in the body is associated only with their excess and accumulation in tissues with age. The essential trace metals, however, are harmful when present in excess or are deficient.

In view of the environmental hazards associated with the abnormal metals, one must consider a variety of technical aspects, including tests to detect subtle, or apparently asymptomatic levels in the population. However, a purely technical solution does not suffice to protect the population. In other words, even if scientists solved all the technical and scientific problems involving environmental pollution, the genetic and toxicological risks to the population from environmental chemicals would not be significantly reduced. To achieve this goal it is also necessary to implement innovative and punitive legislative-administrative approaches (S13).

In view of the above, the following topics are included in this report on heavy metals:

1. Environmental and toxicological effects on the population and occupational groups.

2. Mutagenic and genetic actions.

3. Synergism, i.e., the biological effects of metals in combination with each other and with other agents, not necessarily inorganic, to which man is exposed.

4. Protocols for evaluating the toxicity and mutagenic effects.

5. Chelation therapy for metal poisoning and the use of conditional or effective constants for anticipating the *in vivo* behavior of metal chelates.

6. Provocative chelation for evaluating relative levels of heavy metals in the body and as a means of uncovering possible relationships between trace metals and disease.

7. Control, reduction, or elimination of heavy metal exposure to the population.

A. Definitions and Permissible Levels

The concept of "toxicity" has changed considerably during the past few decades because of the realization and experience that usually employed criteria such as pathology, weight changes, carcinogenicity, fertility, and hematology do not necessarily reflect more subtle effects, e.g., teratogenic (producing birth defects), mutagenic (producing genetic aberrations), and damage to the brain or nervous system. Further, experience has shown that "levels" of substances, e.g., Be (S7; S24, p. 36) presumably considered at present to be safe for an occupational group, are regarded not safe for the general population. However, the latter problem often reflects the size and health status of the population sampled and the nature and extent of medical surveillance. In other words, if the incidence of disease is, say, one in a thousand for a given level of exposure, then it would be difficult to find one case in a sample of 50 or 100 individuals.

In the 1972 edition of the list of toxic substances prepared by the National Institute for Occupational Safety and Health (C6, p. xiii), a toxic substance, relative to chemicals, exclusive of radioactive substances, is defined as "one that demonstrates the potential to induce cancer, tumors, or neoplastic effects in man or experimental animals; to induce a permanent transmissible change in the characteristics of an offspring from those of its human or experimental animal parents; to cause the production of physical defects in the developing human or experimental animal embryo; to produce irritation; to diminish mental alertness or motivation; or to endanger the life of man when he is exposed to the substance via the respiratory tract, skin, eye, mouth, or other routes in any quantity reported . . . ."

H. Druckrey (A1) proposed the term "genotoxic" to include lethal or transmissible hereditary changes in somatic as well as germinal cells.

Extensions and modifications of the above definitions are required when dealing with the general population, particularly for mixtures of hazardous substances, and for quantitative evaluation of risk-benefit.

In practice, industrial workers (as well as the general population) are exposed to a mixture of hazardous chemicals. Obviously, an individual should not be permitted to be exposed to the maximum level of a chemical independent of other potentially toxic substances. Hence, for air contaminants, for example, the equivalent exposure, E_m, for occupational exposure, should utilize the following equation (F1):

$$E_m = \frac{C_1}{L_1} + \frac{C_2}{L_2} + \ . \ . \ . + \frac{C_n}{L_n}$$

where C is the concentration of a particular contaminant and L is the exposure limit as given in Table IV. The value of E_m is not supposed to exceed unity.

The possible synergistic action of potentially hazardous

chemicals to the general population as well as the industrial population is rarely considered in experimental design and infrequently investigated. However, the genotoxic effects of mixtures of chemicals will have to be extensively studied in the coming decade.

Single or divided doses of a chemical which produce death in a "relatively" short but defined time may be referred to as a "lethal dose" or as "acute toxicity.

Other terms will be defined as they arise in the discussion. It must be emphasized, however, that definitions involve a matter of informed judgment and necessarily become modified with advances in biology and medicine. Definitions may even prove harmful if in their use one is led to overlook health hazards to the general population, especially those in which harm does not become evident, unfortunately, until many years have elapsed when the magnitude of the problem has become so great that it becomes readily apparent.

II. OVERT METAL TOXICITY AND ACCEPTABLE LEVELS

Doses of heavy metals which have proved fatal to man are summarized in Table III. The chemical and physical form of the metal actually determines the toxic dose, all other factors being equal. The methylmercury salt is far more toxic than the dimethylmercury compound. The lowest human daily intake known to produce toxic symptoms is about 0.3 mg per day (P1). There is little doubt that the daily consumption of food containing 5-20 ppm of Hg in the form of methylmercury will eventually prove lethal, or disabling, or congenitally harmful (L6, P2).

The numbers given in Table III must be considered as rough guides only because they depend on the individual including his state of health, age, diet, etc. In any event, the doses necessary to produce acute death as the endpoint are fairly high, actually higher than most people would assume.

Obviously, in setting "safe" levels for people, it is necessary to include safety factors which, in the most optimistic case, permit indefinite exposure without harm. This is the basis for the threshold limit values of metals tabulated for occupational groups given in Table IV (F1, I2).

A. Occupational Exposures--Recognition of Biological Effects

Recommendations for the safe handling of metals are promulgated by state and federal agencies. The recommendations include air levels, use of environmental control measures (ventilators, respirators, etc.), personnel protective equipment, and medical control.

For technical, administrative, and economic reasons, the degree of control attained in practice leaves much to be desired. For example, the Chicago urban area in 1968 had 14,453 plants which had been repeatedly surveyed for industrial health and hygiene. At that time, an extensive occupational health survey was conducted by sixteen industrial hygienists on loan from the Occupational Health Program of the United States Public Health Service. The majority of workers were exposed to potentially harmful levels of organic and inorganic chemical agents. In fact, in the majority of all plants, regardless of size, about two-thirds were reported to be poorly controlled (M5). These included thousands of workers exposed to unacceptably high levels of lead and metal fumes.

In the following overview, the nature and effectiveness of industrial health resources by government, medical, industry and labor are considered.

Government. The nature and extent of the impact of the micro-chemical environment from hundreds of thousands of industries in the United States has never been established for the industrial population nor for the general population.

a) The direction, interest, and concern of Federal Agencies, reflected throughout the states for many decades, has been on the micro-biological concept of disease causation rather than the micro-chemical etiological spectrum. This has created an extensive area of scientific unknowns, of the toxicological, carcinogenic, teratogenic, and mutagenic effects, of the thousands of industrial chemicals individually and in combination.

b) No effective thorough system of industrial, medical, and environmental surveillance exists for those employed in the work environments throughout the nation. The extent of the biological effects of the silent daily seeding of the industrial population, of these chemicals in varying doses over the years, cannot be detected with any reasonable degree of accuracy in the absence of adequate, competent, and continuous medical and environmental surveillance throughout each industry.

c) The adoption of the Federal Occupational Safety and Health Act, noble of purpose, but in reality negated by policy and operations, provides at present the illusion of environmental surveillance and protection. The laws on occupational health and safety--federal, state, and city--have been rendered ineffective in the past by a combination of forces which reflect the continued struggle of economic versus human values.

d) Nevertheless, the federal legislation has legally recognized the extent of the accumulation of occupational hazards over the years which require control. Eventually, this will lead to better identification and quantification of biological effects among the industrial workers and a broadening of the span of knowledge of the impact of specific chemicals.

Medical. Occupational diseases and related biological effects resulting from exposure to toxic substances in the work environment have never been adequately recognized or reported in the United States.

a) A basic reason has been the lack of training in industrial medicine for the majority of physicians who serve the industries throughout the country, particularly the small plants. Both the small and large plants recruit physicians from general practice to serve on part- or full-time basis and provide little or no subsequent orientation in industrial medicine or on the hazards of the work environment.

b) The situation is compounded for the part-time physician busy with physical examinations, on a specified time schedule, who does not study the actual work environment of the particular manufacturing processes and occupations and, therefore, is in no position to identify nor recognize the relationship of symptoms or illness of the employee to the work environment.

c) Workers in industry, in turn, unfortunately do not know the specific chemicals and compounds to which they have been exposed and are ignorant of the toxic or injurious effects of these chemicals. Consequently, the worker cannot convey this information to the physician, nor can the worker relate his health problem to the work environment, particularly if it is a delayed biological effect.

Industry. Factors which influence industries' decision processes have been predicated in varying degrees on socio-economic rather than human considerations.

a) The continuation through the decades of the concept of expendability of human life and impairment of various organs of the body in exchange for a "job," the economic lever equated to man's timeless and innate need to support and raise a family.

b) The general lack of recognition of social accountability as an integral and responsible function of management in the utilization of human and community resources of air, water, and land. (Profit must respect and not violate the rights of the community and its resources.)

c) Unfortunately, the concept prevails that management's responsibility relates solely to immediate effects within their executive time span rather than concern with the long term human and social consequences of their actions.

d) No adequate monitoring system prevails in the respective industries to determine what is happening, what are the precise biological effects on the employees exposed to combinations of different chemicals and environmental stresses, in different plants, engaged in the manufacture of different products, in different geographical areas of the country, and, in particular, for those employees who have separated from the company and developed adverse health effects many years later.

e) Workers in small plants may be particularly vulnerable to industrial health hazards because of financial costs for control measures. Nevertheless, equally hazardous conditions exist in large industries which can readily afford necessary corrective measures for airborne contaminants.

Labor. In general, industry and labor have been primarily concerned with safety and neglected the micro-chemical environment and the impact of these stresses on the industrial worker.

a) The international unions have never properly recognized the inadequacies and limitations of their own resources and, therefore, of their responsibilities for the identification of occupational health problems. They had placed their reliance on ineffective governmental resources.

b) Improvement in collective bargaining contracts on health and safety, now in focus by labor organizations to provide necessary information on occupational hazards and the participation in their control, has been restrained by industry for reasons of compensation, potential labor-management disputes, and legal liability.

c) Labor leadership, arising from the ranks over decades, is uninformed as to the technical aspects, the significance and extent of potential risk to the workers of the hundreds of airborne contaminants; in essence, the toxicological, carcinogenic, and other biological effects. This is reflected in the acceptance of "hazard pay" in lieu of pressure for control measures.

d) The economic factors of depression and inflation act in cyclic pressures to remind the workers of the economic tribute for their labor and mutes the strength of motivation for control measures of unwarranted occupational risks.

B. Population Exposures

When the public is exposed to potentially harmful chemicals, far more critical and stringent criteria must be applied to minimize biological damage. The acceptable levels, unlike that in a small and medically prescreened working population, must be based on the most susceptible part of the population if they constitute a reasonably large percentage. For many reasons, the acceptable levels to which the public may be exposed are set at 10-100X less than those set for occupational groups.

Generally, there are two groups of the population who respond most sensitively to toxic chemicals--the fetus and the very old. It is difficult and sometimes practically impossible to detect other than acute effects in a small population, i.e., if the carcinogenic effect of a metal at a given level has an incidence of one in a thousand, with a latent period of ten to twenty years, it can and usually is overlooked, especially in small occupational groups. Hence, the oft-repeated cliche that people have been exposed to a given chemical for decades with no observed harmful effects. Such a statement is meaningless without follow-up observations, sophisticated diagnostic and laboratory examinations, and suitable statistical evaluations.

A good example of the relative consequences of occupational and non-occupational exposure to a metal is illustrated by beryllium (S7, S24). The acceptable occupational level of Be of $2 \mu g/m^3$ (Table IV) was first adopted in 1955, and adherence to this limit has been cited not to have produced any cases of beryllium poisoning as manifested by berylliosis--a chronic disease involving the lungs which occurs years after exposure and characterized by shortness of breath, lowered vitality, weight loss, and reduction in respiratory capacity. About a third of those afflicted die as a direct result of the disease. However, the reported absence of any such cases, under these concentrations, must be held with considerable reservation in view of the inadequate medical surveillance of workers exposed to beryllium in different occupations throughout the country and lack of necessary continuous environmental monitoring.

In 1948, people living within two miles of commercial beryllium processing facilities were reported to have developed berylliosis. By 1949, eleven cases were identified (S24, pp. 35-37). It appeared that as little as $0.1 \mu g/m^3$ of Be was sufficient to produce berylliosis. The beryllium processing plants discharged about 5 kg/day of the metal through the stacks, roof exhaust outlets, and windows--a sad commentary on the working conditions.

The concept that the occupational limit of 2μg/m^3 has not yet produced a significant number of cases of berylliosis and, therefore, has been considered safe is subject to challenge. In fact, the distribution of beryllium products to consumers in small plants since the 1950's without adequate monitoring of beryllium or appropriate medical surveillance raises serious questions as to what has not been recognized. The reputed harmlessness may be due simply to a relatively low incidence of the disease since it takes years to be manifested and that employees may have moved from that community and developed the disease elsewhere which was not detected. More frequently, the employees' death of heart disease and berylliosis was not recorded (M2). If the incidence of the chronic disease is about 0.3% (S24, p. 36) with a latent period of a decade or more, then the probability is too low that it would be detected in a relatively small working population with a finite turnover. In other words, I question the safety of 2μg/m^3 of Be for occupational groups. The established limit of 2μg/m^3 appears safe because of the long latent period, inadequate observation period, incomplete diagnostic information, incorrect reporting, and inadequate epidemiological studies (M2).

III. GENETIC AND MUTAGENIC CONSIDERATIONS

Changes in the genes can lead to an increase in hereditary diseases if the germ cells are involved, or to cancer and congenital malformations if somatic cells are affected. The population is already heavily afflicted with what are called genetic diseases, that is, those with a partial or complete genetic component. According to the geneticist Curt Stern (S23), genetic diseases account for at least 25 per cent of hospital and institutional beds occupied by patients with physical or mental illnesses or defects. Obviously, it would be prudent to avoid an increase in genetic diseases induced by exposure to unnecessarily high levels of chemical agents.

As with other chemicals, metals are potentially capable of teratogenesis and of mutagenesis resulting in visible chromosomal aberrations or invisible chromosomal aberrations (point mutations). An example of a consequence of chromosomal aberrations is shown by the fact that at least 20 per cent of aborted fetuses have chromosome abnormalities. Most of these die *in utero* since only about 0.5 per cent of liveborn infants have chromosomal abnormalities (C3).

The metabolism of essential trace metals is influenced by genetic factors. For example, one inherited metabolic disorder is hemochromatosis, an iron storage disease in which massive amounts of iron are deposited in the tissues because of increased gastrointestinal absorption (P3). In Wilson's disease excessive amounts of copper are deposited in the tissues and results in cirrhosis of the liver and degenerative changes in the brain. Wilson's disease is also inherited in a recessive manner as shown by the high incidence in the offspring of first cousin marriages (B1).

Humans whose red blood cells are deficient in glucose-6-phosphate dehydrogenase have significantly higher concentrations of lead in their red blood cells and lower serum lead than the corresponding nondeficient subjects (M6). Evidence also exists for a genetically induced abnormality in the absorption of manganese in the gut of mice (C9).

A. Mutagenicity Tests

A battery of mutagenicity tests are available which can supplement the usual toxological tests. While no one test result by itself can be considered conclusive, the results from a series of tests can be used togehter with the toxological data for evaluation of potential toxicity to man. Further, if these tests are performed before the long-term toxicity tests, it may eliminate the time and expense for the latter (S13). Naturally, more credence is given to _in vivo_ than to _in vitro_ tests when extrapolation of the findings to man are made.

Three recommended mutagenicity tests now being utilized by several government and private institutions are (S13):

Dominant lethal: Male animals are treated with a given chemical and then paired with suitable females. After different periods, the mated females are killed and examined for the number of living embryos and dead implantations.

Cytogenetics: The chromosomes in ciriculating blood cells and germ cells are examined. Viable cells which show chromosomal aberrations capable of transmitting mutations to the progeny are very important. Cells with multiple breaks are usually non-viable and, hence, do not constitute a genetic threat to future progeny.

Host-mediated assay: An indicator microorganism is incorporated within a mammalian host. Direct administration of the compound to the host allows the animal to activate or detoxify the potential mutagen before it encounters the microorganism in the peritoneal cavity. Afterwards, the microorganisms are removed and tested for mutant frequency. The test, while flexible and applicable to a wide range of animal species and microbial indicators, is an indirect test for the detection of point mutations.

B. Mutagenic Effects of Metals

Metals are mutagenic in a variety of *in vitro* test systems (Table V). Several scattered reports exist in the literature on the *in vivo* effects of metals in humans and other mammals. More clastogenic (chromosome-breaking) investigations are needed, especially with mixtures of metals.

Chromosomal studies were made on nine individuals who had consumed fish contaminated with methyl mercury three times a week for more than five years (S19). The fish consumed had mercury levels of 1 to 7 ppm. Chromosomal abnormalities were observed in the lymphocytes grown *in vitro*. The incidence of abnormalities observed was independent of the time the cell cultures were incubated but were correlated with mercury concentration (S19). Reviews of the clastogenic effects of mercury compounds in non-human test systems are summarized elsewhere (M1, R1).

Lead has also been shown to produce chromosome damage in leukocyte cultures of mice fed a diet containing 1% lead acetate for two weeks. The observed chromosome abnormalities involved single chromatids mainly, suggesting that the damage occurred after DNA replication (M11). Chromosomal abnormalities were observed in the lymphocytes of lead-poisoned men (L1, p. 166). In this work, G. Lehnert (ref. 344, L1) found that workers with blood lead from 62μg to 89μg/150 ml showed a generally positive correlation between an increase in urinary δ-aminolevulinic acid (ALA) and the percentage of abnormal mitoses. Lead adetate solutions (10^{-4}-5×10^{-6}) added to normal leukocyte cultures produced the same chromosomal abnormalities.

Several metals have been tested for ability to produce dominant lethals in the mouse (E2). The metals subjected to the screening procedure were cadmium chloride, manganous chloride, triphenyltin acetate, and triphenyltin hydroxide. Results were negative.

It should be pointed out that, aside from the chemical forms involved, the mutagenic investigations involving mammals described here were short-term and not necessarily equivalent to the human cases in which exposure to the metals takes place over periods of months and years. It is clear that much more work needs to be done on the mutagenic effects in mammalian systems. The effects of mercury and other metals on producing congenital abnormalities or poisoning in the fetus as a result of maternal intake are discussed later.

IV. GENOTOXICITY OF INDIVIDUAL TOXIC METALS

The genotoxic properties of metals must enter into an evaluation of their potential or actual environmental risks. Such information is needed for decision-making relative to the amounts which can prudently be released to the environment, and for the setting of safe levels of exposure by the general population. Specifically, I will devote most of the discussion to lead, mercury, and cadmium for reasons cited in the introduction to this paper. Further, in order to be as brief as possible, I will often relate levels and effects to children and the fetus rather than adults for the obvious reason that they are most susceptible and have a longer life-expectancy, thus increasing the probability that harmful effects will be manifested. An estimate of toxic metal levels in urban and non-urban areas is given in Table IX.

A. Lead

1. Introduction. Several reviews on the medical, ecological, geochemical, and other aspects related to lead toxicity are available (B6; H1, H3; L1; L21, Chapter 3; P1; S26). This discussion will deal with a few selected aspects--especially those which relate to levels of lead in tissues which may constitute a hazard to the population but do not produce obvious symptoms.

More than 98% of the 400 million pounds of lead released to the environment in the U.S.A. each year is derived from combustion of leaded gasoline (Table I) containing about 1.9 to 2.6 grams of lead per gallon in the U.S. (S26, p. 116). The concentrations of lead in Greenland snows increased dramatically beginning about 1950, and are now about 400 times natural levels (L1, p. 6). Lead is spewed into the water of lakes in the U.S.A. used by motorboats to the extent of 0.1 to 0.5 gram per liter of gasoline or a total (in 1963) of about 4×10^8 grams/year (S26).

The lead in the soil of cities is much higher than that of the average content of the earth's crust (10-15 ppm) (S26, p. 7). In fact, measurements of the lead content of soils and air near highways increases with closeness to the road, often reaching levels of more than 1,000 ppm in the soils (Table VI) and 17μgm/m^3 in the air (Table VII). Monthly dustfall samples from 77 midwestern cities gave mean lead levels of: residential--1,636μg/g; commercial--2,413μg/g; and industrial--1,512μg/g (L1, p. 139). A child who swallows 1g of such dust would exceed by more than a factor of 10, the estimated mean daily intake of lead from normal food and drink. Thus, a child with pica--abnormal craving for unnatural foods including dirt--could become acutely poisoned by the eating of a few chips of paint (1-50 mg of lead) augmented by contaminated street dusts. The lead in soils is derived for the most part from gasoline engine fallout and the washing of lead into the soil from exterior paint used on buildings as well as industrial discharges. Therefore, soils can constitute a serious hazard to children in many outdoor play areas (H3).

Canned milk is another source of lead contamination to babies. It has been reported that lead concentrations in canned evaporated milk ranged from 0.3 to 3.2 ppm compared to 0.012 ppm in human milk and no more than 0.28 ppm in raw whole cow's milk (S17). The World Health Organization's standard is 0.05 ppm for lead in drinking water. A child who drinks four 8-ounce bottles of canned milk a day containing 1 ppm of lead would have ingested 400µg of lead--a level which could lead to about 55µg/100 ml of blood after four months of exposure. A Committee of the Department of Health, Education, and Welfare (HEW) in 1970 recommended that children one year or older should receive no more than 300µg of lead daily from all sources (K3). Roughly an additional 10% of lead or about 30µg could enter the body by inhalation.

The reason for the reported high lead content in canned milk is due to the fact that the canned milk producers continue to use an old style can sealed with solder containing lead. As Shea concluded (S17): "In some respects, therefore, beer is now a safer diet for your infant than--canned milk." Practically, it would be better to substitute inexpensive powdered milk even though it appears that current levels of lead in canned milk on the average are not as high as those cited in Shea's article. However, the issue remains open until better sampling techniques are used and the reliability of analytical procedures employed by different laboratories are cross-checked.

2. *Lead poisoning--effects and symptoms*. The effects of lead poisoning or of any other metal can be classified into three broad categories (H3): (1) acute manifest illness; (2) symptomatic chronic illness; (3) "asymptomatic" chronic poisoning. Acute lead poisoning symptoms in children include lassitude, vomiting, and loss of appetite. This may progress rapidly to uncoordinated bodily movements, convulsions, and stupor. Without treatment the children go into a coma and die. Other symptoms include anemia and kidney damage.

According to Hardy, *et al*. (H3), the symptomatic chronic illness which is nonspecific and may be overlooked include those of the initial stages of acute poisoning--irritability, decreased appetite, lassitude, and vomiting. Long-term effects include kidney malfunction, brain damage with resulting behavior problems, intellectual impairment, and hyperactivity. Finally, the symptoms of "asymptomatic" chronic poisoning may not be attributed to lead and include what is called psychological impairment, low-grade anemia, and liver cirrhosis. It should be noted that 90% of children with blood lead levels of 60µg/100 ml or higher are asymptomatic (H3).

From the occupational standpoint, the usual limits of variation of lead concentrations in blood, urine, and feces are given in Table VIII. The blood levels in children which were earlier considered to require medical attention were 80µg/100 ml but were later reduced to 60µg/100 ml and, most recently, to 40µg/100 ml (K3). The Surgeon General in 1971 (K3) considered children with blood lead levels of 40µg/100 ml as suggestive of undue absorption of lead, those with 50-79µg/100 ml as possible cases of lead poisoning, and those with levels of 80µg/100 ml as unequivocal cases of lead poisoning requiring emergency treatment. It has been reported (C4, C5) that children with a blood lead level as little as 63µg/100 ml exhibited clinical symptoms of lead poisoning and as little as 100µg/100 ml for those with severe encephalopathy (inflamations of the brain). As little as 138µg/100 ml of blood lead has proved fatal.

In New York City a lead poisoning case is defined as one who has a blood lead level of 60µg/100 ml or more--the mean blood lead of New York City residents is 20µg/100 ml (B7). It is estimated that about 120,000 children are exposed to the risk of lead poisoning in New York City with as many as 6,000 actual cases (B7).

Occasional measurements of blood lead levels cannot be relied upon to give more than a crude index of exposure except when the levels are very high. Repeated samples of blood from the same individual may show appreciable differences during a period when the total body burden has not changed (B6, p. 10). The *in vitro* activity of aminolevulinic dehydratase (ALAD) is inversely related to blood lead levels over a range of 5-95µg/100 ml of blood (K3). Excretion of aminolevulinic acid (ALA) for young children with blood lead levels between 25 and 75µg/100 ml blood and without clinically evident lead poisoning appeared to follow an exponential relationship. While no firm quantitative relationships can be depended upon between blood lead and ALA excretion, which reflect disturbances in heme synthesis, it does appear that blood lead levels above approximately 40µg/100 ml indicates physiologically significant *in vivo* inhibition of ALAD (L1, pp. 101-110). However, as a screening procedure, urinary ALA excretion measurements are unreliable and a high proportion (33%) of children with high blood lead values can be missed (B7). It should be noted, however, that an elevated blood lead level indicates recent exposure, while urinary ALA levels may represent high lead exposure several years previously (B7, V2). The ALAD activity remains decreased in persons who had chronic lead poisoning but had no subsequent exposure for periods of up to 17 years (P4). A good discussion of blood lead levels in relationship to possible clinical and sub-clinical effects is given by Lin-Fu (L5) and Bryce-Smith (B11).

Later in this paper I will discuss the use of provocative or challenging chelation for the purpose of evaluating exposures to lead and other metals which may not be revealed by blood or urine lead levels. At this stage, I would like to focus attention on some subtle, possibly harmful effects of lead in individuals with apparently low blood or urine lead levels who may, nevertheless, constitute a relatively high risk population.

In 1934, one of the pioneers of industrial toxicology, Alice Hamilton, pointed out that lead enters the fetal blood during gestation and that "it is a matter of common knowledge among women who work in lead industries . . . that lead has an abortifacient action . . . ." She cited evidence for injury to the male germ-cell including that of a Japanese study in which a group of workmen in non-lead occupations were matched and compared to a group exposed to lead in Japanese storage-battery plants. Sterile marriages constituted 24.7% for the lead group and 14.8% for the non-lead group. The percentage of pregnancies ending prematurely or in stillbirth was 8.2 for the lead group and 0.2 for the control group (S26, p. 15). In this connection I recommend the provocative article by Gilfillan (GI) on lead poisoning and the fall of Rome.

"Excessive" exposure to lead by pregnant women has resulted in neurologic disorders in children (LI, p. 117). However, most of the reported effects of lead on reproduction are due to relatively high levels of lead. The important question arises, however, whether relatively low levels of lead produce detectable impairment or harm. Unfortunately, most of the literature and studies to date cannot answer this question simply because adequate long-term epidemiological studies have not been properly carried out. Hence, most statements as to the safety of current lead levels in the population consist more of propaganda than substance. Thus, one finds unsupported statements to the effect that current lead levels in the population have shown ". . . no evidence of any current threat to the health of representative urban populations in the United States" (KI). The same author (I2, p. 122) concluded from two volunteers that exposure to absorbable airborne lead of 150μg/m^3 for two years was safe because the volunteers had no symptoms of lead poisoning. It is difficult to imagine in this day and age that conclusions as to the safety of lead levels on a whole population would be based on such short-term, statistically inadequate experimental data as cited above. Unfortunately, that this represents the current situation is shown by the examination of the monograph on lead issued by the National Academy of Sciences in 1972 (LI), but more on this later.

The World Health Organization has also criticized the inadequacy of views on the long-term effects of lead because of the lack of valid data (I2, p. 123).

One long-term British study (D2) examined a group of pensioners retired from an electric accumulator works. It was found that those who had experienced the greatest exposure to lead (Group C), as shown by a mean urine lead content of 0.1-0.25 mg/liter, suffered a significant ($p < 0.001$) excess of deaths from cerebrovascular catastrophes and at an earlier age than the general population. All these workers had been employed for many years, had been the subject of continuous medical observation, and had worked in environmental conditions better than average for the time. Note that a urinary output of 0.2 mg/liter of urine is considered to be safe in the U.S.A. (I2, p. 123). The National Academy of Sciences' evaluation of the British study (LI, p. 149) concluded that the lead exposures of Group C represented some sort of a threshold, i.e., that lower levels were safe! Such a view is untenable unless a long-term prospective epidemiological study under precise conditions with appropriate environmental and medical data of larger populations covering all age groups is made.

3. Subtle effects. Subtle or insidious effects of lead have come under sporadic study and been discussed by many investigators (B6, B7, H3, K3, L5, M8, NI). Blood or urine lead levels alone cannot be used to detect the so-called "asymptomatic" chronic poisoning, though a semi-log plot of ALAD activity vs. blood lead in children decreases in linear fashion so that even the presumed "safe" blood lead of 20μg/100 ml shows definite inhibition of ALAD activity.

It is well known that at least 25% of children who survived acute encephalopathy from lead poisoning have permanent central nervous system injury (LI, p. 159). A recent study utilized provocative chelation as a means of ascertaining the body lead levels in children classified as hyperactive compared to controls (D2). In these investigations, children were given 250 mg of penicillamine before bedtime and the first urine voided in the morning were collected, brought to the hospital, and analyzed for lead. Penicillamine is a chelating agent which combines with available tissue lead forming a readily diffusible metal chelate resulting in increased excretion. In my opinion, the evaluation of these findings would have been improved if the urine had been collected until lead excretion had dropped to near control levels, at which time the area of the curve, i.e., Pb levels vs. time, of those treated could be compared to controls.

It was found that hyperactive children had higher blood lead (26.2μg/100 ml) than the controls (22.2μg/100 ml) and that the post-penicillamine urine lead levels were 77±92μg/l in the controls and 146±144μg/l in the hyperactive group ($p < 0.025$), while children with known history of lead poisoning had 41.1μg/100 ml blood lead and 325±167μg/l in post-penicillamine urine. The authors concluded that lead elevations above 24.5μg/100 ml of blood be considered dangerous and receive serious attention.

The investigation cited above is obviously incomplete and

deficient in several respects, but does indicate that evaluation of the health effects of current lead levels in the population be examined with great care.

B. Mercury

1. Introduction. Methylmercury salts were introduced in 1914 in Germany for seed dressing and their use became widespread. Within a decade of their introduction in Sweden in the 1940's, conservationists claimed that the mercury-treated seeds poisoned seed-eating birds. These allegations were dismissed by the manufacturers and their agricultural experts (L6, p. 6). Human fatalities traced to the mercury-treated seeds were first noted in Iraq in 1956, and later in West Pakistan, Guatemala, Japan, and the United States (D5, H4, L6, M9, P2, W4). The high levels of mercury in Swedish hen eggs, fish, and other samples eventually led to a complete ban in Sweden of the use of methyl and ethyl mercury compounds in agriculture as of February 1, 1966. At that time, farmers in the United States were using about 40 times more mercury per acre than Swedish farmers!

A large number of human fatalities took place in Minamata, Japan, beginning in 1952 when large quantities of mercury were discharged into the bay and river. About that time it was noticed that cats became ill, screamed incessantly, and some killed themselves by diving from the cliffs into the ocean (W4). At the present time, it has been determined that 397 persons were victimized, of whom 68 died (H2) as a result of eating fish and shellfish from the contaminated waters. Among the Minamata victims were at least 22 cases of congenital mercury poisoning in which the mercury from the mother passed to the fetus. As with most poisons, the fetus and children may show symptoms at levels far less than adults. In a U.S.A. case in New Mexico, the mother was asymptomatic but gave birth to a brain damaged infant with typically severe mercury poisoning--cerebral palsy, mental retardation, convulsions, involuntary movements, and defective vision (P2, S21).

Similarly, in Nigata, Japan, 330 victims of mercury poisoning were found, of whom 13 have died (H2). Several hundred deaths from mercury poisoning were reported to have occurred in 1972 in Iraq from flour prepared from mercury-treated wheat seeds. When the government issued a decree stipulating death to those who deal in or sell the poisonous seeds, some of the farmers panicked and dumped the seeds into the Tigris River resulting in pollution of the river bed (*New York Times*, March 9, 1972).

During late 1969, a family of nine in Alamogordo, New Mexico, ate a hog which had been fed with a maxiture of garbage and mercury-treated grain. Three of the family became ill from eating the pork exhibited typical symptoms of methylmercury poisoning. No suitable treatment was given since mercury poisoning was not suspected. Subsequently, chelation treatment was administered when the correct diagnosis was made (P2). Patient 1 is now in a long-term care facility--unable to sit, blind, unable to talk. Patient 2 can now feed himself with difficulty but he is blind, while Patient 3 is now able to walk with crutches but her sight is markedly impaired (P2). The infant born to the asymptomatic mother was severely poisoned as described above.

Detailed information on the pharmacology and toxicology of mercury is available in several reviews including C7; H4; L2, Chapter 2; L6.

2. <u>Mercury pollution</u>. More than 163 million pounds of mercury have been used in the United States since 1900 (E1, p. 30). By 1970, U. S. industry was releasing nearly a million pounds of mercury yearly into the environment including lakes and rivers. According to Klein (E1, p. 90), the amount of mercury dumped into the St. Clair River in the last 20 years is nearly high enough for the sediments to be commercially mined. He also pointed out that the fish in Lake St. Clair--which connects Lake Huron with Lake Erie--average about 0.5 ppm of mercury. Since the usual fish catch is about 6×10^6 pounds a year, normal fishing removes about 3 pounds of mercury a year. However, the sediments contain roughly 2×10^5 pounds, hence, the mercury reservoir in the sediments constitutes a long-term hazard which can be brought under control by elimination of industrial dumping of mercury and by costly procedures such as dredging, possibly in combination with other procedures.

The biggest source of mercury pollution in the U.S.A. stems from the manufacture of chlorine, production of which amounts to about 26,000 tons of chlorine daily, of which about a fourth is produced by the electrolysis of brine using mercury metal electrodes. The discharge of mercury metal and salts has been widespread in the inland and coastal waterways of the U.S.A. and Canada. Elevated mercury levels have been found in the waterways of 33 states (D5). The plastics industries use mercury in the manufacture of vinyl chloride and the paper industry uses methylmercury slimicides. The so-called long life alkaline batteries contain about 8% of mercury and are often burned in municipal incinerators. The principal sources of mercury consumption are shown in Table X.

Mercury is released to the air during the combustion of coal which contains anywhere from 0.012-33 ppm, but averaging about 1 ppm. It is estimated that about 10^3 metric tons of mercury are emitted into the U. S. atmosphere from the combustion of coal, while about 10^4 metric tons is released from the worldwide combustion of coal (B5b). A 700-megawatt coal-burning unit releases about 2.5 kg of mercury per day (B5a).

Another source of mercury exposure to the general population stems from latex-based interior paints containing diphenyl mercury dodecenyl succinate to prevent fungus growth. Measurements of elemental gaseous mercury in the air in several homes, offices, and laboratories in the Dallas, Texas area showed that the mercury levels depended on the type of paint used and the length of time since the room was painted (F4). In freshly painted rooms values as high as $3 \mu g/m^3$ were found compared to an average of $0.07 \mu g/m^3$ three years after painting. Wood-panelled and unpainted rooms were down to about $0.01 \mu g/m^3$. Doctors' and dentists' examination and treatment rooms showed mercury concentration of more than $5 \mu g/m^3$. The values given here should be compared to suggested ambient air levels of mercury for 24-hour exposure of less than $1 \mu g/m^3$ while the U.S.S.R. recommends not more than $0.3 \mu g/m^3$ (F4).

The health hazard to dentists could be serious since they are exposed for several decades to high levels of mercury vapor ensuing from drops of metallic mercury collected in corners, along walls, and in rugs. About 14% of dental offices have excess mercury levels and a high percentage of dentists excrete above normal levels of mercury according to the American Medical Association's Department of Environmental, Public, and Occupational Health as of April, 1971. Actually, the greatest potential hazards may result from contamination of hands after working with mercury metal or fresh amalgam (B12). In fact, Buchwald found (B12) most dental assistants and many dentists are not aware that mercury is hazardous, hence, safety measures were nearly nonexistent.

A review of a large body of data indicates that the natural mercury concentration in fish is less than 0.2 mg/kg (L6, p. 15) and the level in water about a thousandfold less (E1, p. 90). The mercury concentration in oceanic fish is less than 0.15 mg/kg and shows no significant change since the 1930's. It is estimated that the Japanese victims in Minamata and Nigata who ate about 200g of fish and shellfish daily took in less than 2 mg of methylmercury per day, assuming that fish contained about 5-10 mg/kg <u>fresh</u> weight (L6, p. 3; W4, p. 49).

Fish in Lake St. Clair in 1935 had about 0.07-0.11 mg/kg of methylmercury, but by 1970 fish with levels of 7 mg/kg were found--about a hundredfold increase (W4, p. 51). Some lakes in North America, e.g., Lake Wabigoon, have fish containing 24 mg/kg of methylmercury.

Why was no action taken in the U.S.A. to halt the dumping of mercury into lakes and rivers until 1970?

In 1970, the numerous companies who had dumped mercury for

decades claimed "informed ignorance" to the possibility that inorganic and metallic mercury could be converted to diffusible organic forms (D5). Consequently, it seemed reasonable to continue to dump about 250,000 pounds of mercury a year for decades into the Great Lakes. While mercury is an expensive substance, cost-accounting analysis undoubtedly showed that it was cheaper to dump mercury into the environment rather than reclaim it.

The facts of the matter are that sufficient information was available to reasonably well-read industrial hygienists and chemists and government officials to have halted the dumping of mercury several years previously. A brief summary of some of the facts prior to 1970 which might have been expected to have led to preventive action on dumping and restrictions on the use of mercury follow:

1956 - Fishing banned in Minamata Bay because of 42 cases of mercury poisoning from the daily eating of fish (L6, p. 2).

1960 - Kurland (K4; L6, p. 3) suggested that inorganic mercury could be methylated biologically. No further action was taken because methylmercury was a component of the mercury discharged from the Minamata factory.

1961-1966 - Mass poisonings by mercury (> 30 cases)--Iraq (1961); Pakistan (1963); Nigata (1965); Guatemala (1966).

1963 - When methyl thiomethylmercury was isolated from shellfish in the Minamata Bay, Fujiki (F6) suggested that mercury could be alkylated by marine life.

1965 - A reduction of the mercury content of treated seeds was made compulsory in Sweden (L6, p. 7).

1965 - A scientific conference was held in September in Sweden to provide advice to the government. Several publications at this time proved that the use of methylmercury in agriculture was responsible for the poisoning of wild bird populations (Refs. 18-23 in L6).

1966 - An international conference on "The Mercury Problem" was held in Stockholm on January 24. The United States was represented by five official delegates including a representative of the White House Office of Science and Technology.

1967-1969 - Jensen and Jernalöv (J1) reported that anaerobic microorganisms in sludge (i.e., organisms such as those found in lake-bottom mud) can methylate inorganic mercury and suggested a mechanism for the methylation reaction.

1967 - Methylation of inorganic mercury in the liver was demonstrated by Westöö (W4, p. 47).

1967 - In May, 1967, a symposium on mercury poisoning was sponsored by the International Atomic Energy Agency.

1967 - In July, the Batelle Memorial Institute in Columbus, Ohio issued a final report under its contract with the U. S. Public Health Service. The report was entitled "Design of an Overview System for Evaluating Health Hazards of Chemicals in the Environment" (D5). Commenting on the Minamata and Swedish experience, they concluded that ". . . mercury from environmental sources can accumulate in the food chain and be transferred to man. Mercury contamination in chlorine production is especially important . . . significant quantities of mercury are released to the environment each year [from chlorine-caustic production] which apparently cannot be traced and which to date cannot be accounted for. There exists a striking lack of fundamental information on national levels of mercury in our air, water, and food . . . ."

1968 - Wood, Kennedy, and Rosen at the University of Illinois published an article in Nature (W3) confirming the fact that the methylation reaction can proceed enzymatically by the use of cell extracts of a methanogenic bacterium. The latter was originally isolated from a symbiotic mixed culture from canal mud. They also suggested the mercury can be methylated nonenzymatically involving methylcobalamin, a vitamin B_{12} analog. It should be noted that the richest sources of vitamin B_{12} include estuarine mud (3μg/g dry weight) and activated sewage sludge (50μg/g).

Wood, et al. (W3) explicitly pointed out the consequences of the bioconversion of mercury to the food chain:

> Apparently transfer of methyl groups from Co^{+++} to Hg^{++} in biological systems may also occur as a nonenzymatic process. If this methyl-transfer reaction is significant in biological systems, then it will be enhanced by anaerobic conditions and by increasing numbers of bacteria capable of synthesizing alkylcobalamines. This means, for example, that pollution of a body of water with nutrients (that is, sewage) will increase the rate of formation of methylmercury at a certain concentration of Hg^{++}. Methylmercury could be formed by both

enzymatic and non-enzymatic reactions, thus making this cumulative poison available for incorporation into various organisms in the aquatic environment, and secondarily into terrestrial predators. The cumulative nature of mercury poisoning can be titrated in fish.

The paper by Wood, et al. (W3) which eventually appeared in Nature should have appeared about six months earlier, but Science, the journal to which it was originally submitted, declined to publish it (M9).

Professor Wood also alerted "all the people I thought might be able to influence developments in this area in Washington," and he also contacted a few companies but was obviously unsuccessful in impressing upon them the gravity of the problem (D5).

1969 - In May, a series of three articles in which the mercury data from Japan and Sweden were summarized and interpreted appeared in Environment. The articles discussed the biotransformation of mercury and accumulation in the food chain and the ecological implications.

1969 - As a result of press coverage of the three articles in Environment, the FDA in July adopted an interim maximum limit of 0.5 ppm of mercury in a daily diet.

1969 - The New Mexico mercury poisoning patients described earlier (Section IV-B-1) reached an acute phase requiring hospitalization (P2), however, realization that they were suffering from mercury poisoning was made a month later or several months after they were first stricken.

1970 - On February 17, the New Mexico tragedy was the subject of a nationwide broadcast on NBC. Within 24 hours the Chief of the U. S. Department of Agriculture's Pesticide Regulation Division announced that as a result of ten years of research on the environmental danger of mercury pesticides, he was cancelling registration of 17 alkylmercury fungicides--four years after a complete ban by Sweden had been announced.

1970 - Ontario (Canada) banned all fishing in the St. Clair River, Lake St. Clair, and the Detroit River because of high mercury levels in the fish.

Industrial representatives have repeated testified that they were unaware of the possibility that mercury could be methylated and, thus, enter the food chain. For example, the then president of Dow Chemical stated that not until March, 1970, when a report on mercury contamination of fish from Lake St. Clair was made public, was his company aware that inorganic mercury could be methylated and biologically concentrated in biological systems. To quote part of his statement made at the annual stockholders meeting, May 6, 1970 (EI, p. 54): "No one realized until recent research in Sweden brought it to light that mercury itself or inorganic mercury could be biologically converted to organic dimethylmercury. When I say no one, I mean no one in Dow, no one in industry, no one in the government or the universities."

In a similar vein, consider testimony on May 8, 1970, by a vice-president of Wyandotte Chemicals Corp. of Wyandotte, Michigan (EI, pp. 20-21):

> Wyandotte does not pose, nor do I, as an expert with respect to the biochemical effects of the emission of elemental or inorganic mercury compounds. However, our scientists are now making preliminary studies of this field. They find that harmful ecological effects of mercury have come to light only relatively recently in incidents of exposure of humans to heavy concentrations of specific methylmercury compounds. Studies only very recently conducted suggest possible difficulties arising from the concentration of methylmercury compounds in specific fish, which have a tendency to preserve the compounds and concentrate them.
>
> Our men have also found that recent studies in these fields have for the first time suggested the possibility that these concentrations of mercury compounds in fish may arise from the action on mercury by marine organisms in presently unknown and previously unsuspected ways.

Senator Hart then asked (EI, p. 21): "You suggest that the possibility that concentrations of mercury compounds in the fish may arise from the action of mercury compounds on marine organisms is a matter of recent discovery?" The vice-president's reply was "Yes, sir." He went on to say that he personally became aware of the bioconversion phenomenon only a few weeks earlier.

3. _Methylation of mercury_. Inorganic mercury, as mentioned above, may be converted into alkylmercury by enzymatic and nonenzymatic pathways. The methyl radical donor stems from methylcobalamin, the vitamin B_{12} derivative, which is produced in large quantities as an intermediate of methane biosynthesis. Many routes exist for the biological methylation of mercury (D5, W4). The relative amounts of dimethylmercury and monomethylmercury formed depend on such factors as the concentration of the mercury ion and pH.

Theoretically, one pathway for the methylation of mercury

would proceed as follows (B4):

$$CH_3^+ + Hg^0 \longrightarrow CH_3Hg^+$$

$$CH_3\cdot + Hg^+ \longrightarrow CH_3Hg^+$$

$$CH_3:^- + Hg^{2+} \longrightarrow CH_3Hg^+$$

$$CH_3:^- + RHg^+ \longrightarrow CH_3HgR$$

where RHg^+ is an organic mercury cation. Monomethylmercury is the toxic form in contrast to dimethylmercury and the higher monoalkyl (butyl and higher) salts of mercury (M7).

The formation of monomethylmercury from dimethylmercury takes place under mild acid conditions (W4):

$$R\text{-}Hg\text{-}R' + HX \longrightarrow R\text{-}Hg\text{-}X + R'H$$

Further details and elaboration of the methylation reactions of mercury are given in references B4, D5, I1, ref. 140 in L6, W4, among others, some of which have already been cited. Recently, it has been shown that long-term incubation of Hg^{2+} with sediments taken from the delta area of the St. Clair River resulted in a build-up of methylmercury (S22). Four bacterial isolates were obtained which were capable of degrading methylmercury to methane and Hg. Klein (E1, p. 90) points out that one pound of methylmercury introduced into an aquatic environment is the ecological equivalent of 10,000 pounds of inorganic mercury.

Löfroth (L6) has summarized the findings of a number of reports on the metabolism of mercury in man. In adults ingesting methylmercury present in fish, the retention of CH_3Hg^+ was about 10% and the biological half-life was 76±10 days. The methylmercury is excreted mainly via the feces. While urinary excretion is low, it increase exponentially with time.

The average mercury concentration in the blood corpuscles was measured in mothers and infants at the time of birth. The mothers had 8.6ng/g and the children 11.5ng/g while the plasma values were 2.5 and 1.9ng/g, respectively.

The suggested maximum-acceptable concentrations (MAC) for methyl- and ethylmercury compounds are 10μg/100 ml of whole blood in adult males corresponding to a daily intake of 100μg of Hg per day. The Swedish government recommended (L6, p. 53) that fish with elevated mercury levels of 0.2-1 ppm should not be consumed more than once a week.

C. Cadmium

Tables II-V and IX provide a variety of data on cadmium levels, including those in the human body and environment. Aside from its general toxicity, interest in cadmium stems originally from animal experimentation which would suggest that cadmium in the kidneys is one cause of hypertension and cardiovascular disease in man (H6; L2, Chapter 4; S2, S3). The feeding of cadmium to experimental animals does lead to hypertension as shown by the extensive investigations of Schroeder (S2, S3). Zinc also appears to play a role since a high Cd:Zn ratio in the kidneys has been associated with arterial hypertension (S3).

At birth practically no Cd is present in the body of man. Subsequently, the levels increase. About 5% of ingested cadmium (∿100μg/day--Table IX) is absorbed (Hammer, _et al_. in ref. H6). The amount of Cd in the body of man, without classification into smokers, non-smokers, _et al_., is given roughly as 30 mg in Table IX but according to Lewis, _et al_ (L4), the average adult American non-smoker at age 60 has a _total_ body burden of about 13 mg of Cd while the corresponding smoker has about 30-40 mg. Anywhere from 0.75-3.0μg of Cd could be absorbed per pack of cigarettes. Foods which contain relatively high amounts of Cd are beef kidney (12 ppm) and anchovies (5 ppm) whereas most other foods contain 0.1-1 ppm (M4).

Acute effects following oral ingestion of as little as 15 mg of cadmium salts include nausea and vomiting--a symptom which reached epidemic levels when cadmium plated articles were used as food containers (L2, p. 111). Excess occupational exposure has resulted from inhalation of cadmium oxide fume or dust leading to delayed pneumonitis--lung inflammation, cough, and chest pains. Deaths have occurred from exposure to cadmium fume levels of 3-100 mg/m^3. Chronic exposure can lead to emphysema.

For decades, the Mitsui Mining and Smelting Company in Japan dumped cadmium waste into waters used for rice-paddy irrigation. For about 50 years the wives of farmers living along the Juntsu River, 150 miles west of Tokyo, complained of a disease in which the victim's bones and skin became painful to the touch. The disease was called Itai-Itai disease (itai means "it hurts") by Dr. Noboru Ogino, a Toyama physician who first concluded that there was a connection between the disease and cadmium (_New York Times_, July 11, 1971). The disease was particularly prevalent among elderly women 50-60 years of age who had borne several children (L2, p. 116). At present, the Japanese Government has identified 265 victims of whom 47 have died (_New York Times_, March 21, 1973). The mining company which recently was ordered to pay $400,000 compensation refused to pay earlier when 14 victims brought suit in 1968, because they asserted that no causal relationship could be proved with certainty (_New York Times_, July 11, 1971). Some believe that cadmium alone did not produce the disease but that other factors such as nutritional were essential (L2, p. 117).

Whether cadmium is a causal factor in human hypertension and cardiovascular disease is far from resolved (H6; L2, Chapter 4). Recent data indicate that the presumed association of cadmium with arterial hypertension is more apparent than real. In persons not exposed to cadmium in their occupations, an association with higher body cadmium seems to exist. However, it turns out that smoking contributes a major source of cadmium to man. In fact, it contributes more to the total body burden than the amount derived from other sources (L4). According to Lewis, _et al_. (L4): "The diseases, chronic bronchitis and emphysema . . . cancer of the bronchus or lung . . . and arterial hypertension . . . are associated with high kidney or liver levels of cadmium in man. The present study indicates that when control is made for smoking habits, subjects with these diseases do not have more cadmium in their tissues than those without. An increased prevalence of the above mentioned diseases has long been known to occur in subjects who smoke cigarettes. This relationship constitutes the most likely explanation for the association of elevated cadmium levels with these illnesses . . . ."

Admittedly in this retrospective study, information on lifetime work exposure to cadmium of these individuals were not available and, indeed, would be difficult to reconstruct. Consequently, we do not know the time contribution from the work environment.

The relationships between trace metals and a given disease are of intense interest (M3) and, as I will discuss in the Section VII, cannot be resolved by occasional measurements of trace metal concentrations in tissues and excreta.

Many of the problems involved in evaluating levels of Cd which do not produce obvious symptoms are similar to those discussed for lead. Prudency again would necessitate strict controls in the release of Cd to the environment. All together about 5×10^6 pounds of Cd are emitted each year in the U.S.A. into the air, water, and soils, especially in urban areas where metallurgical plants are located. The article by McCaull (M4) provides a good summary of the sources of environmental contamination from cadmium. One example he cites is a battery industry which dumped cadmium-nickel wastes into a stream feeding the Hudson River. The mud in the stream contains 16.2% of Cd and 22.6% of Ni on a dry-weight basis--a level well worth mining! Mud dredged from Lake Erie contained as much as 130 ppm of Cd.

Chronic poisoning from cadmium in industrial workers has been reported (M4). These include plants in Sweden, Japan, England, and the U.S.A. It is obvious that much more has to be done from the industrial hygiene standpoint, medical diagnosis, and long-term epidemiology.

D. Miscellaneous

Only a very brief discussion will be given here regarding the toxic effects of metals other than Pb, Hg, and Cd. Brief mention of Be and others have already been made.

An outbreak of fulminating heart failure was reported in 1967-1968, among heavy beer drinkers in Quebec, Canada and Omaha, Nebraska, and in Minneapolis, Minnesota from 1964-1967 (Aa). Cobalt was incriminated when it was found that the particular brands of beer involved contained cobalt salts as an additive in order to stabilize the foam. Heart failure appeared some time after addition of cobalt and disappeared when the use of cobalt was discontinued.

Large doses of cobalt salts to rats are cardiotoxic (G6). A number of factors affected cobalt toxicity including routes of administration, pre-existing heart damage, length of exposure, nutritional status, and diet. The feeding of large quantities (equivalent to 24-48 bottles a day in humans) of cobalt-containing beer to rats did not produce toxicity after 60 days of treatment (W1). Yet beer containing 1.1 to 1.2 ppm of cobalt proved fatal to humans (S25). The levels of cobalt in the heart muscle in the hearts of patients with heart disease were about ten times higher than found in control tissues. However, the cardiotoxic effect of cobalt in beer is difficult to explain since the amount ingested of up to 10 mg/day is far less than the amount used to treat refractory anemia (up to 50 mg/day). Alexander (Aa) suggests that the heart is rendered more sensitive to cobalt because of inadequate protein and vitamin intake, especially thiamine, zinc depletion, and prior heart damage from alcohol. Other possible explanations include the chemical form of cobalt in beer in contrast to that in the pills.

The fatalities caused by cobalt in beer illustrate the hazard of adding untested additives to food and the uncertainties of relating effects in man to animals and vice versa. However, the animal evidence unambiguously did reveal damage to the heart in at least one species--though several species would have had to be used. When the toxicological data, safety factors, and the fact that beer can and is manufactured without cobalt, then prudency considerations alone could have banned its use. This approach, of course, is easier said than done when potentially carcinogenic dyes used purely for cosmetic purposes are still permitted in foodstuffs (S13).

Numerous papers have appeared relating trace elements to cardiovascular diseases (M3). It appears that cardiovascular disease is more prevalent in soft water than in hard water areas. Several reasons have been suggested for the apparent correlation including a higher sodium intake in soft water areas (D1) and higher lead levels (B2). Much more work has to be done before definitive conclusions can be reached as to the reasons for the observations, and, in fact, whether a true correlation does exist. Similar conclusions pertain to correlations between trace metals and cancer (A2); infant mortality and water hardness (C10); and zinc deficiency and congenital malformations (S14).

V. SYNERGISM INVOLVING TOXIC METALS

The setting of acceptable levels (e.g., Table IV) does not take into account the fact that the population is exposed simultaneously to a multiplicity of potentially harmful agents. Consequently, one must consider the possibility that the actual toxicity of a given toxic metal is greater (i.e., synergistic) than that deduced by investigation of its effect in the absence of other metals or stress. It is apparent that people contain body burdens of several potentially genotoxic metals (Table II). The possible adverse effects of two or more metals already present in the body such as Pb, Hg, and Cd are, therefore, an important and urgent problem for evaluating occupational and environmental hazards. The degree to which they are synergistic, additive, or antagonistic using various criteria of genotoxicity is a relatively unexplored area. Consideration of additive effects were discussed earlier (Section I-A).

Bridges (B8) has emphasized that each toxic chemical or mutagen cannot be treated as an isolated insult.

> What is unlikely, however, is that informed people would tolerate more than an overall doubling (mutation rate), whatever the benefits from any environmental mutagen. It follows that what is a suitable recommendation for one mutagen (i.e., radiation) will not suffice when each of a number of mutagens is considered. It has been estimated that about a thousand new chemicals are introduced into the environment each year, of which no more than a minute fraction are tested for mutagenic activity. If a thousand mutagens were each allowed at population doses which doubled the spontaneous rate, then the overall rate might go up a thousandfold, quite apart from any synergistic interaction which might occur.

The import of Bridges' statement cannot be overemphasized, especially when it is realized that workers exposed to, say, lead in a foundry, already have mercury and cadmium in their tissues stemming from various sources, such as the industrial environment, ingestion of food, breathing of urban air, smoking, etc.

Some examples of synergistic or possible synergistic effects involving toxic metals are cited here. For example, cadmium is often found to be elevated in blood along with lead in adults and some children (C2). It is suggested that cadmium may be a synergistic agent in lead poisoning. There exist considerable data (U1) regarding interaction of a toxic metal with a given essential trace metal such as zinc, iron, and cadmium, e.g., cadmium and zinc (S3). Synergistic effects between calcium or iron levels and lead toxicity are well known (G4). Many factors, in fact, influence lead (and other metal) toxicity such as age, season of the year, calcium-phosphorus levels, iron deficiency, Vitamin D, alcohol, pregnancy, coexistent disease, etc. (G4, H3). An illness such as pneumonia can produce an acidosis causing a release of lead stored in the bones and, thus, produce an acute attack of lead encephalitis. Experimentally, lead and cadmium in combination have been found to cause teratogenic effects (F2).

Individual metals reduce resistance to infections, as has been demonstrated experimentally in mice treated with subclinical doses of lead nitrate and subsequently challenged with Salmonella typhimurium (H5). Similarly, an alteration of the immune response in cadmium-exposed rats has been noted (J2). Non-toxic levels of lead have been claimed to multiply the lethal effects of non-toxic levels of bacterial endotoxins in rats and primates by a factor of 10^5 (L7). Hence, it has been concluded that many children's deaths attributed to lead poisoning may actually result from acute sensitivity to bacterial endotoxins.

The chromosome-breaking effects in plants of the well-known mutagen, ethyl methane sulfonate, are enhanced by copper and zinc salts (M10).

The toxicity and absorption of metals sometimes depend on the gastric acidity. For example, the toxicity of zinc phosphide is enhanced by gastric HCl, hence, cats and dogs are more resistant than rats, rabbits, or fowl because the former secrete HCl intermittently (O1). Alcohol which increases gastric HCl output results in higher absorption of iron salts. Since wine contains high levels of iron, along with alcohol, it serves as a pleasant source of iron, especially for anemic women as noted by Sigmund Freud who prescribed it for his fiancee.

VI. THERAPY OF HEAVY METAL POISONING

The harmful effects of heavy metals in humans can be minimized or prevented by chemically modifying the biochemically active forms of the metal ion. This can be accomplished by the administration of chemicals which form insoluble compounds with the metal, or chelating agents which form diffusible, readily excreted metal-chelates (S11). An intriguing example of the former is provided by the larvae of the clothes-moth (Tiniola biselliella) which detoxifies metal ions lethal to other insects by converting them to insoluble sulfides (W2, p. 42). The moth produces H_2S from the digestion of wool cystine in the midget. We have used a lake-forming dye, aurintricarboxylic acid, to protect mice acutely poisoned with soluble beryllium salts (S9).

In this section I will focus on chelating agents for treatment of heavy metal poisoning since it is the most widely used and successful form of therapy. The chemical principles involved are instructive as well since they bear a close relationship to the mechanisms underlying the toxicity of heavy metals. After all, the body contains numerous endogenous chelating agents influencing the retention and transport of heavy metals, e.g., the blood plasma is 10^{-4}M in citrate.

Finally, I will describe the theoretical and practical advantages of using combinations of chelating agents as well as mixed ligand complexes. For discussion of certain aspects not covered here, several sources are available as the text will make clear. In any event, the monograph by Catsch is a useful introduction (C1).

A. Mixed Chelates and Mixed Ligand Complexes

The efficacy of treatment of metal intoxication by a single chelating agent appears to have reached a plateau. Those commonly employed--$CaNa_2EDTA$, BAL (2,3-Dimercaptopropanol-1), $CaNa_3DTPA$, D-penicillamine--are about as effective from the standpoint of both binding constants and toxicity as we are likely to attain. Further, the use of a single chelating agent for therapy may be and has, in fact, been proven dangerous. Several clinical cases are available, for example, in which it was found that chelation treatment with single chelants aggravated toxic symptoms. These include arsenic and BAL (S1), lead and EDTA (G3, p. 980; B10a; C4), lead tetraethyl and BAL (V1), thallum and BAL (G5), mercury and BAL in subacute cases (B9, p. 256), mercury and EDTA (B9, p. 238), while cadmium and BAL are not recommended because of experimental demonstration of nephrotoxic action (D3, p. 166) as is also known experimentally for cadmium and $CaNa_2EDTA$ (B10), cadmium and cysteine (G7, K2), and beryllium and citrate (S9).

In view of the foregoing, and for theoretical reasons, it would appear reasonable to employ mixtures of chelating agents and of mixed ligand complexes since it could lead to significant improvement in the treatment of chronic and acute metal poisoning. In fact, the use of two chelating agents in combination has been used by Chisolm (C5) with marked success in the clinical treatment of lead poisoning, e.g., EDTA plus BAL.

The rationale for using combinations of chelating agents can simplistically be ascribed to two factors. These are--assuming that the highest practical chelant-metal ratio is maintained--: (1) the relative degree of hydrophobic (extracellular) and hydrophilic (intracellular) properties and/or net charge possessed by the chelating agent and the corresponding metal chelate, and (2) the relative stability of the metal-chelates under *in vivo* conditions.

The first factor relates to the fact that minimal contact often occurs between tissue deposits of a given metal and that of a strongly hydrophilic chelating agent, i.e., chelants possessing water-seeking functional groups such as sulfonic and carboxylic. For example, two chelating agents, salicylic and sulfosalicylic acid, having equal binding affinity for Be were equally effective against acute Be poisoning when administered during the period that most of the Be was in the circulating blood (S9). However, later when significant fractions of the Be were deposited in the tissues, only salicylic acid was effective. In fact, the most effective antidote was the dye, aurintricarboxylic acid, which did not modify the excretion or tissue distribution of Be. In the case of lead poisoned rats, a significant reduction of mortality was obtained by inducing elevated levels of citrate endogenously in the kidney and other organs by vluorocitrate administration, while injected citrate was ineffective (F5).

A striking example of the differences in selective decorporation involving intracellular and extracellular chelating agents is the greater effectiveness of DTPA plus the penta-ethyl ester of DTPA for removing Pu from the soft tissues of mice (C1). Unfortunately, the toxicity of such esterified chelants is usually high, i.e., have low therapeutic index.

The second interrelated factor, stability, is critical because the metal-chelate must remain largely intact to permit excretion, or, if deposited in tissues, to remain intact. Otherwise, the chelating agent may shift additional amounts of the metal as the metal-chelate to sensitive organs, where dissociation of the weak metal-chelate permits direct reaction of the metal ions with receptors in the organs, e.g., -SH groups. Consequently, the chelating agent may aggravate the toxic effects of the metal. Thus, the administration of a chelating agent may increase excretion of a toxic metal but, at the same time, accelerate the transfer of some of the mobilized metal into the brain or kidney. If these organs already contain levels of the metal nearly sufficient to produce encephalitis or kidney damage, a small increase in metal concentration could suffice to produce acute damage. Experimentally, for example, the administration to rats of sulfur-containing chelating agents, including penicillamine, produced a large increase in the amount of cadmium deposited in the kidneys (N2). Similarly, BAL increased the uptake of mercury by the brain (B3).

A drop in the chelant/metal ratio can be critical and, because of mass action equilibria, amounts to a decrease in stability of the metal chelate. In this connection it is of interest to note the observation by Chisolm (C5) that no obvious clinical complications arise from the use of EDTA in patients with mild degree of lead poisoning but do in those with severe intoxication. It would appear that the *in vivo* stability of the lead-EDTA chelate is borderline (see Section VI-B) and/or the concentration of available EDTA is low, especially that of the fully dissociated form, L^{-4}, which is responsible for most of the chelating action at physiological pH.

In the cases of severe intoxication with lead it is especially important that the chelating agent form a particularly stable complex because the blood cells become saturated with lead at relatively low levels permitting lead to "spill" into the plasma. In the blood cells of the rat lead has a half-time of 30 hours but only 42 minutes in the plasma (S6). When as little as 1 mg of lead acetate was injected, most of the lead was in the plasma whereas with the carrier-free lead practically all of the lead was in the blood cells. The plasma lead is immediately available to the chelating agent. However, unless a very stable chelate is formed, the brain and kidney are liable to be insulted with increased levels of lead in a form able to react with sensitive receptors.

A related approach to mixed chelate therapy is to employ mixed ligand complexes in which the chelating agents are part of the same coordination sphere of the metal (S19), e.g., MAB, MAB_2, MABC, etc., where A, B, and C represent different chelating agents. In this way, one obtains an enhanced binding because of statistical factors and, depending on the nature of the chelating agents occupying the same coordination sphere, an additional enhancement of binding. The deliberate and systematic application of mixed ligand complexes is an unexplored phenomenon in the treatment of toxic metal poisoning. In some cases, mixed ligand complexes are inadvertently formed when combinations of chelating agents are employed or form spontaneously when a single chelating agent is used by participation of endogenous ligands such as citrate.

The advantage of mixed ligand complexes stems from two factors. For example, MAB is favored over that of MA_2 or MB_2 where M is the metal ion and A and B are complexing or chelating ligands. Aside from this favorable statistical effect (S15), the stability of the mixed complex is often enhanced above and beyond the statistical contribution, especially when the ligands are dissimilar types (S15) and involve an aromatic ligand.

From the purely statistical standpoint, a mixed chelate of, for instance, a divalent cation, M^{2+}, or two monovalent ligands, A and B, forming MAB would be twice as stable as MA_2 or MB_2, while

MABC would be six times more stable than either MA_3, MB_3, or MC_3, and MA_2B would be three times as stable as the saturated simple complexes. If the complexing ligands are dissimilar, the stability would be enhanced over and beyond the statistical factor by factors as high as five or more. The statistical factors are calculated by the expression (S15):

$$\frac{n!}{A!\ B!\ C!\ D!\ E!\ F!} \quad (1)$$

where n is the total number of ligands in the complex species. The equation gives the number of ways in which these species A, B, etc. can be formed.

Mixed ligand complex formation does not necessarily take place if a given chelating agent, A, occupies, very strongly, all of the available coordination sites of the metal ion. However, if chelating agent B is present at much higher concentration than A and possesses at least as strong binding sites, mixed complex formation can occur. If ligand A cannot occupy all of the coordination positions of the metal ion, then ligand B could occupy the remaining sites to form a mixed complex. Another possibility is the situation where ligand A has six sites but used only two of them, for example, while ligand B fills the remaining positions. Then a second metal ion can be accommodated in the four available sites of A, forming a mixed metal-mixed ligand complex.

In summary, some of the reasons for anticipating and explaining the greater effectiveness of mixed chelant therapy include: (1) A higher fraction of metal can be bound since a higher chelant/metal ratio can be maintained by the use of compatible and complimentary chelating agents as in the use of BAL plus EDTA in lead poisoning (C9); (2) Each chelating agent may remove and chelate with metals bound to different tissue sites; (3) As the metal ion from one metal-chelate is released, it may be bound by the other chelating agent; (4) One of the metal-chelates may be more hydrophobic, i.e., more non-polar or fat soluble than the other, and may remove metal ions deposited at sites unavailable to another chelant; (5) Mixed ligand complexes may be formed.

It cannot be overemphasiced that the effectiveness of chelation therapy for metal poisoning cannot be gaged simply by the degree of decorporation. Concomitantly, it is necessary to examine the effects on the acute and more subtle manifestations produced by metal poisonings. Ideally, these would include the clastogenic, mutagenic, teratogenic, and carcinogenic manifestations.

B. *In vivo* Stability of Metal-Chelates and the Effective or Conditional Constant

A quantitative formulation of the reaction between a chelating agent and a radioelement under physiological conditions becomes quite unwieldy, since it is necessary to include the effect of interfering ions, e.g., Ca^{++}, OH^-, H^+, carrier, and endogenous complexing agents, on the stability of the chelate formed between the metal ions and the chelating agent (SIO). In other words, one cannot rely on tabulated values of stability constants alone to evaluate the *in vivo* effectiveness of a chelating agent. Other factors of a pharmacological nature include the perturbations caused by the lower pH in the kidney, the ability of the chelated metal ions to enter intracellular compartments, and the metabolism of the administered chelating agent. For these reasons, it is well to outline some of the quantitative aspects involved in the chelating reactions.

The relative fraction of metal chelated--hence, solubilized--by the chelating agent can be characterized to a first approximation by simple mass action:

$$M + L \rightleftharpoons ML \qquad (2)$$

where M represents the concentration of the free metal ion, and L represents the corresponding concentration of the chelating ligand, and ML is the chelate. It follows that the fraction of M present as ML is:

$$\frac{(ML)}{(M)} = K(L) \qquad (3)$$

where K is the formation constant as discussed shortly. Consequently, for a given chelating agent, the fraction of M present as ML approaches a constant value depending on the concentration of L.

Most useful chelating agents under biological conditions form a stable 1:1 complex or chelate with a metal ion. The anionic state of the chelating agent involved is usually the most higher dissociated form. Consider ethylenediaminetetraacetic acid, EDTA, which is represented by H_4L. As the pH of the system is increased, the protons dissociate:

$$H_4L \overset{-H^+}{\rightleftharpoons} H_3L^- \overset{-H^+}{\rightleftharpoons} H_2L^{-2} \overset{-H^+}{\rightleftharpoons} HL^{-3} \overset{-H^+}{\rightleftharpoons} L^{-4} \qquad (4)$$

The form which reacts with a metal ion most strongly is L^{-4} and the contribution of the other anionic forms, at neutral pH, can, to a first approximation, be disregarded.

The effective concentration of EDTA for chelation is not the stoichiometric concentration but the actual concentration of the chelating species. At pH7, for example, L^{-4}, the reacting species is less than a thousandth of the stoichiometric concentration of H_4L. Consequently, at the peak concentration attainable in blood, $\sim 10^{-4}$M, the "effective" concentration in relation to competing ions is only 10^{-7}M.

When a chelating agent is administered to a mammal previously exposed to a metal, the information desired is the ratio of the metal in chelate form relative to the "free" or unchelated metal. On the assumption--not always true--that the rate of formation of the chelate is rapid, then the reaction can be expressed by equation (3).

Chelating agents of the EDTA type are injected as a calcium chelate or become converted to the calcium form upon entrance into the blood stream. In other words, one is actually dealing with an exchange reaction:

$$CaL + M \rightleftharpoons ML + Ca \tag{5}$$

Hence, the ratio:

$$\frac{(ML)}{(M)} = \frac{K_{ML}}{K_{CaL}} \times \frac{(CaL)}{(Ca)} \tag{6}$$

This expression permits the relative effectiveness of chelating agents to be made more reliably by emphasizing the importance of the formation constant of the calcium chelate in biological interactions. Nevertheless, the above expression (6) is too limited and doesn't take into account competition from other ions. More general expressions have been introduced, but are not convenient for taking into account the effects of several interfering ions on the interaction of a metal with a specific chelating agent.

A convenient and simply applied procedure (S10) for evaluating *in vivo* stability of metal-chelates employs an approach from analytical chemistry (R2). In brief, an effective constant is calculated which becomes identical to the formation constant tabulated in the literature when no interfering ions are present. Consider first the formation constant for ML as defined in equation (3).

In point of fact, the concentration of each of the components is rarely represented by the stoichiometric concentrations. For example, the actual concentration of M may be diminished by the formation of hydroxide complexes such as M(OH) and $M(OH)_2$, while the concentration of L may be diminished because of incomplete dissociation of the acid H_nL; that is to say, because of competition from H^+. Furthermore, the concentration of 1:1 chelates may be greater when the possibility of acid or basic complexes such as MHL and M(OH)L is considered. In the body, other ions, such as calcium ions, compete against M for L, while citrate, phosphate, proteins, and others may compete for M.

The effective constant is defined as:

$$K_{eff} = K_{(ML')\ M'L'} = \frac{[ML']}{[M'][L']} \tag{7}$$

where ML' represents the sum of concentration of all species containing M and L in the molar ratio 1:1. Similarly, [M'] is the total concentration of the metal ion in all combinations excluding that fraction combined with L. Finally, [L'] is the total concentration of the ligand in all forms excluding that fraction not specifically combined with the given metal.

The effective constant is handled in the same way as ordinary equilibrium constants, and one does not need to know the exact nature of the species contributing to M and L and ML. For this purpose, alpha coefficients or the so-called "side-reaction" coefficients are employed, which afford a measure of the extent of side reactions. For ML, M, and L these are defined as follows:

$$\alpha_{ML} = \frac{[ML']}{[ML]} \tag{8}$$

$$\alpha_{M} = \frac{[M']}{[M]} \tag{9}$$

$$\alpha_{L} = \frac{[L']}{[L]} \tag{10}$$

If the reactions take place without interference or competing reactions, then the alpha values are unity. The effective constant is calculated from the alpha values from the simple definition:

$$K_{eff} = K_{(ML')M'L'} = \frac{\alpha_{ML}}{\alpha_M \alpha_L} K_{ML} \tag{11}$$

so that if no interference or competition reactions are involved:

$$K_{eff} = K_{ML} \tag{12}$$

There exists an important simplifying rule for calculating overall alpha coefficients when at least one alpha term is a large one, namely: <u>the overall alpha coefficients can be taken to be approximately equal to the sum of the individual coefficients</u>. For example, assuming that in a system there are, apart from L, several ligands, L_1, L_2, etc., competing for the metallic ion M, this means that

$$\alpha_M = \alpha_{M(L_1)} + \alpha_{M(L_2)} + \cdots \quad (13)$$

and, likewise, if several cations interfere, B_1, B_2, which may include H^+, then:

$$\alpha_L = \alpha_{L(B_1)} + \alpha_{L(B_2)} + \cdots \quad (14)$$

Usually one of the alpha terms predominates, so that the other terms can be dropped.

Numerical values for alpha are obtained from known formation constants as compiled in many general references and, in particular, in the comprehensive tables of Sillen and Martell (S18). Typical equations for calculating alpha are as follows:

1) For a disturbing ligand, A, that forms complexes with M:

$$\alpha_{M(A)} = 1 + [A]K_{MA} + [A]^2K_{MA_2} + \cdots \quad (15)$$

2) For disturbing cations, B, where B can also be H^+:

$$\alpha_{L(B)} = 1 = [B]K_{BL} + [B]^2K_{B_2L} + \cdots \quad (16)$$

3) When acid or basic complexes are formed:

$$\alpha_{ML(H)} = 1 + [H]K^H_{MHL} \quad (17)$$

$$\alpha_{ML(OH)} = 1 + [OH]K^{OH}_{MOHL} \quad (17a)$$

where K^H_{MHL} is the constant for the formation of the acid complex, MHL, from ML and H^+, while K^{OH}_{MOHL} is the constant for the formation of MOHL from ML and OH.

A typical example of the application of the relations described above is one involving the calculation of K_{eff} for the reaction of Sr^{++} with EDTA in the blood. In this case, we assume that the peak concentration of EDTA in the blood stream is 10^{-4}M corresponding to the dosages of $CaNa_2EDTA$ usually administered. Consider the interferences from Ca^{++}, citrate, Cit^{-3}, OH^-, H^+, and serum albumin (P):

$$\alpha_{EDTA} = \alpha_{EDTA(H)} + \alpha_{EDTA(Ca)}$$

and

$$\alpha_{Sr} = \alpha_{Sr(OH^-)} + \alpha_{Sr(Cit)} + \alpha_{Sr(P)}$$

The formation of hydrogen and hydroxy chelates of Sr, namely, SrHCit and SrOHCit, can be disregarded.

From the concentrations in blood we have $(H^+) = (OH^-) \cong 10^{-7}M$, $(Cit) \approx 10^{-4}M$, $(Ca^{++}) \approx 10^{-3}M$, and further, $K_{CaEDTA} = 10^{10.7}$, $K_{SrEDTA} = 10^{8.6}$, $K_{SrCit} = 10^{2.8}$. From equations (15) and (16) and the known proton dissociation constants of EDTA, we have:

$$\alpha_{EDTA(Ca)} = 1 + (10^{-3})10^{10.7} = 10^{7.7}$$

In similar fashion we obtain that $\alpha_{EDTA(H)} = 10^{3.4}$. From the rule given in equation (14):

$$\alpha_{EDTA} = 10^{7.7} + 10^{3.4} \approx 10^{7.7}$$

In other words, we can disregard the competition of H^+ for EDTA in this pH range.

In this case of Sr we find that the effects of OH^-, Cit, and serum albumin ($K_{SrP} \sim 10^3$) are too small, so that $\alpha_{Sr} = 1$. Inserting the proper values in equation (11), we have:

$$K_{eff(SrEDTA)} = \frac{\alpha_{SrL}}{\alpha_{Sr}\alpha_L} K_{SrEDTA} = \frac{1}{1 \times 10^{7.7}} \times 10^{8.6} = 10^{0.9}$$

Consequently, from equation (3):

$$\frac{(SrEDTA)}{(Sr)} = 10^{0.9} \times 10^{-4} = 10^{-3.1}$$

which means that less than 0.1% of blood strontium is bound by injected EDTA.

With polyvalent, easily hydrolyzable cations such as iron, the hydroxide competition becomes very important. For example, $Fe^{III}EDTA$ is about a million times stronger than $Cu^{II}EDTA$ in terms of the formation constants as found in the tables. Yet, at pH around 7, the copper-EDTA chelate is more than a hundred times stronger than the iron chelate. One reason is the fact that the $\alpha_{Fe(OH)} = 10^8$ while $\alpha_{Cu(OH)} \cong 1$.

The rate at which insoluble or polymerized species of metal ions become solubilized in body fluids and by selected concentrations of chelating agents is best determined experimentally. In a general way, the larger the effective constant and the lower the concentration of carrier, the more rapid is the rate of solubilization.

The effects of side reactions on mixed complexes can also be handled by the concepts described above (see R2, p. 54). Finally, the formation of mixed metal complexes must be considered. For example, the decorporation of lead and mercury in the same person by penicillamine (Pen) may proceed by the formation of Pb-Hg-Pen complexes.

I. _Ultrafiltration screening technique_. The testing of chelating agents by administration to animals burdened with toxic metals is a time-consuming and expensive procedure. For example, the _in vivo_ evaluation of a chelating agent for decorporation therapy involves several animals, tissue analyses, numerous urine and fecal analyses, effectiveness as a function of the time of administration of the chelating agent relative to that of the metal ion, the effect of the physico-chemical state of the metal ion, etc. However, such _in vivo_ studies are unnecessary except for that chelating agent selected as most promising from the _in vitro_ screening investigations (S12).

A low pressure ultrafiltration technique can be employed. In this technique (S12) the metal is administered to the animal, following which various tissues are removed. The tissues are homogenized in the presence of different concentrations of chelating agents, and the fraction of metal found in the ultrafiltrate or dialysate determined. Alternatively, or, in addition, the chelating agents can be administered later, and the degree of ultrafiltrability measured as a function of time. The latter method has the advantage that it simulates the _in vivo_ situation most closely, e.g., the results also reflect the actual _maximum_ amount (since we are not considering here the intracellular depots) of chelating agent reaching a target organ and available to react with the deposited metal.

VII. PROVOCATIVE OR CHALLENGING CHELATION

I mentioned earlier (IV-A-3) that blood and urine levels of a metal do not necessarily reflect body burdens or the hazard potential. However, the use of "provocative" chelation for diagnosis can and should be applied more widely (see IV-A-3 for an example). Teisinger (T1) demonstrated that the increased excretion of lead in workmen in the first 24 hours following a test dose of $CaNa_2EDTA$ was a good basis for evaluating the degree of exposure to lead. If the workmen excreted more than 1 mg of lead in their urine during this time, they were removed from further exposure, at least until the working conditions were presumably improved. The assumption is that "the amount of lead so mobilized bears a direct relation to the active deposit of lead in the soft tissues of the body (including the trabecular bone)." Results of a similar study with children is shown in Table XI.

Hardy and co-workers (H3) employed penicillamine for diagnosis of over-exposure to lead. Penicillamine has the advantage that it is given orally. They consider the test of unique value.

Prerovska and Teisinger (P4) showed that provocative chelation, given 3 to 17 years after the termination of exposure to lead in previously chronic lead poisoning cases, led to an average increase of urinary lead excretion of nearly twentyfold (P4).

The technique of provocative chelation can and should be extended for evaluating the possible role of trace metals in a variety of diseases in which they have been implicated or suggested. These run the gamut from cardiovascular to cancer to schizophrenia.

VIII. COMMITTEES AND PUBLIC HEALTH

I have discussed elsewhere some of the regulatory and administrative problems involved in protecting the population from toxic chemicals (S13). In this concluding section I would like to amplify briefly certain aspects.

Our economy and technology can function without many toxic or potentially toxic agents, e.g., lead in gasoline (see Section IV-A). The concept of prudency regarding exposure of the general population to toxic metals should be the overriding factor in decisions to eliminate or to discontinue use. It is often stated that until evidence of actual harm to the population is demonstrated, a substance employed for years may continue to be used. This approach--not uncommon--amounts to human experimentation. In fact, if a substance did produce an obvious, readily detectable epidemic of disease, injury, or death, the effect would be catastrophic in that millions in the population would have been needlessly hurt. Further, when it is considered that often the "evidence" of no harm is not based on long-term follow-up studies, then the misjudgments are compounded. Consider the following statement in a monograph produced by the Subcommittee on Environmental Improvement, Committee on Chemistry and Public Affairs of the American Chemical Society (C8, p. 55):

> The argument against removing lead from gasoline is that the economic penalties involved would be unacceptable without the development of much more definitive evidence than now exists of the effects of lead emission, particularly its effect on human health . . . .

In a similar vein, consider the recommendation regarding lead issued by the Committee on Biologic Effects of Atmospheric Pollutants of the Division of Medical Sciences, National Research Council of the National Academy of Sciences (L1). The panel primarily responsible for the report made a series of recommendations for research which amounted to a do-nothing policy on minimizing lead pollution. Their recommendations included the gathering of more data regarding lead levels, lead chemistry, monitoring of food and drink, dose-response relations, search for possible subtle effects, etc. However, no recommendation for immediate action to remove lead from gasoline and other sources was made. One can literally be studied to death with such recommendations.

The NAC committee did refer to lead removal from gasoline and wrote (L1, p. 206):

> . . . Any proposal for the removal of lead from automotive fuels to rid the environment of lead pollution must take into consideration the fact that the fate of other lead products, such as paints and manufactured items, is largely unknown. The Panel can do no more than call attention to the lack of information on this point.

It would appear that the Panel recommends that nothing be done until all information is in. Since automotive fuels account for more than 98% of lead (IV-A) spewed into the atmosphere, why the delay? After all, gasoline engines can be designed to operate on lead-free gasoline.

The committees of the ACS and NAC cited above illustrate a problem associated with the staffing of such committees (S8). Too often, the members are chosen by the crony system, so that not only a homogeneous point of view is given, but the committee as a whole is not qualified because it lacks a representative group of disciplines.

While a committee may seem to have representatives from universities, industry, and government, in fact, this is more apparent than real. For example, to what extent is a university representative actually an industrial representative?

I would like to suggest some criteria for appointment of committees, especially those involving public health:

1. Apply a simple financial criteria to ascertain whether a given member really represents the public, industry, government, or is an impartial university representative:

   <u>What fraction of his total earned income comes from outside sources and what is the source of these funds?</u>

2. Many decisions involve an interplay of technical and non-technical, even political factors. Hence:

Appointment of committee members should deliberately include those with different social and political viewpoints as long as they are technically competent individuals.

Industry would be better served if they followed the same rule in choosing consultants.

3. Most public health problems require individuals in different disciplines to reach well-rounded conclusions of benefit to society, e.g., biostatistics, genetics, biochemistry, etc. I am referring here to primary committees, since the impact of subcommittees, except for technical data, does not often get reflected in the final recommendations of main committees. In the NAC report on lead the views of many of the consultants and contributors were not expressed in the final report. Hence:

   a) The report of committees or panels which involve public health such as the NAC panel on lead (L1) should also include a minority report deliberately solicited as such from members and contributors whose conclusions differ.

   b) Members of continuing committees should have limited terms allowing for gradual but complete turnover of the membership.

I believe if recommendations such as the above were in force, many unfortunate situations as have occurred in areas such as foods (cyclamates), ionizing radiation (S8), and others (S13) might have been averted.

IX. TABLES

Table I. Lead Emission in the United States in 1968 (LI, p. 13).

Emission Source	Lead Emitted, Tons/Year
Gasoline combustion	181,000
Coal combustion	920
Fuel oil combustion	24
Lead alkyl manufacturing	810
Primary lead smelting	174
Secondary lead smelting	811
Brass manufacturing	521
Lead oxide manufacturing	20
Gasoline transfer	36
Total	184,316

Table II. Concentration of Abnormal Trace Metals in the Human Body[1,2].

Metal	Daily Intake in Diet mg	Approximate Body Content			Comments
		mg	μg/g	μmol/g×10^{3}	
Cadmium	0.22	50	0.7	6.2	28% in kidney and liver
Lead	0.45	120	1.7	8.2	92% in bone. Brain is important target organ.
Mercury	0.02	13	0.2	1	69% in fat and muscle. Brain is important target organ.
Beryllium	0.013	0.03	4×10^{-4}	0.04	75% in bone. Toxic action usually involves lung.
Arsenic	1.0	~18	~0.3	~4	

[1]Adapted with additions from reference Schroeder and Nason (S4).

[2]It must be noted that the chemical forms of the metal present in the diet and body are not identified--a point discussed in the body of this report.

Table III. Approximate Fatal Human Doses of Some Metallic Poisons[1].

Metal	~Total Dose (Milligrams)
Antimony (trivalent inorganic salts)	100-200
Arsenic trioxide	120
Arsenic (organic)	7,000-35,000
Cadmium "compounds"	>10
Chromium (chromates)	5,000
Lead "compounds" (absorbed)	>500
Mercuric salts (including mercurial diuretics)	1,000
Mercury (organic mercurials)	3,000-5,000
Mercury (diethyl & dimethyl)	100
Mercury (monomethyl) (2-4 mg/day)[2,3]	200-400

[1]Derived from R. H. Dreisbach (1971), Handbook of Poisoning, 7th ed., Los Altos, Calif.: Lange Medical Publications.

[2]Reference P2.

[3]Reference L6.

Table IV. Threshold Limit Values[1] for Airborne Metals and Their Salts for Workplaces in the U.S.A.[2,3].

Substance	ppm	mg/M^3
Antimony and compounds (as Sb)[3]	-	0.5
Arsenic and compounds (as As)[3]	-	0.5
Arsine[3]	0.05	0.2
Barium (soluble compounds)[3]	-	0.5
Beryllium	-	0.002
Cadmium (metal dust and soluble salts)	-	0.2
Cadmium oxide fume	-	0.1
Calcium arsenate	-	1
Calcium oxide[3]	-	5
Chromic acid and chromates (as CrO_3)	-	0.1
Chromium-metal & insoluble salts[3]	-	1
Cobalt (metal fume and dust)	-	0.1
Copper (fume)	-	0.1
Hafnium	-	0.5
Iron salts, soluble (as Fe)	-	1
Iron oxide fume[3]	-	10
Lead	-	0.2
Lead arsenate	-	0.15
Lead tetraethyl (as Pb)-(s)[3,5,6]	-	0.075
Lead tetramethyl (as Pb)-(s)[3,5]	-	0.075
Lithium hydride	-	0.025
Magnesium (oxide fume)	-	15
C Manganese[4]	-	5
Mercury (s)[5]	-	0.1
Mercury (organic compounds) (s)[5]	-	0.01
Molybdenum (soluble compounds)	-	5
Molybdenum (insoluble compounds)	-	15
Nickel carbonyl	0.001	0.007
Nickel, metal and soluble compounds	-	1

Table IV. Continued.

Substance	ppm	mg/M^3
Osmium tetroxide	-	0.002
Rhodium (metal fume and dusts)	-	0.1
- Soluble salts	-	0.001
Selenium compounds (as Se)	-	0.2
Selenium hexafluoride	0.05	0.4
Silver, metal and soluble compounds	-	0.01
Tantalum	-	5
Tellurium	-	0.1
Tellurium hexafluoride	0.02	0.2
Tetramethyl lead (TML) (as Lead) (s)[5]	-	0.075
Thallium (soluble compounds) (s)[5]	-	0.1
Thallium (soluble compounds) (as Tl) (s)[3,6]	-	0.1
Tin (inorganic compounds, except oxide)	-	2
Tin (organic compounds)	-	0.1
Titanium dioxide	-	15
Tungsten and compounds (as W)		
- soluble	-	1
- insoluble	-	5
Uranium (soluble compounds)	-	0.05
Uranium (insoluble compounds)	-	0.25
Uranium (soluble and insoluble compounds) (as U)	-	0.2
C Vanadium (V_2O_2 dust)[4]	-	0.5
C Vanadium (V_2O_2 fume)[4]	-	0.1
Yttrium	-	1
Zinc chloride (fume)	-	1
Zinc oxide (fume)	-	5
Zirconium compounds (as Zr)	-	5

[1] "The Threshold Limit Values refer to airborne concentrations of substances and represent conditions under which it is believed that nearly all workers may be repeatedly exposed, day after day,

Table IV. Continued.

without adverse effect . . . ." (ILO, p. 297). The values given (except for those with a "C" listing) refer to time-weighted average concentrations for a conventional 7- or 8-hour workday (ILO, p. 297).

[2]Except where noted, values listed are those given in: Permissible Levels of Toxic Substances in the Working Environment (1970), International Labour Office (ILO), Geneva, pp. 297ff.

[3]These substances, not listed in the ILO report, are given in Table G-1, Federal Register, Chapter XVII, Part 1910--Occupational Safety and Health Standards, Vol. 36, No. 157, August 13, 1971. In nearly all cases, the values for substances listed in both the ILO report and the Federal Register here are identical.

[4]A listed value bearing a "C" designation refers to "ceiling" value that should not be exceeded; all values should fluctuate below the listed value (ILO, p. 318).

[5]The "s" designation refers to the potential contribution to the overall exposure by the cutaneous route including mucous membranes and eye, either by airborne, or more particularly, by direct contact with the substance (ILO, p. 299).

[6]American Industrial Hygiene Association as recommended by the Ethyl Corporation (ILO, p. 328).

Table V. Mutagenesis by Metal Salts[1].

Metal	Salt	Point Mutations	Chromosome Aberrations	
		Microorganisms	Plant	Mammalian
Aluminum	Chloride		+	
Arsenic	Arsenate			+
Cadmium	Nitrate		+	
Calcium Deficiency				+
Ferrous	Chloride	+		
Lead	Acetate			+
Magnesium Deficiency				+
Manganous	Acetate	+		
Manganous	Chloride	+		
Mercury	Methyl		+	+
Mercury	Methoxyethyl		+	
Mercury	Phenyl Hydroxide		+	
Potassium Excess		+		

[1]Data obtained principally from references F3 and S16.

Table VI. Lead Content of Soils[1].

Rt. 93, Wilmington, Mass.		Memorial Drive Cambridge, Mass.	
Location	Amount (ppm)	Location	Amount (ppm)
Roadside	380	Roadside	1,300
10 yards	85	10 feet	440
45 yards	35	20 feet	320
440 yards	20	60 feet	270

[1]Data from reference H3.

Table VII. Air Lead Concentrations Versus Distance From Road[1].

Location	Micrograms lead/ cubic meter
Roadside	17.5
10 feet from road	6.5
50 feet from road	3.0
100 feet from road	2.7

[1]Data from reference H3.

Table VIII. Usual Variations in Lead Levels in "Normal" Adults[1].

Site	Lead Levels	Lead Levels in 95% of Population[2]
Blood	20-80 µg/100 ml	<40 µg/100 ml
Urine	20-80 µg/l	<65 µg/l
Feces	200-400 mg	--

[1]Data from reference I2, p. 95.

[2]Non-exposed population.

Table IX. Toxic Metal Levels in U. S. Cities (58) and Nonurban Areas (29)[1].

Item Evaluated	Metal						
	Nickel	Beryllium	Cadmium	Tin	Antimony	Lead	Mercury
Cities (58)-μg/m^3	0.001-0.118	0.001-0.002	0.002-0.370	0.01-0.03	0.042-0.085	0.1-2.3	****
Nonurban-μg/m^3	0.0006-0.012	0.00013	0.004-0.026	0.0002-0.0018	0.001-0.002	0.0001-0.0004	0.003-0.009
Intake (μg/day) Food and Water	600	12	100	7,300	<100	300	20
Air	2.36	0.04	7.4	0.6	1.7	46	0.108
Total Body (μg)	10,000	30	50,000	5,800	7,900	120,000	13,000
Soil (ppm)	40	6	0.06	10	4	10	0.03-1.3
Principal Source	Coal Petroleum	Coal?	Zinc	?	?	Gasoline Additives	Coal Petroleum

[1]Adapted from H. A. Schroeder (S5).

Table X. Consumption of Mercury in the U.S.A. in 1969[1].

Source	Amount (10^3 pounds)
Chlor-alkali production	1575
Electrical apparatus	1417
Paints	739
Industrial and control instruments	531
Dental preparations	232
Catalysts	225
Agriculture[2]	204
General laboratory use	155
Pharmaceuticals	55
Pulp- and papermaking	42
Other uses[3]	736
Total	5,911

[1] U. S. Bureau of Mines.

[2] Includes fungicides and bactericides for industrial purposes.

[3] Includes university and other research, and government allocation for military and scientific purposes.

Table XI. Provacative Chelation in Children--Excretion of Lead in Urine Before and After Administration of $CaNa_2EDTA$[1].

Patient Group	Patients (No.)	Urinary Lead Excretion (μg/liter)	
		Average Before Administration	Average After Administration
Controls	24	15	165
Suspected Lead Poisoning	11	24	995
Clinical poisoning	8	146	1966

[1] Adapted from reference G3, p. 951.

X. REFERENCES

Aa Alexander, C. S. (1972), Amer. J. Med., 53, 395.

A1 Ambio, Special Report on the "Evaluation of Genetic Risks of Environmental Chemicals" (in press).

A2 Ann. N. Y. Acad. Sci. (1972), Geochemical Environment in Relation to Health and Disease, H. C. Hopps and H. L. Cannon, eds., June 28, Vol. 199.

B1 Bearn, A. G. (1966), "Wilson's Disease," The Metabolic Basis of Inherited Disease, Chapter 34, J. B. Stanbury, J. B. Wyngaarden, and D. S. Frederickson, eds., New York: McGraw-Hill.

B2 Beattie, A. D., et al. (1972), Brit. Med. J., May 27, 491.

B3 Berlin, M., Jerksell, L. G., and Nordberg, G. (1965), Acta Pharmacol. et Toxicol., 23, 312.

B4 Bertilsson, L. and Neujahr, H. Y. (1971), Biochemistry, 10, 2805.

B5 (a) Billings, C. E. and Matson, W. R. (1972), Science, 176, 1232.
(b) Friedman, I. and Peterson, N. (1971), Science, 172, 1027.

B6 Blokker, P. C. (1972), Atm. Env., 6, 1.

B7 Blumenthal, S., Davidow, B., Harris, D., and Oliver-Smith, F. (1972), Amer. J. Public Health, 62, 1060.

B8 Bridges, B. (1971), Newsletter Environ. Mutagen Soc., #5, November, 13-15.

B9 Browning, E. (1969), Toxicity of Industrial Metals, New York: Appleton-Century-Crofts, 383 pp.

B10 (a) Brugsch, H. G. (1959), A.M.A. Archiv. Indust. Health, 20, 285.
(b) Brugsch, H. G. (1965), J. Occup. Med., 7, 394.

B11 Bryce-Smith, D. (1972), Lancet, Oct. 14, 817.

B12 Buchwald, H. (1972), Amer. Indust. Hyg. Assoc. J., July, 492.

B13 Burdé, B. and Choate, Jr., M. S. (1972), J. Pediatrics, 81, 1088.

C1 Catsch, A. (1964), Radioactive Metal Mobilization in Medicine, Springfield, Ill.: C. C. Thomas, 170 pp.; ibid. (1968), Dekorporierung Radioakfiver und Stabiler Metallionen, München: Karl Thiemig Kg, 176 pp.

C2 Challop, R. S. (1971), New Engl. J. Med., 285, 970.

C3 Childs, B., Miller, S. M., and Bearn, A. G. (1972), "Gene Mutation as a Cause of Human Disease," Mutagenic Effects of Environmental Contaminants, H. E. Sutton and M. F. Harris, eds., New York: Academic Press, pp. 3-14.

C4 Chisolm, Jr., J. J. and Harrison, H. E. (1956), Pediatrics, 18, 943.

C5 Chisolm, Jr., J. J. (1968), J. Pediatr., 73, 1.

C6 Christensen, H. E., ed. (1972), The Toxic Substances List, Rockville, Md.: U. S. Department of Health, Education, and Welfare, Health Sciences and Mental Health Administration, National Institute for Occupational Safety and Health.

C7 Clarkson, T. W. (1972), Ann. Rev. Pharmacology, 12, 375.

C8 Cleaning Our Environment--The Chemical Basis for Action (1969), Washington, D. C.: American Chemical Society, 249 pp.

C9 Cotzias, G. C. (1967), in Proceedings University of Missouri 1st Annual Conference on Trace Substances in Environmental Health, Columbia, Mo.: Univ. Mo. Environmental Health Center and Extension Division, pp. 5-19; Cotzias, G. C., et al. (1972), Science, 176, 412.

C10 Crawford, M. D., Gardner, M. J., and Sedgwick, P. A. (1972), Lancet, May 6, 988.

D1 Dauncey, M. J. and Widdowson, E. M. (1972), Lancet, April 1, 711.

D2 David, O., Clark, J., and Voeller, K. (1972), Lancet, Oct. 28, 900.

D3 Deichmann, W. B. and Gerarde, H. W. (1964), Sympomalology and Therapy of Toxicological Emergencies, New York: Academic Press, 605 pp.

D4 Dingwall-Fordyce, I. and Lane, R. E. (1963), Brit. J. Ind. Med., 20, 313.

D5 Dunlap, L. (1971), Chem. & Eng. News, July 5, 22-34.

E1 "Effects of Mercury on Man and the Environment," Hearings of the Committee on Commerce, United States Senate, 91st Congress, May 8, 1970, Washington, D. C.: Government Printing Office.

E2 Epstein, S. S., Arnold, E., Andrea, J., Bass, W., and Bishop, Y. (1972), Toxicol. and Appl. Pharmacology, 23, 288.

F1 Federal Register (1971), Occupational Safety and Health Administration, Department of Labor, Chapter XVII, Occupational Safety and Health Standards, Part 1910, Vol. 36, No. 157, Friday, August 13.

F2 Ferm, V. H. (1969), Experientia, 25, 56.

F3 Fishbein, L., Flamm, W. G., and Falk, H. L. (1970), Chemical Mutagens, New York: Academic Press.

F4 Foote, R. S. (1972), Science, 177, 513.

F5 Fried, J. F., Rosenthal, M. W., and Schubert, J. (1956), Proc. Soc. Exptl. Biol. Med., 92, 331.

F6 Fujiki, M. (1963), Kumamoto Igk, Z., 37, 10.

G1 Gilfillan, S. C. (1965), J. Occup. Med., 7, 53.

G2 Goldstein, A., Aronow, L., and Kalman, S. M. (1968), Principles of Drug Action, New York: Harper and Row, 1884 pp.

G3 Goodman, L. S. and Gilman, A., eds. (1970), The Pharmacological Basis of Therapeutics, 4th ed., New York: Macmillan Co., 1785 pp.

G4 Goyer, R. S. and Mahaffey, K. R. (1972), Environ. Health Perspectives, October, 73-80.

G5 Greenfield, O. (1963), New Engl. J. Med., 269, 1138.

G6 Grice, H. C., et al. (1969), Clin. Toxicol., 2, 273.

G7 Gunn, S. A., Gould, T. C., and Anderson, W. A. D. (1968), Soc. Exptl. Biol. Med., 128, 591.

H1 Hall, S. K. (1972), Environ. Sci. & Technol., 6, 31.

H2 Halloran, R. (1973), New York Times, March 21.

H3 Hardy, H. L., et al. (1971), Clin. Pharmacol. and Therapeutics, 12, 982.

H4 Hartung, R. and Dinman, B. D., eds. (1972), Environmental Mercury Contamination, Ann Arbor Sci. Publ., 349 pp.

H5 Hemphill, F. E., Kaeberle, M. L., and Buck, W. B. (1971),

Science, 172, 1031.

H6 Hemphill, D. D., ed. (1972), Trace Substances in Environmental Health, Columbia, Mo.: University of Missouri.

I1 Imura, N., et al. (1971), Science, 172, 1248.

I2 International Labour Office, Sixth Session of the Joint ILO/WHO Committee on Occupational Health, Geneva (1968).

J1 Jensen, S. and Jernelöv, A. (1967), Nordforsk Biolidinformation, 10, 4; ibid. (1968), 14, 3; ibid. (1969), Nature, 223, 753.

J2 Jones, R. H., Williams, R. L., and Jones, A. M. (1971), Proc. Exptl. Biol. Med., 137, 1231.

K1 Kehoe, R. A. (1965), Arch. Environ. Health, 11, 736.

K2 Kennedy, A. (1968), Brit. J. Exp. Path., 49, 360.

K3 King, B. G., et al. (1972), Amer. J. Public Health, 62, 1056.

K4 Kurland, L. T., et al. (1960), World Neurology, 1, 370.

L1 Lead, Committee on Biologic Effects of Atmospheric Pollutants (1972), Washington, D. C.: National Academy of Sciences, 330 pp.

L2 Lee, D. H. K., ed. (1972), Metallic Contaminants and Human Health, New York: Academic Press, 244 pp.

L3 Lenz, W. (1961), Deutsche med. Wochenschr., 86, 2555.

L4 Lewis, G. P., Jusko, W. J., and Coughlin, L. L. (1972), J. Chron. Dis., 25, 717.

L5 Lin-Fu, J. S. (1972), New Engl. J. Med., 286, 702.

L6 Löfroth, G. (1970), "Methylmercury," Ecological Research Committee, Bulletin No. 4, 2nd ed., Stockholm, Sweden: Swedish National Science Research Council.

L7 Luzio, N. di (1972), Chem. & Eng. News, February 28, 56.

M1 Malling, H. V., Wassom, J. S., and Epstein, S. S. (1970), Environmental Mutagen Soc. Newsletter, June, 7.

M2 Mancuso, T. F. and El-Attar, A. A. (1969), J. Occup. Med., 11, 422.

M3 Masironi, A. T., et al. (1972), Bull. Wld. Hlth. Org., 47, 139.

M4 McCaull, J. (1971), Environment, 13, 3.

M5 McClure, C. D. (1970), Trans. N. Y. Acad. Sci., 32, 204.

M6 McIntire, M. S. and Angle, C. R. (1972), Science, 177, 520.

M7 Miettinen, J. K. (1972), Science, 176, 1072.

M8 Millar, J. A., et al. (1970), Lancet, Oct. 3, 695.

M9 Montague, P. and Montague, K. (1971), Saturday Review, Feb. 6, 50-55.

M10 Moutschen, J., Moës, A., and Gilot, J. (1964), Experientia, 20, 494.

M11 Muro, L. A. and Goyer, R. A. (1969), Arch. Path., 87, 660.

N1 Needleman, H. L. and Scanlon, J. (1973), New Engl. J. Med., 288, 466.

N2 Niemeier, B. (1967), Int. Archiv f. Gewerbepathologie u. Gewerbehygiene, 24, 160.

O1 Oehme, F. W. (1970), Clin. Toxicol., 3, 5.

P1 Patterson, C. C. (1965), Arch. Environ. Hlth., 11, 344.

P2 Pierce, P. E., et al. (1972), J. Amer. Med. Assoc., 220, 1439.

P3 Pollycove, M. (1966), "Hemochromatosis," The Metabolic Basis of Inherited Disease, Chapter 35, J. B. Stanbury, J. B. Wyngaarden, and D. S. Frederickson, eds., New York: McGraw-Hill.

P4 Prerovska, I. and Teisinger, J. (1970), Brit. J. Med., 27, 352.

R1 Ramel, C. (1969), Hereditas, 61, 208; Ramel, C. and Magnusson, J. (1969), Hereditas, 61, 231.

R2 Ringbom, A. (1963), Complexation in Analytical Chemistry, New York: Interscience, 395 pp.

S1 Sands, J. H., Berris, B., and Scherer, L. R. (1950), New Engl. J. Med., 243, 559.

S2 Schroeder, H. A. (1965), J. Chron. Dis., 18, 647.

S3 Schroeder, H. A., et al. (1967), J. Chron. Dis., 20, 179.

S4 Schroeder, H. A. and Nason, A. P. (1971), Clin. Chem., 17, 461.

S5 Schroeder, H. A. (1971), Environment, 13, 18.

S6 Schubert, J. and White, M. R. (1952), J. Lab. Clin. Med., 39, 260.

S7 Schubert, J. (1958), Scient. Amer., 199, 27.

S8 Schubert, J. and Lapp, R. E. (1958), Bulletin of the Atomic Scientists, 14, 23.

S9 Schubert, J. and Rosenthal, M. W. (1959), A.M.A. Arch. Ind. Health, 19, 169.

S10 Schubert, J. (1964), "The Chemical Basis of Chelation," Iron Metabolism, F. Gross, ed., Springer-Verlag, pp. 466-494.

S11 Schubert, J. (1966), Scient. Amer., 214, 40.

S12 Schubert, J. (1972), Radiobiology of Plutonium, W. S. S. Jee and B. J. Stover, eds., University of Utah, 552 pp.

S13 Schubert, J. (1972), Ambio, 1, 79.

S14 Sever, L. E. and Emanual, I. (1972), Teratology, 7, 117.

S15 Sharma, V. J. and Schubert, J. (1969), J. Chem. Ed., 46, 506.

S16 Shaw, M. (1970), Ann. Rev. Med., 21, 409.

S17 Shea, K. P. (1973), Environment, 15, 6.

S18 Sillen, L. G. and Martell, A. E., eds. (1964), Stability Constants of Metal-Ion Complexes, Special Publication No. 17, London: The Chemical Society, Burlington House, W. 1; ibid. (1971), Supplement No. 1, Special Publication No. 25.

S19 Skerfving, S., Hansson, K., and Lindsten, J. (1970), Arch. Environ. Health, 21, 133.

S20 Skoryna, S. C. and Waldron-Edward, D., eds. (1971), Intestinal Absorption of Metal Ions, Trace Elements and Radionuclides, Pergamon Press, 431 pp.

S21 Snyder, R. D. (1971), New Engl. J. Med., 284, 1014.

S22 Spangler, W. J., Spigarelli, J. L., Rose, J. M., and Miller, H. M. (1973), Science, 180, 192.

S23 Stern, C. (1971), Ann. Int. Med., 75, 623.

S24 Stokinger, H. E., ed. (1966), Beryllium--Its Industrial Hygiene Aspects, New York: Academic Press.

S25 Sullivan, J., Parker, M., and Carson, S. B. (1968), J. Lab. Clin. Med., 71, 893.

S26 Symposium on Environmental Lead Contamination (1966), U. S. Public Health Service Publication No. 1440, March, Washington, D. C.: Superintendent of Documents, U. S. Government Printing Office, 177 pp.

T1 Teisinger, J. (1971), Arch. Environ. Health, 23, 280.

U1 Underwood, E. J. (1971), Trace Elements in Human and Animal Nutrition, 3rd ed., New York: Academic Press, 543 pp.

V1 Vigliani, E. C. and Zurlo, N. (1951), Brit. J. Ind. Med., 8, 218.

V2 Vincent, W. F., et al. (1970), Amer. J. Clin. Path., 53, 963.

W1 Wiberg, et al. (1969), Clin. Toxicol., 2, 257.

W2 Williams, R. T. (1959), Detoxication Mechanisms, 2nd ed., New York: John Wiley, 796 pp.

W3 Wood, J. M., Kennedy, F. S., and Rosen, C. G. (1968), Nature, 220, 173.

W4 Wood, J. M. (1970), Adv. Environ. Sciences, 2, 39.

SUBJECT INDEX

GPSR Compliance
The European Union's (EU) General Product Safety Regulation (GPSR) is a set of rules that requires consumer products to be safe and our obligations to ensure this.

If you have any concerns about our products, you can contact us on

ProductSafety@springernature.com

In case Publisher is established outside the EU, the EU authorized representative is:

Springer Nature Customer Service Center GmbH
Europaplatz 3
69115 Heidelberg, Germany

www.ingramcontent.com/pod-product-compliance
Ingram Content Group UK Ltd.
Pitfield, Milton Keynes, MK11 3LW, UK
UKHW051127260726
13967UKWH00010B/2903
* 9 7 8 1 4 6 8 4 3 2 4 1 1 *